W0018640

Unit Operations Methods in Environmental Engineering

Unit Operations Methods in Environmental Engineering

Contributors

Manuel Ferreira Rebelo, Gilberto Santos et al.

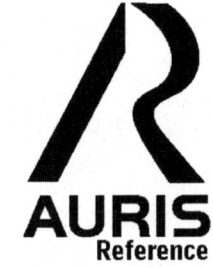

AURIS
Reference

www.aurisreference.com

Unit Operations Methods in Environmental Engineering

Contributors: Manuel Ferreira Rebelo, Gilberto Santos et al.

Published by Auris Reference Limited

www.aurisreference.com

United Kingdom

Copyright 2016
Printed in 2017 for Sale in the Indian Subcontinent

The information in this book has been obtained from highly regarded resources. The copyrights for individual articles remain with the authors, as indicated. All chapters are distributed under the terms of the Creative Commons Attribution License, which permit unrestricted use, distribution, and reproduction in any medium, provided the original author and source are credited.

Notice

Contributors, whose names have been given on the book cover, are not associated with the Publisher. The editors and the Publisher have attempted to trace the copyright holders of all material reproduced in this publication and apologise to copyright holders if permission has not been obtained. If any copyright holder has not been acknowledged, please write to us so we may rectify.

Reasonable efforts have been made to publish reliable data. The views articulated in the chapters are those of the individual contributors, and not necessarily those of the editors or the Publisher. Editors and/or the Publisher are not responsible for the accuracy of the information in the published chapters or consequences from their use. The Publisher accepts no responsibility for any damage or grievance to individual(s) or property arising out of the use of any material(s), instruction(s), methods or thoughts in the book.

Unit Operations Methods in Environmental Engineering

ISBN: 978-1-78154-968-1

British Library Cataloguing in Publication Data
A CIP record for this book is available from the British Library

Printed in the United Kingdom

Exclusively distributed by CBS Publishers & Distributors Pvt. Ltd.

Sales & Distribution Rights only for India, Pakistan, Bangladesh, Sri Lanka, Nepal and Bhutan. This book is not to be sold outside these territories.

Contents

List of Abbreviations

AMAP	Arctic Monitoring.Assessment Program
ARMA	Auto regressive moving average
CCS	Capture and geo-sequestration
DNP	Dinitrophenol
EDCs	Endocrine disrupting chemicals
EESA	Environmental and Economic Sustainability Assessment
ER	Estrogen receptor
FAZ	Frankfurter Allgemeine Zeitung
DPG	German Physical Society
GHG	Greenhouse gases
IWM	Institute of Water Modelling
IOS	Fisheries and Oceans Canada, DFO
LNG	Liquefied natural gas
LNG	Liquefied natural gas plant
MS	Mass spectrometry
MALDI	Matrix-assisted laser deionization
MT	Metallothionein
MCM	Million cubic meters
MW	Molecular weights
NERC	Natural Environment Research Council
PCP	Pentachlorophenol
PFCAs	perfluorinated carboxylates
PFCs	Perfluorinated contaminants
FOSEs	perfluorinated sulfonamide alcohols
PFSAs	perfluorinated sulfonates
PFCs	perfluorinated.contaminants
PFBA	perfluorobutanoic acid
PFHxA	perfluorohexanoic acid
FOSAs	perfluorooctane sulfonamides
PFOS	perfluorooctane sulfonate
PFOA	perfluorooctanoate
POPs	Persistent organic pollutants
PVP	Polyvinylpyrrolidone
PCA	Principal component analysis
PES	Protein expression signatures
SELDI	ProteinChip, surface-enhanced laser desorption ionization
RISC	RNA-induced silencing complexes
SPD	Social democratic party
SDF	Spectral density function
SAMS	Stochastic Analysis Modeling and Simulation

SZ	Süddeutsche Zeitung
THCs	total hydrocarbons
UNEP	United Nations Environmental Program
UPLC	Using ultra-performance liquid chromatography

List of Contributors

Manuel Ferreira Rebelo
CLEGI, Lusíada University, Vila Nova de Famalicão, Portugal

Gilberto Santos
CLEGI, Lusíada University, Vila Nova de Famalicão, Portugal
College of Technology, Polytechnic Institute of Cávado and Ave, Barcelos, Portugal

Rui Silva
CLEGI, Lusíada University, Vila Nova de Famalicão, Portugal

Paul E. Hardisty
WorleyParsons EcoNomics™, Perth, WA 6000, Australia
Imperial College London, London SW7 2AZ, UK

Mayuran Sivapalan
WorleyParsons EcoNomics™, Houston, TX 77079, USA

Peter Brooks
WorleyParsons, Brisbane, QLD 7000, Australia

Juan José Alava
School of Resource & Environmental Management, Faculty of Environment, Simon Fraser University, Burnaby, British Columbia, Canada
Facultad de Ingeniería Marítima, Ciencias Biológicas, Oceánicas y Recursos Naturales, Escuela Superior Politécnica del Litoral (ESPOL), Guayaquil, Ecuador

Mandy R.R. McDougall
School of Resource & Environmental Management, Faculty of Environment, Simon Fraser University, Burnaby, British Columbia, Canada

Mercy J. Borbor-Córdova
Facultad de Ingeniería Marítima, Ciencias Biológicas, Oceánicas y Recursos Naturales, Escuela Superior Politécnica del Litoral (ESPOL), Guayaquil, Ecuador

K. Paola Calle
Facultad de Ingeniería Marítima, Ciencias Biológicas, Oceánicas y Recursos

Naturales, Escuela Superior Politécnica del Litoral (ESPOL), Guayaquil, Ecuador

Mónica Riofrio
Instituto Antártico Ecuatoriano (INAE), Guayaquil, Ecuador

Nastenka Calle
Pacific Institute for Climate Solution (PICS), Simon Fraser University, Burnaby, British Columbia, Canada

Michael G. Ikonomou
Institute of Ocean Sciences, Fisheries and Oceans Canada, Sidney, BC, Canada

Frank A.P.C. Gobas
School of Resource & Environmental Management, Faculty of Environment, Simon Fraser University, Burnaby, British Columbia, Canada

Masachika Suzuki
Faculty of Commerce, Kansai University, Suita-shi, Osaka, Japan

Volker Schneider
Department of Politics and Public Administration, University of Konstanz, Germany

Jana K. Ollmann
Department of Politics and Public Administration, University of Konstanz, Germany

Saad Merayyan
Department of Civil Engineering, California State University, Sacramento, USA

Salwa Mrayyan
AL-Balqa Applied University, Al-Huson, Jordan

Misbah Uddin
Department of Civil and Environmental Engineering, Shahjalal University of Science and Technology, Sylhet, Bangladesh

Jahir Bin Alam
Department of Civil and Environmental Engineering, Shahjalal University of Science and Technology, Sylhet, Bangladesh

Zahirul Haque Khan
Coast, Port and Estuary Division, Institute of Water Modelling, Dhaka, Bangladesh

G. M. Jahid Hasan
Department of Civil and Environmental Engineering, Shahjalal University of Science and Technology, Sylhet, Bangladesh

Tauhidur Rahman
Department of Civil and Environmental Engineering, Shahjalal University of Science and Technology, Sylhet, Bangladesh

Ahmad Shakir Mohd Saudi
East Coast Environmental Research Institute, University Sultan Zainal Abidin, Kuala Terengganu, Malaysia
Science and Technology, Open University Malaysia, Shah Alam, Malaysia

Hafizan Juahir
East Coast Environmental Research Institute, University Sultan Zainal Abidin, Kuala Terengganu, Malaysia

Azman Azid
East Coast Environmental Research Institute, University Sultan Zainal Abidin, Kuala Terengganu, Malaysia

Mohd Khairul Amri Kamarudin
East Coast Environmental Research Institute, University Sultan Zainal Abidin, Kuala Terengganu, Malaysia

Mohd Ekhwan Toriman
East Coast Environmental Research Institute, University Sultan Zainal Abidin, Kuala Terengganu, Malaysia

Nor Azlina Abdul Aziz
East Coast Environmental Research Institute, University Sultan Zainal Abidin, Kuala Terengganu, Malaysia

Nidhal Saada
Civil Engineering Department, AL Ahliyya Amman University, Amman, Jordan

Golden Makaka
University of Fort Hare, Alice, South Africa

Letensie Tseggai Hadgu
Department of Civil, Construction, and Environmental Engineering, Jomo Kenyatta University of Agriculture and Technology, Nairobi, Kenya

Maurice Omondi Nyadawa
Jaramogi Oginga Odinga University of Science and Technology, Bondo, Kenya

John Kimani Mwangi
Department of Civil, Construction, and Environmental Engineering, Jomo Kenyatta University of Agriculture and Technology, Nairobi, Kenya

Purity Muthoni Kibetu
Department of Civil, Construction, and Environmental Engineering, Jomo Kenyatta University of Agriculture and Technology, Nairobi, Kenya

Beraki Bahre Mehari
Department of Civil, Construction, and Environmental Engineering, Jomo Kenyatta University of Agriculture and Technology, Nairobi, Kenya

Manvendra Singh Chauhan
Department of Civil Engineering, Indian Institute of Technology, IIT (BHU), Varanasi, India

Prabhat Kumar Singh Dikshit
Department of Civil Engineering, Indian Institute of Technology, IIT (BHU), Varanasi, India

Shyam Bihari Dwivedi
Department of Civil Engineering, Indian Institute of Technology, IIT (BHU), Varanasi, India

Ramesh Prasad Tripathi
Department of Land Resources and Environment, Hamelmalo Agricultural College, Keren

Woldeselassie Ogbazghi
Department of Land Resources and Environment, Hamelmalo Agricultural College, Keren

Semere Amlsom
Department of Land Resources and Environment, Hamelmalo Agricultural College, Keren

Simon Measho
Department of Land Resources and Environment, Hamelmalo Agricultural College, Keren

Paul S. Okweye
College of Engineering, Technology & Physical Sciences, Department of Physics, Chemistry and Mathematics, Alabama A&M University, Normal, AL, USA

Karnita G. Garner
College of Agricultural, Life and Natural Sciences, Department of Biology and Environmental Sciences, Alabama A&M University, Normal, AL, USA

Anthony S. Overton
College of Agricultural, Life and Natural Sciences, Department of Biology and Environmental Sciences, Alabama A&M University, Normal, AL, USA

Elica M. Moss
College of Agricultural, Life and Natural Sciences, Department of Biology and Environmental Sciences, Alabama A&M University, Normal, AL, USA

Hyo Jeong Kim
Institute of Environmental Medicine for Green Chemistry, Dongguk University Biomedi Campus 32, Dongguk-ro, Ilsandong-gu, Goyang-si, Gyeonggi-do 410-820, Korea
Department of Life Science, Dongguk University Biomedi Campus 32, Dongguk-ro, Ilsandong-gu, Goyang-si, Gyeonggi-do 410- 820, Korea

Preeyaporn Koedrith
Institute of Environmental Medicine for Green Chemistry, Dongguk University Biomedi Campus 32, Dongguk-ro, Ilsandong-gu, Goyang-si, Gyeonggi-do 410-820, Korea
Faculty of Environment and Resource Studies, Mahidol University, 999 Phuttamonthon 4 Rd., Phuttamonthon District, Nakhon Pathom 73170, Thailand

Young Rok Seo
Institute of Environmental Medicine for Green Chemistry, Dongguk University Biomedi Campus 32, Dongguk-ro, Ilsandong-gu, Goyang-si, Gyeonggi-do 410-820, Korea

Department of Life Science, Dongguk University Biomedi Campus 32, Dong-guk-ro, Ilsandong-gu, Goyang-si, Gyeonggi-do 410- 820, Korea

Preface

The text *Unit Operations Methods in Environmental Engineering* provides thorough, updated coverage of unit operations in environmental engineering. It presents a practical approach to the design of water and wastewater treatment plants, carefully explaining new technologies that affect the design of such facilities. A methodology to develop the integration of the environmental management system with other standardized management systems has been presented in first chapter. The environmental and economic sustainability of carbon capture and storage has been discussed in second chapter. Third chapter focuses on perfluorinated chemicals in sediments, lichens, and seabirds from the Antarctic Peninsula. Fourth chapter shows a broad landscape of barriers in technology diffusion in the developing countries by addressing two levels of barriers. In fifth chapter, we apply a specific form of quantitative discourse analysis to the debate on global warming and related policy decisions. Sixth chapter addresses the water supply challenges that Jordan faces and what has been accomplished to improve supply and/or reduce demand. Two dimensional hydrodynamic modelling of northern Bay of Bengal coastal waters has been outlined in seventh chapter. Eighth chapter constructs downscaling statistical model in analyzing the hydrological modeling in the study area which faces the risk of flood occurrence as the impact of climate change. The objective of ninth chapter is to investigate the use of autoregressive moving average (ARMA) models in modeling and simulation of annual rainfall data in Saudi Arabia and their ability to capture the long term statistics observed in the historical records. Tenth chapter investigates the effects of fly ash on brick properties and the thermal comfort of a passive solar fly ash brick house. The aim of eleventh chapter is to model the water quality of the polluted segment of Ndarugu River by the comprehensive application of water quality model QUAL2K and evaluate the performance of the model using statistics based on correlation coefficient (R2) and standard error (SE). The purpose of twelfth chapter is to develop a correlation among parameters of Varanasi bend of holy River Ganga which are directly related to the physical parameters of river i.e. discharge, depth and velocity. The objective of thirteenth chapter is to design and develop runoff harvesting system in agricultural watersheds associated with nonagricultural lands to facilitate runoff farming of rice in semiarid environments of Hamelmalo, Anseba region of Eritrea. Fourteenth chapter evaluates spatial and compositional patterns in the FC and FR sediment contaminant data using the multivariate techniques—principal components analysis (PCA) and cluster analysis (CA). Ecotoxicogenomic approaches for understanding molecular mechanisms of environmental chemical toxicity using aquatic invertebrate, daphnia model organism have been proposed in last chapter.

Chapter 1

A METHODOLOGY TO DEVELOP THE INTEGRATION OF THE ENVIRONMENTAL MANAGEMENT SYSTEM WITH OTHER STANDARDIZED MANAGEMENT SYSTEMS

Manuel Ferreira Rebelo[1*], Gilberto Santos[1,2], Rui Silva[1]

[1]CLEGI, Lusíada University, Vila Nova de Famalicão, Portugal

[2]College of Technology, Polytechnic Institute of Cávado and Ave, Barcelos, Portugal

ABSTRACT

Traditionally the global management system of an organization is frequently split into a number of individual management systems that are defined and implemented according to specific man- agement systems standards (MSSs) as well as managed independently. The individual implemen- tation of MSSs is an option that leads to several inefficiencies and sub-optimization of the global management system of an organization. As referred by ISO [1] the interested parties' require- ments increase. A more effective and efficient option for an organization is to integrate, into an integrated management system (IMS), the implementation and management of requirements of multiple MSSs. Certain difficulties are associated to the structuring process, implementation, verification, evaluation, improvement and progressive development of an IMS in the organizations. Several scholars have proposed various theoretical approaches regarding the integration of individual management systems (MSs) leading to the conclusion that there is not a common practice for all organizations as they encompass different characteristics. This paper aims to present and justify a designed methodology to be used by organizations to support the integration of various MSs. Among them are highlighted: the Environmental Management System (EMS) according ISO 14001 [2] , the Quality Management System (QMS) according ISO 9001 [3] , and the Occupational Health and Safety Management System (OH & SMS) according OHSAS 18001 [4] . The methodology was designed in the context of a Portuguese company, on sequence of an organizational diagnosis and a research that was performed through a questionnaire. The strategy and the research methods took into consideration the case study.

INTRODUCTION

Due to the demand of the market itself or by other internal reasons, there are many organizations that implement different standardized MSs [5] . Standards arise through the development of detailed descriptions of particular characteristics of a product or service by experts from companies and scientific institutions [6] . According to ISO [7] , the domain of standardized management systems (MSs) has expanded greatly over the last years and nowadays there exist a relevant number of MSSs for individual MSs, which apply to any type of organization independently of its external and internal context. The objective of the development of standards is to support both individuals and companies when procuring products and services [6] . According to ISO publication [1] a common objective of MSSs is to assist organizations to manage the risks associated with providing products and services to customers and other interested parties.

As the number of MSSs versus standardized MSs increases, their integration becomes a necessity [1] [8] . In the literature and in MSSs are presented several definitions/descriptions of the concept of management system (MS) and integrated management system (IMS). Empirical studies were conducted with the aim of the integra- tion of individualized MSs around the world [9] and several tangible and intangible gains for organizations, as well as to their internal and external interested parties, are achieved with the integration of the individual stan- dardized MSs [10] - [15] , among others. Integration of MSs promotes synergies and cost savings, as well as a re- duction of the time spent when managing the systems [16] . Olaru [17] summarized forty benefits that an or- ganization can gain from the implementation of an IMS. Organizations therefore need a framework to integrate these MSs and facilitate their contribution to the operation of the overall business MS [18] . On the other hand an essential element in the strategy of any organization is the minimization of business risk to a level that ensures the security market [19] . According to Suditu [20] for organizations that want to survive and compete in the ac- tual market, it is necessary to continually improve performance of their business and MSs in a sustainable way, taking in consideration the necessities of the interested parties. In turn, according to Oliveira [21] , integration is justified as a function of the benefits that it provides; certifiable management systems that work separately are more bureaucratic and costly, and generate poorer results than those obtained employing integration. Study conducted by Simon [22] concluded that Organizations prefer integration of MSs to managing them separately and the integration of systems is one of the major strategies for ensuring survival and savings for the organizations of the sample.

Quality is no longer, as formerly, a redundant and restricted concept and must be managed in a global per- spective and of sustainability not only focused on satisfying customers, but on a whole range of interested par- ties [11] . Are example, those identified in the ISO 9004 [7] , several others exist. On the other hand, the increas- ing global competition potentiates, therefore, an increase in the expectations of all the interested parties of or- ganizations. On Table 1 there are listed several interested parties and associated needs and expectations to be satisfied by organizations according to the requirements of related MSSs.

So, more than ever, business sustainability gains increased importance and focus is shifting from their finan- cial results. These results will not verify if that focus does not prioritize also, the satisfaction in a balanced and integrated fashion of customers and others relevant interested parts, that are clearly and objectively the employees for example. In this context of real and new paradigms of management—the Global Quality Management—it is required a constant search for Business Excellence [11] .

Hence, in a not distant past, some organizations in Portugal and other countries, although in a small percent- age, began to integrate their individual standardized management systems like: EMS; QMS; OH & SMS; CSRMS-Corporate Social Responsibility Management System, among others. For this purpose, organizations began to conceive integrated procedures in order to make the integration of two systems (QMS & EMS or EMS & OHSMS) and whenever possible, the three standardized Management Systems: EMS, QMS, and OHSMS [23] . This reveals the growing interest that has been demonstrated by organizations in the adoption of the MSSs ISO 14001, ISO 9001 and OHSAS 18001.

On the other hand, the integration of MSs, supported by those MSSs in a single system, taking into account the correspondence and the level of compatibility between them and potential tangible and intangible gains re-sulting from this integration will be an added value that organizations cannot ignore [11] . On the other hand, regulations based on ISO 9000 have been created to guide companies in developing systems for management and prevention of worker risks. Annex A and B of ISO 9001 [3] gives various clauses and subclauses related to the necessary elements of this standard [24] .

Table 1: Examples of interested parts and their needs and expectations

Interested Parties	Needs and Expectations
Customers	Quality; price; delivery performance of products
Owners/shareholders	Sustained profitability; transparency
People in the organization	Good work environment; job security; recognition and reward
Suppliers and partners	Mutual benefits and continuity
Society	Environmental protection; ethical behaviour; compliance with statutory and regulatory requirements
Competitors	Ethical behaviour; fair competition; zero ethical faults
Government Labor unions Regulators	Attractive employer; business continuity; compliance with statutory and regulatory requirements; energy efficiency; mutual benefits; on time payment of taxes and others fees; risk management; sustained profitability; transparency

Note: Adapted and upgraded from ISO [7] .

Human resources are the main subject of an organizations activity. Quality improvement and the efficiency of the organization activity depend greatly on the quality of human resources [25] . Top management support and commitment are thus essential for the initiation of the integration process, completed and subsequently main- tained. Managers consequently need to recognise that for the IMS to be implemented and maintained, they must continuously push it forward [26] . One interesting finding in research conducted by Alolayan [27] was the fairly strong correlation between top management commitment and the adoption of continuous improvement programs in the work organization. Sustainable management is the combination of management theory and the concept of sustainable development [28] , cited by Tsai and Chou [29] and according to Salomone [30] , a cultural shift is underway and the number of companies with more than one certification is constantly increasing. Many of them are advancing towards integration.

Within the past decade, the application of certification has spread from documenting quality standards to ad- ditional areas, including the management of occupational health and safety (OHS) [31] . To satisfy the require- ments of each standardized MSs like: EMS, QMS, and OHSMS organizations have to assure a lot of docu- mented procedures and other documentation, checking processes and associated records forms, among other several paperwork [32] . As stated by Zeng [33] major problems for enterprises to operate multiple parallel MSs include: it causes complexity of internal management, it lowers management efficiency, it incurs in cultural in- compatibility and it causes employee hostility and increases management costs. In a recent past, in Portugal and other countries, some companies have begun to integrate their individual MSs [34] . Figure 1 [9] presents the evolution on the number of certifications of integrated MSs (Quality, Environment, and OH & S) for 2007, 2010, and 2011 in Portugal.

IMSs and its certification are increasingly used by organizations namely to document and develop optimized conformance and business risk management, in a lean and sustainable way, in a variety of different management areas considering, in general, its internal and external context and the needs and expectations of interested par- ties, in particular. According to Jørgensen [37] the third and most ambitious level, the integration, concerns the creation of a culture of learning, focus on interested parties, continuous improvements, and synergies between the subject areas. This integration creates a sound basis for working towards a more sustainable MS.

Approach to Structuring a Methodology to Develop an IMS

The standards ISO 14001 [2] , ISO 9001 [3] and OHSAS 18001 [4] , and the identification of common areas and requirements versus correspondences between them allowed to structure from the existing individual MSs in organizations, a methodology to develop the integration of the EMS with others standardized.

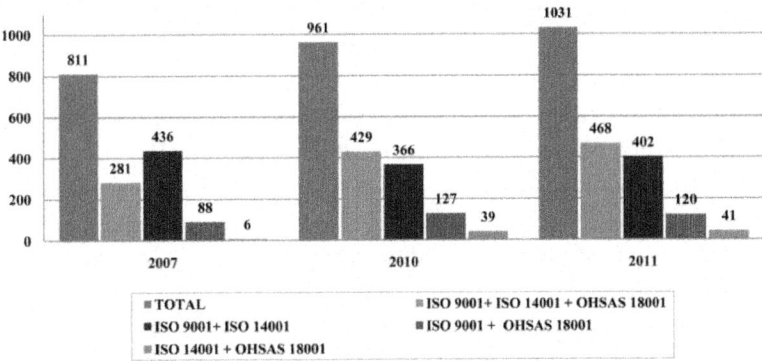

Figure 1: Evolution on the number of certified companies in Portugal related to inte-grated MSs (Quality, Environment, and OH & S) for 2007, 2010, and 2011. Source: Data from Portuguese Certified Companies Guide [35] [36].

MSs like QMS and OH & SMS. There is a convergence in terms of the management model of the standards: ISO 14001 [2] , ISO 9001 [3] , OHSAS 18001 [4] , specifically in terms of the objectives associated with process efficiency, and to the fact that the three MSSs are supported on the continuous improvement principle of the Deming's Cycle-PDCA (Plan/Do/Check/Act), as described in Table 2.

The ISO 14001 [2] specifies requirements for an EMS to enable an organization to develop and implement a policy and objectives which take into account legal requirements and other requirements to which organizations subscribe, and information about significant environmental aspects. It applies to those environmental aspects that organizations identify as those which can be controlled and those which it can be influenced; the ISO 9001 [3] specifies requirements for a QMS where an organization: a) needs to demonstrate its ability to consistently provide products that meet customer and applicable statutory and regulatory requirements; and b) aims to en- hance customer satisfaction through the effective application of the system, including processes for continual improvement of the system and the assurance of conformity to customer and applicable statutory and regulatory requirements and the OHSAS 18,001 [4] specifies requirements for an OH & SMS to enable an organization to develop and implement a policy and objectives which take into account legal requirements and information about OH & S risks. So, there are several MSSs domains with potential for the effective integration of EMS with QMS and OH&SMS and integration gives a true usefulness and added value to the organization's business, more easier manageable and securely enhances the improvement of conditions in organizations in terms of management, the prevention component of EMS; QMS and OH & SMS [11] . According to Santos [23] , the com- patibility among different MSSs for an effective integration of the individual standardized MSs should be done in moderation and Almeida [38] [39] states that the success of the integration of the MSs is significantly related to the true motivations that leads organizations to integration. To achieve sustained success, top management should establish and maintain a mission, a vision and values for the organization. These should be clearly under- stood, accepted and supported by people in the organization and, appropriate to other interested parties [7]

MATERIALS AND METHOD

A preliminary investigation was conducted in the business environment, in a company, localized in the northern region of Portugal. Over the years the company—a SME, has been progressively adopting, in whole or in part, individualized MSSs and others specifications to implement independent MSs. Relevance to the ISO 14001 [2] for the EMS; ISO 9001 [3] for the QMS; OHSAS 18001 [4] for the OH & SMS, and ISO/International Electrotech- nical Commission (IEC) 17025 [40] for Laboratories MS and Accreditation.

While it was imperative to assess the perception of employees of the Company on the methodology, structur- ing, implementation and evaluation of the integration model of IMS and its validation in a real business envi- ronment, it was developed an internal investigation by questionnaire, previously tested and validated, to a repre- sentative sample, of employees. Figure 2 shows the distribution of collaborators surveyed from the different le- vels of the organizational structure.

Number (percentage) of Collaborators by Hierarchical Level

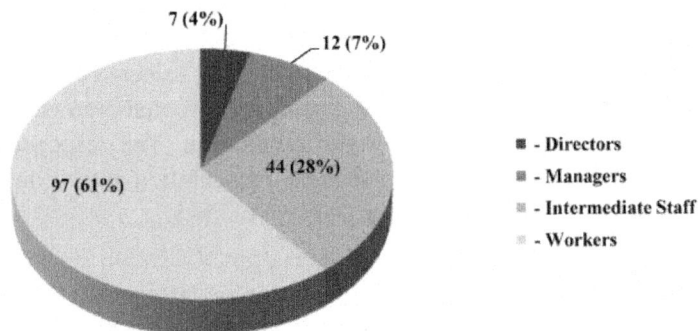

Figure 2: Distribution of collaborators by hierarchical level. Source: [15]

Table 2: PDCA (Plan/Do/Check/Act) convergence at the level of the management model by the correspondent MSSs-ISO 14001 [2] , ISO 9001 [3] , and OHSAS 18001 [4]

PDCA1 Cycle	Management Systems Standards		
	ISO 14001 [2]	ISO 9001 [3]	OHSAS 18001 [4]
PLAN To understanding the context of the organizations, to establish the necessary objectives and plans to achieve them, the processes, and their sequences and interconnections, as well as the required criteria and methods to guarantee their control and effectiveness	PLANNING Environmental targets and objectives; identification of environmental aspects and assessment of potential impacts; extent to which material and financial resources are affected; competence, training and awareness of the Human Resources; planning of the activities	RESOURCE MANAGEMENT Personnel competence, training and awareness; material resources; infrastructure and work environment	PLANNING OH & S objectives; hazard identification and risk assessment; extent to which material and financial resources are affected; training, awareness and competence of the Human Resources; communication, participation and consultation, planning of the activities
DO Implement the processes, ensuring the availability of support resources and appropriate documentation and the required documented information for realization and monitoring	IMPLEMENTATION AND OPERATION Environmental management programs; control of documents; operational control; emergency preparedness and response	PRODUCT AND/OR SERVICE REALIZATION Key processes; purchasing; service provision; measurement and monitoring equipment	IMPLEMENTATION AND OPERATION OH & S management programs; control of documents; operational control; emergency plans
CHECK Perform the measurements of the processes and monitoring their performance. Analyze and assess on a regular basis the obtained results	CHECKS AND CORRECTIVE ACTIONS Monitoring and measurement; compliance-nonconformities assessment, corrective and preventive actions; records control; internal audits	MEASUREMENT, ANALYSIS AND IMPROVEMENT Client and employee satisfaction; corrective actions for non-conformities; preventive actions; internal audits	CHECKS AND CORRECTIVE ACTIONS Monitoring and measurement; accidents, nonconformities and corrective and preventive actions; records control, internal audits
ACT Take the needed actions to meet the objectives, achieve the expected results and to encourage improvements	POLICY Environmental aspects; legal and other requirements; environmental policy	MANAGEMENT RESPONSIBILITY Quality policy Decentralized management; communication and information	POLICY Hazards and risks; legal and other requirements; OH & S policy
IMPROVE Continual improvement and innovation of MSs processes	Continual improvement and innovation of the MSs, with the main objectives of its optimization and satisfaction of all interested parties with focus on them for consequent development and sustained success of the organizations		

Note: Adapted and upgraded from [11] [13] .

It was considered a Likert scale on the questionnaire with five levels: 1) irrelevant; 2) not so relevant; 3) rele- vant; 4) ery relevant; 5) determinant. The population was the total of the collaborators of the company from whom the objective was to draw conclusions about the project objectives and the issues that are being re- searched. After the questionnaire had been tested and improved in some of its questions, it was sent by e-mail to each one the company collaborators of the sample that had been carefully selected according to their position in the hierarchy. The sample considered 49 collaborators, representing 30.62% of the total collaborators—the population. The responses rate was 86%. In the data collection, analyses and presentation were considered the guidelines of the Portuguese standard—NP 4463 [41] .

There were considered four main questions: Question 1—importance of the twelve factors identified as moti- vation for the implementation of the IMS; Question 2—influence of nine identified interested parties on the performance and evolution of the IMS; Question 3—main difficulties in a group of seven potentials, in the con- text of the development and implementation of the IMS model. Question 4—potential benefits with the imple- mentation of the IMS-QES.

The main final objective of this preliminary research was to contribute to the integration of EMS, QMS, and OH&SMS in a specific organizational context, supported on a structured methodology of development of the integration, implementation and evaluation of the designed integration model and its validation. The survey re- sults, by them self, justify, validate and prioritize enormously the structure of the designed methodology and model of IMS.

MATRIX OF COMPATIBILITY OF REQUIREMENTS-SUPPORT TO INTEGRATION

One of the activities that forms part of the scope and objectives of this preliminary research to which we have paid particular attention is the compatibility of the requirements of the MSSs, in context and framework of the characterization of the company's situation, backed up by an analysis of these MSSs. This compatibility, as pre- sented in Table 3, represent a starting point for consequents activities of integration, simplification and optimi- zation, to achieve a level of the strictly necessary and consequently the three sub MSs— EMS, QMS, and OH & SMS are integrated to the maximum extent possible.

On the matrix of Table 3 it is shown the requirements of the ISO 14001 [2] , ISO 9001 [3] , and OHSAS 18001 [4] , as well as the established correspondences, made them compatible with each other and associated with the phases of the cycle PDCA-"Plan-Do-Check-Act". This matrix orientate and align the organizational structure of the enterprise in the same direction and in addition creates a structured and useful referential methodology of work to support an effective alignment and correspondences of the sub management systems of Environment, Quality, and Safety with consequent compatibilities between each other, for consequent design and implementa- tion of the IMS. From this matrix it can also be depicted a correspondence with the PDCA cycle, in this circum- stance for the IMS, as well as a set of stages (1.1; 2.1...2.4; 3.1...3.7; 4.1...4.6 and 5.1) associated with each of the phases of the PDCA cycle.

MODEL OF DEVELOPMENT OF THE IMS

The continuous improvement of the global performance of an organization is an objective always present in the development of an IMS. The organization should therefore potentiate for each stage: Plan, Do, Check, Act, to be carefully and methodically analyzed in their differences that effectively can be observed in terms of MSSs re- quirements under clauses equivalent involved and for each phase and each stage of development of the IMS, according to the model of Figure 3 [11] [12] , to ensure its compliance and evidence of it, in full conformity.

Organizational diagnosis is an exercise done to check an organization's current health. A complete diagnosis not only checks the current health, but also suggests corrective measures [42] . So, first of all, the understanding of context of an organization, and definition, approval and communication of the integrated management policy, attentive that is common requirement to the different normative references. The leadership of top management is extremely important for the improvement of management quality. If the quality of top management is bad even if the management system of organization is good, it cannot continue to supply good outcomes such as products or services. As a result, it is thought that consequent profit is not ensured [25] . So, top management should demonstrate a strong commitment leadership and personal involvement through a defined Strategy, Policy, Ob- jectives and Targets for Quality, Environment and Safety, as well as, to make available all the needed resources to achieve the objectives [14] . The policy has to be coherent with the Mission, Vision, and Values of the com- pany, these supported on a strategy and specific objectives which in turn, support the implementation of that policy and its consequent effectiveness and continual improvement. The planning of activities in the aim of the Integrated Management System-Phase I—PLAN, is perhaps the most important of all [11] [12] . In fact, a neglected planning, including the non-identification of critical success factors, will lead to inefficiencies that can be translated into potential deviations to the objectives and consequent unsuccessful implementation of IMS.

Table 3: Matrix of compatibility of MSSs requirements and of support to the integration of the individual standardized MSs: EMS, QMS, and OH & SMS

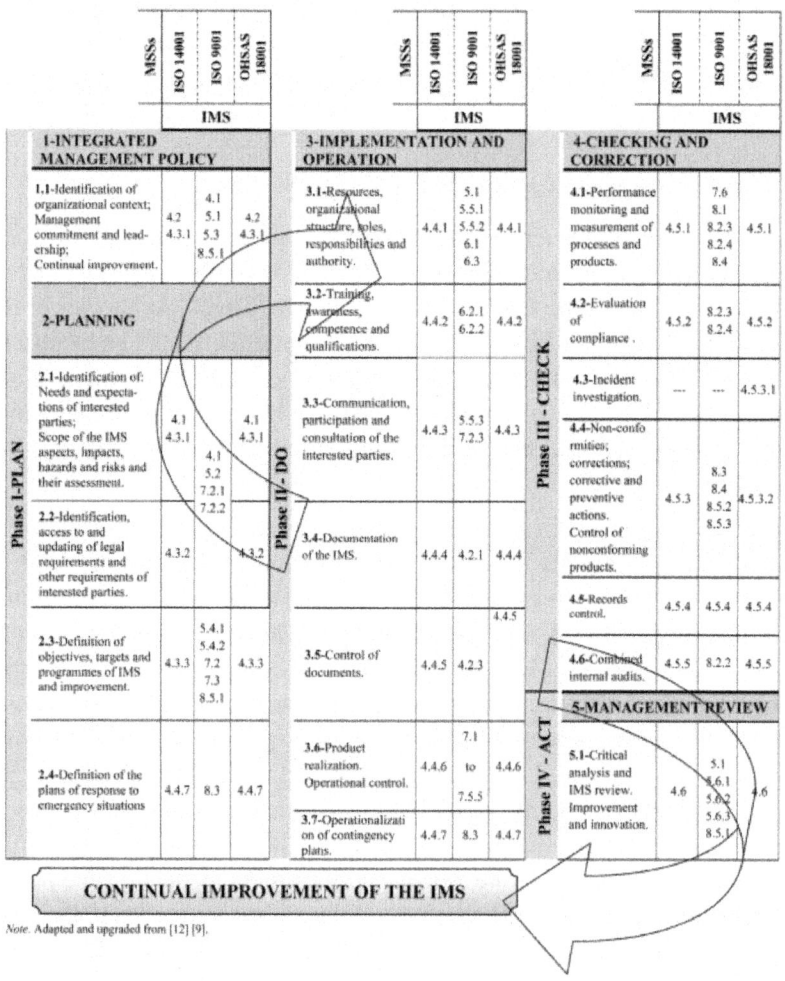

Phase I-PLAN

MSSs	ISO 14001	ISO 9001	OHSAS 18001
1-INTEGRATED MANAGEMENT POLICY		**IMS**	
1.1-Identification of organizational context; Management commitment and leadership; Continual improvement.	4.2 4.3.1	4.1 5.1 5.3 8.5.1	4.2 4.3.1
2-PLANNING			
2.1-Identification of: Needs and expectations of interested parties; Scope of the IMS aspects, impacts, hazards and risks and their assessment.	4.1 4.3.1	4.1 5.2 7.2.1 7.2.2	4.1 4.3.1
2.2-Identification, access to and updating of legal requirements and other requirements of interested parties.	4.3.2		4.3.2
2.3-Definition of objectives, targets and programmes of IMS and improvement.	4.3.3	5.4.1 5.4.2 7.2 7.3 8.5.1	4.3.3
2.4-Definition of the plans of response to emergency situations	4.4.7	8.3	4.4.7

Phase II - DO

MSSs	ISO 14001	ISO 9001	OHSAS 18001
3-IMPLEMENTATION AND OPERATION		**IMS**	
3.1-Resources, organizational structure, roles, responsibilities and authority.	4.4.1	5.1 5.5.1 5.5.2 6.1 6.3	4.4.1
3.2-Training, awareness, competence and qualifications.	4.4.2	6.2.1 6.2.2	4.4.2
3.3-Communication, participation and consultation of the interested parties.	4.4.3	5.5.3 7.2.3	4.4.3
3.4-Documentation of the IMS.	4.4.4	4.2.1	4.4.4
			4.4.5
3.5-Control of documents.	4.4.5	4.2.3	
3.6-Product realization. Operational control.	4.4.6	7.1 to 7.5.5	4.4.6
3.7-Operationalization of contingency plans.	4.4.7	8.3	4.4.7

Phase III - CHECK

MSSs	ISO 14001	ISO 9001	OHSAS 18001
4-CHECKING AND CORRECTION		**IMS**	
4.1-Performance monitoring and measurement of processes and products.	4.5.1	7.6 8.1 8.2.3 8.2.4 8.4	4.5.1
4.2-Evaluation of compliance.	4.5.2	8.2.3 8.2.4	4.5.2
4.3-Incident investigation.	---	---	4.5.3.1
4.4-Non-conformities; corrections; corrective and preventive actions. Control of nonconforming products.	4.5.3	8.3 8.4 8.5.2 8.5.3	4.5.3.2
4.5-Records control.	4.5.4	4.5.4	4.5.4
4.6-Combined internal audits.	4.5.5	8.2.2	4.5.5

Phase IV - ACT

MSSs	ISO 14001	ISO 9001	OHSAS 18001
5-MANAGEMENT REVIEW			
5.1-Critical analysis and IMS review. Improvement and innovation.	4.6	5.1 5.6.1 5.6.2 5.6.3 8.5.1	4.6

CONTINUAL IMPROVEMENT OF THE IMS

Note. Adapted and upgraded from [12] [9].

It is therefore fundamental to invest resources and expertise at this stage, via a thorough and careful work, in order to respond effectively, within an integrated approach, to all interested parties requirements arising from the in- volved standards and others applicable specifications, including legal requirements, in this phase of the planning of the IMS with particular focus on environmental issues, customer satisfaction, and occupational health and safety of the collaborators and their families [11] [12] . Following is the Implementation and Operation—"Do", the organization should, in this Phase II—DO, pro- mote the "Make/Do" in coherence with what was previously planned, attentive the scope of the IMS. Corresponds mainly to clauses: 4.4—Implementation and operation of ISO 14001 [2] , 7—Product Realization, of ISO 9001 [3] , and 4.4—Implementation and operation of OHSAS 18001 [4] .

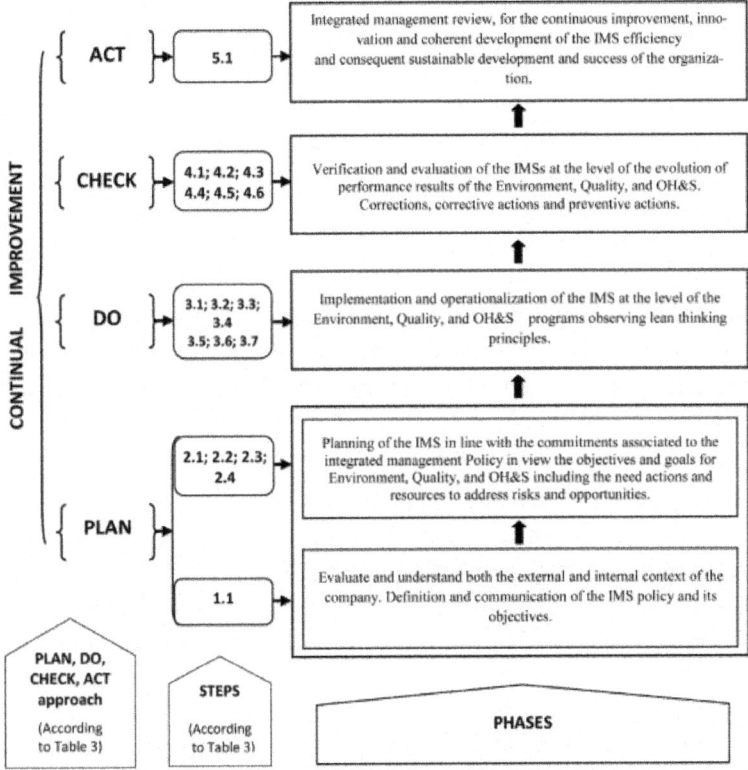

Figure 3: Model of development of the IMS Source: Adapted and upgraded from [11] [12] .

In the case of ISO 9001 [3] it should be consid- ered associated with product realization, other complementary clauses, particularly in the context of resource al- location (6.1, 6.2, 6.3, 6.4) and management commitment (5.1, 5.5.1). Different activities of the organization should be set out as processes and put into operation, as per the applicable requirements, in order to strictly comply with the policies and instructions, whether they are documented or not, such as, to ensure that the or- ganization's objectives and targets are achieved and the different Stakeholders are satisfied [14] . In the Phase III—CHECK, are considered six steps (4.1 to 4.6) designed to meet requirements of clauses: 4.5—Checking of the ISO 14001 [2] , 8—Measurement, analysis and improvement of ISO 9001 [3] , and 4.5—Checking of OHSAS 18001 [4] . With the exception of step 4.3—Investigation of incidents resulting from a specific sub-section, the 4.5.3.1—Incident investigation, the OHSAS 18001 [4] has no correspondence in the ISO 9001 [3] and ISO 14001 [2] . Critical information in the aim of the IMS should be identified, collected and analysed. Consistent with the integrated policy and the commitment to compliance, organization should estab- lish, implement and maintain documented records to demonstrate the improvements and compliance with all the applicable internal and external requirements. The use of KPIs—key process indicators, to monitor the processes, their control and continual improvement should be made systematically and guaranteed by the process owners, through the active involvement and participation of collaborators [14] . Particular relevance for the internal and external combined audits of the different components and areas of the IMS, and suppliers and subcontractors. Audits shall be scheduled at a frequency that takes into account the risk of the business. It must be conducted to assess the level of implementation and compliance of the IMS, its evolution, effectiveness and potentiate the identification of the necessary corrections and opportunities for improvement, which must be listed and priori- tized on its evaluation and implementation [14] [15] .

Finally, in the Phase IV—ACT, it was identified step 5.1—Critical analysis and review of the integrated management system, which refers to requirements of clauses: 4.6—Management review of ISO 14001 [2] , 5.6— Management review of ISO 9001 [3] , and 4.6—Management review of OHSAS 18001 [4] . Top management should ensure that processes of assessment, improvement and innovation on the different components of the management system. Management reviews should be conducted at planned intervals, to identify opportunities for improvements based namely in the lean philosophy, to assess the need for modifications to the IMS and to the integrated management policy, to ensure that it continues to be appropriate, suitable, effective and efficient [14] . Top management should review the integrated management system, defining the review inputs according to the requirements of each MSSs. Management review requires, in itself, a very careful preparation phase, par- ticularly, to the level of various information that supports the inputs [11] [12] . Records of the management re- view should be retained [2] - [4] .

DISCUSSION AND CONCLUSIONS

Management systems standards have developed in an unprecedented manner in the last few years. There exist at least, one for each interested part and new ones are going to be published. There is no international ISO standard with a specific structural model for Integrated Management Systems for Quality, Environment and Safety, or for other management areas such as: Risk Management; Information Security Management; RDI Management; and Social Responsibility Management, among others. The impact generated by environmental, quality, and safety and other MSSs is demonstrated by the importance of such standards worldwide, ISO 14001 [2] , ISO 9001 [3] , and OSHAS 18001 [4] . Organizations have to understand its context, the needs and expectations of interested parties and associated requirements, and consequently to determine the scope of the integrated management sys- tem, and formalise a policy to be communicated to internal and external interested parties.

The continuous improvement of global performance of organizations must be always a present goal in a per- spective of sustainability [43] . The development methodology to integrate various sub-management systems supported on a model of integration of the EMS, QMS, and OH & SMS at organizations should therefore poten- tiate, for each phase: Plan, Do, Check, Act, a careful and methodical analysis of the differences that effectively are observed at the level of normative requirements under the equivalent clauses and for each step of their de- velopment as the advocated model of integration.

Making MSSs requirements compatible through the analysis of their similarities promotes integration and can be depicted from the compatibility matrix presented—ISO14001 [2] , ISO 9001 [3] , and OHSAS 18001 [4] , es- tablishing correspondences, matching them with each other and associate the following phases of the PDCA cy- cle—Plan, Do, Check, Act: Policy and principles; Planning, Implementation and Operation, Performance Eva- luation, Improvement, Management Review. This is one of the activities that in the aim and objectives of the in- tegration model was given special attention in context of characterization and framework of the situation diag- nosed in the company in which the research was conducted. That compatibilization constitutes, the starting point for subsequent activities of integration, simplification and optimization, to a level of the strictly necessary and consequent integration maximized as desired of the three individual MSs—EMS, QMS, and OH & SMS, in con- text of strong competitiveness and changing of the business environments, which occurs at an accelerated and turbulent manner, enhancing also the sustained development of the business with added value for the relevant interested parts.

The integration of the three individual MSs represent added value both in the present and, fundamentally, for the future, not only for the company, as well as for a whole range of interested parties [43] . Examples are also highlighted by the surveyed respondents: the elimination of conflicts between individual systems with optimiza- tion of Resources; the improvement at the level of the coordinated and integrated management of the risk asso- ciated to the safety of the persons and company assets, environment and quality of the products; the reduction on the number of internal and/or external audits and audits to suppliers, and spent time versus associated costs; the creation of added value for the business through the elimination of several types of waste.

From the statistical analyses, resulting from the responses to the survey, there are shown a set of conclusions that by them self reveal: the importance, presently and for the future, of various "motivating factors" that were evaluated and alone justify and validate the model of implementation of the integration of the EMS, QMS, and OH & SMS in the company, either from internal aspects such as rationalization and optimization of resources, reduction of costs and bureaucracy; otherwise from external aspects, such as increasing competitiveness, to sat- isfy the growing demands of customers and others stakeholders. There were also identified a relevant number of difficulties, as well as a range of potential benefits resulting from the integration of the EMS, QMS, and OH & SMS into an IMS.

One of the major problems that organizations are facing with the integration of several MSSs is regarding the conception of a methodology and the implementation of an adequate structure of IMS to overcome the problems resulting from multiple MSSs. The continuous improvement of the global performance of organizations must be always a present goal in a perspective of sustainability and the route that should be taken by maximizing the in- tegration of the several individual MSs [13] , supported in an model of IMS, flexible, integrator and lean [15] .

ACKNOWLEDGEMENTS

This work had the financial support of the Portuguese Foundation for the Science and Technology (FCT) through the Strategic Project-UI 4005-2014, Project Reference PEst-OE/EME/UI4005/2014.

REFERENCES

1.   International Organization for Standardization (ISO) (2008) The Integrated Use of Management System Standards. In- ternational Organization for Standardization, Geneva.

2.   International Organization for Standardization (ISO) (2004) ISO 14001: Environmental Management Systems. Re- quirements with Guidance for Use. 2nd Edition, ISO Copyright Office, Geneva.

3.   International Organization for Standardization (ISO) (2008) ISO 9001: Quality Management Systems—Requirements. 4th Edition, ISO Copyright Office, Geneva.

4.   British Standards Institution (BSI) (2007) BS OHSAS 18001: Occupational Health and Safety Management Systems—Requirements. 2nd Edition, BSI Limited, London.

5.   Karapetrovic, S. and Casadesus, M. (2009) Implementing Environmental with Other Standardized Management Sys- tems: Scope, Sequence, Time, and Integration. Journal of Cleaner Production, 17, 533-540. http://dx.doi.org/10.1016/j.jclepro.2008.09.006

6.   Disterer, G. (2013) ISO/IEC 27000, 27001 and 27002 for Information Security Management. Journal of Information Security, 4, 92-100.http://dx.doi.org/10.4236/jis.2013.42011

7.   International Organization for Standardization (ISO) (2009) ISO 9004: Managing for the Sustained Success of an Or- ganization—A Quality Management Approach. 3rd Edition, ISO Copyright Office, Geneva.

8.   Bernardo, M., Casadesus, M., Karapetrovic, S. and Heras, I. (2009) How Integrated Are Environmental, Quality, and Other Standardized Management Systems? An Empirical Study. Journal of Cleaner Production, 17, 742-750.http://dx.doi.org/10.1016/j.jclepro.2008.11.003

9.   Rebelo, M.F., Santos, G. and Silva, R. (2014) Integration of Individualized Management Systems (MSs) as an Aggre- gating Factor of Sustainable Value for Organizations: An Overview through a Review of the Literature. Journal of Modern Accounting and Auditing, 10, 357-384.

10.   Khanna, H.S., Laroiya, S.C. and Sharma, D.D. (2010) Integrated Management Systems in Indian Manufacturing Or- ganizations: Some Key Findings from an Empirical Study. The TQM Journal, 22, 670-686. http://dx.doi.org/10.1108/17542731011085339

11.   Rebelo, M.F. (2011) Contribution to the Structuring of a Model of Integrated Management System QES. Master Thesis, Polytechnic Institute of Cávado and Ave, Barcelos.

12.   Santos, G., Rebelo, M.F., Barros, S. and Pereira, M. (2012) Certification and Integration of Environment with Quality and Safety—A Path to Sustained Success. In: Curkovic, S., Ed., Sustainable Development—Authoritative and Leading Edge Content for Environmental Management, InTech, Croatia, 193-218. http://dx.doi.org/10.5772/48414

13.   Rebelo, M.F. and Silva, R.G. (2012) Integration of Individual Management Systems—An Organizational Pillar for the Competitiveness and the Sustainability of Business. The International Conference on Innovation

for Sustainability—IS 2012, Porto, 27-28 September 2012, 1-12.

14. Rebelo, M.F., Santos, G. and Silva, R. (2014) A Generic Model for Integration of Quality, Environment and Safety Management Systems. TQM Journal, 26, 143-159.http://dx.doi.org/10.1108/TQM-08-2012-0055

15. Rebelo, M.F., Santos, G. and Silva, R. (2014) Conception of a Flexible Integrator and Lean Model for Integrated Management Systems. Total Quality Management & Business Excellence, 25, 683-701. http://dx.doi.org/10.1080/14783363.2013.835616

16. Simon, A., Bernardo, M., Karapetrovic, S. and Casadesus, M. (2011) Integration of Standardized Environmental and Quality Management Systems Audits. Journal of Cleaner Production, 19, 2057-2065. http://dx.doi.org/10.1016/j.jclepro.2011.06.028

17. Olaru, M., Maier, D., Nicoara, D. and Maier, A. (2014) Establishing the Basis for Development of an Organization by Adopting the Integrated Management Systems: Comparative Study of Various Models and Concepts of Integration. Procedia-Social and Behavioral Sciences, 109, 693-697. http://dx.doi.org/10.1016/j.sbspro.2013.12.531

18. Asif, M., Bruijn, E.J., Fisscher, O.A.M. and Searcy, C. (2010) Meta-Management of Integration of Management Sys- tems. The TQM Journal, 22, 570-582.http://dx.doi.org/10.1108/17542731011085285

19. Nowicki, P. (2013) Risk Management—An Important Issue in Quality Management Systems. The 7th International Quality Conference, Kragujevac, 24 May 2013, 267-272.

20. Suditu, C. (2007) Positive and Negative Aspects Regarding the Implementation of an Integrated Quality-Environ- mental-Health, and Safety Management System. Annals of the Oradea University, Fascicle of Management and Tech- nological Engineering, 6, 2013-2017.

21. De Oliveira, O.J. (2013) Guidelines for the Integration of Certifiable Management Systems in Industrial Companies. Journal of Cleaner Production, 57, 124-133.http://dx.doi.org/10.1016/j.jclepro.2013.06.037

22. Simon, A., Bernardo, M., Karapetrovic, S. and Casadesus, M. (2013) Implementing Integrated Management Systems in Chemical Firms. Total Quality Management and Business Excellence, 24, 294-309. http://dx.doi.org/10.1080/14783363.2012.669560

23. Santos, G., Mendes, F. and Barbosa, J. (2011) Certification and Integration of Management Systems: The Experience of Portuguese Small and Medium Enterprises. Journal of Cleaner Production, 19, 1965-1974. http://dx.doi.org/10.1016/j.jclepro.2011.06.017

24. Vinodkumar, M.N. and Bhasi, M. (2011) A Study on the Impact of Management System Certification on Safety Management. Safety Science, 49, 498-507.http://dx.doi.org/10.1016/j.ssci.2010.11.009

25. Esaki, K. (2013) General Frame Work of New TQM Based on the ISO/ IEC25000 Series of Standard. Intelligent Information Management, 5, 126-135.http://dx.doi.org/10.4236/iim.2013.54013

26. Zeng, S.X., Xie, X.M., Tam, C.M. and Shen, L.Y. (2011) An Empirical Examination of Benefits from Implementing Integrated Management Systems (IMS). Total Quality Management and Business Excellence, 22, 173-186.http://dx.doi.org/10.1080/14783363.2010.530797

27. Alolayan, S., Hashmi, S., Yilbas, B. and Hamdy, H. (2013) An Empirical Evaluation of the ISO 9001 Quality Man- agement Systems for Certified Work Organizations in Kuwait as Benchmarked against Analogous Swedish Organiza- tions. Journal of Service Science and Management, 6, 80-95. http://dx.doi.org/10.4236/jssm.2013.61009

28. Daub, C.H. and Ergenzinger, R. (2005) Enabling Sustainable Management through a New Multi-Disciplinary Concept of Customer Satisfaction. European Journal of Marketing, 39, 998-1012. http://dx.doi.org/10.1108/03090560510610680

29. Tsai, W.H. and Chou, W.C. (2009) Selecting Management Systems for Sustainable Development in SMEs: A Novel Hybrid Model Based on DEMATEL, ANP, and ZOGP. Expert Systems with Applications, 36, 1444-1458.http://dx.doi.org/10.1016/j.eswa.2007.11.058

30. Salomone, R. (2008) Integrated Management Systems: Experiences in Italian Organizations. Journal of Cleaner Production, 16, 1786-1806. http://dx.doi.org/10.1016/j.jclepro.2007.12.003

31. Granerud, L. and Rocha, R.S. (2011) Organisational Learning and Continuous Improvement of Health and Safety in Certified Manufacturers. Safety Science, 49, 1030-1039. http://dx.doi.org/10.1016/j.ssci.2011.01.009

32. Karapetrovic, S. and Jonker, J. (2003) Integration of Standardized Management Systems: Searching for a Recipe and Ingredients. Total Quality Management & Business Excellence, 14, 451-459. http://dx.doi.org/10.1080/1478336032000047264

33. Zeng, S.X., Shi, J.J. and Lou, G.X. (2007) A Synergetic Model for Implementing an Integrated Management System: An Empirical Study in China. Journal of Cleaner Production, 15, 1760-1767. http://dx.doi.org/10.1016/j.jclepro.2006.03.007

34. Santos, G., Barros, S., Mendes, F. and Lopes, N. (2013) The Main

Benefits Associated with Health and Safety Man- agement Systems Certification in Portuguese Small and Medium Enterprises Post Quality Management System Certi- fication. Safety Science, 51, 29-36.http:// dx.doi.org/10.1016/j.ssci.2012.06.014

35. Cempalavras, Comunicação Empresarial (2012) Portuguese Certified Companies Guide. 7th Edition. http://www.cempalavras.pt/GEC_2012/ EN/index.html

36. Cempalavras, Comunicação Empresarial (2013) Portuguese Certified Companies Guide. 8th Edition. http://www.cempalavras.pt/GEC_2013/ EN/index.html

37. Jørgensen, T.H. (2008) Towards More Sustainable Management Systems: Through Life Cycle Management and Integration. Journal of Cleaner Production, 16, 1071-1080.http://dx.doi.org/10.1016/j.jclepro.2007.06.006

38. Almeida, J., Sampaio, P. and Santos, G. (2012) Integrated Management Systems—Quality, Environment, and Health and Safety: Motivations, Benefits, Difficulties, and Critical Success Factors. Book of Abstracts, The International Symposium on Occupational Safety and Hygiene— SHO 2012, Guimarães, 9-10 March 2012, 13-15.

39. Almeida, J., Domingues, P. and Sampaio, P. (2014) Different Perspectives on Management Systems Integration. Total Quality Management and Business Excellence, 25, 338-351.http://dx.doi.org/10.1080/14783363.2 013.867098

40. International Organization for Standardization/International Electrotechnical Commission [ISO/IEC] (2005) ISO/IEC 17025: General Requirements for the Competence of Testing and Calibration Laboratories. ISO Copyright Office, Geneva.

41. Instituto Português da Qualidade (2009) NP 4463: Linhas de orientação sobre técnicas estatísticas para a ISO 9001: 2000 (ISO/TR 10017:2003). Instituto Português da Qualidade, Caparica.

42. Saeed, B. and Wang, W. (2014) Sustainability Embedded Organizational Diagnostic Model. Modern Economy, 5, 424- 431. http://dx.doi.org/10.4236/me.2014.54041

43. Rebelo, M.F. and Santos, G. (2012) Integration of the Occupational Health and Safety Management System with the Quality Management System and Environmental Management System—From the Theory to the Action. Book of Abstracts, The International Symposium on Occupational Safety and Hygiene—SHO 2012, Guimarães, 9-10 March 2012, 372-374.

Chapter 2

THE ENVIRONMENTAL AND ECONOMIC SUSTAINABILITY OF CARBON CAPTURE AND STORAGE

Paul E. Hardisty [1,2], Mayuran Sivapalan [3] and Peter Brooks [4]

[1] WorleyParsons EcoNomics™, Perth, WA 6000, Australia

[2] Imperial College London, London SW7 2AZ, UK

[3] WorleyParsons EcoNomics™, Houston, TX 77079, USA

[4] WorleyParsons, Brisbane, QLD 7000, Australia

ABSTRACT

For carbon capture and storage (CCS) to be a truly effective option in our efforts to mitigate climate change, it must be sustainable. That means that CCS must deliver consistent environmental and social benefits which exceed its costs of capital, energy and operation; it must be protective of the environment and human health over the long term; and it must be suitable for deployment on a significant scale. CCS is one of the more expensive and technically challenging carbon emissions abatement options available, and CCS must first and foremost be considered in the context of the other things that can be done to reduce emissions, as a part of an overall optimally efficient, sustainable and economic mitigation plan. This elevates the analysis beyond a simple comparison of the cost per tonne of CO_2 abated—there are inherent tradeoffs with a range of other factors (such as water, NO_x, SO_x, biodiversity, energy, and human health and safety, among others) which must also be considered if we are to achieve truly sustainable mitigation. The full life-cycle cost of CCS must be considered in the context of the overall social, environmental and economic benefits which it creates, and the costs associated with environmental and social risks it presents. Such analysis reveals that all CCS is not created equal. There is a wide range of technological options available which can be used in a variety of industries and applications—indeed CCS is not applicable to every industry. Stationary fossil-fuel powered energy and large scale petroleum industry operations are two examples of industries which could benefit from CCS. Capturing and geo-sequestering CO_2 entrained in natural gas can be economic and sustainable at relatively low carbon prices, and in many jurisdictions makes financial sense for operators to deploy now, if suitable secure disposal reservoirs are available close by. Retrofitting existing coal-fired power plants, however, is more expensive and technically challenging, and the economic sustainability of post-combustion capture retrofit needs to be compared on a portfolio basis to the relative overall net benefit of CCS on new-build plants, where energy efficiency can be optimised as a first step, and locations can be selected with sequestration sites in mind. Examples from the natural gas processing, liquefied natural gas (LNG), and coal-fired power generation sectors, illustrate that there is currently a wide range of financial costs for CCS, depending on how and where it is applied, but equally, environmental and social benefits of emissions reduction can be considerable. Some CCS applications are far more economic and sustainable than others. CCS must be considered in the context of the other things that a business can do to eliminate emissions, such as far-reaching efforts to improve energy efficiency.

INTRODUCTION

Background—Technology, Policy, Economics

Carbon capture and storage (CCS) is one of a host of technical solutions that are currently available for reducing global emissions of greenhouse gases (GHG) to the atmosphere, and thus curb the longer term effects of anthropogenic climate change. However, the barriers to solving the world's climate change challenge are not, in the main, technical. There is now widespread scientific consensus on the major causes of climate change [1], the overall planetary risks of inaction, and even on the combinations of measures which will be required to deliver the emissions reductions required to eliminate the worst of the future risks posed by an earth system in flux [2]. Reductions in emissions will be required in a wide range of sectors, including from land-use changes (forest clearing and agricultural practices), building design and operation, transport, and notably electrical power generation [3]. Research and technical development has been underway in all of these areas for many years now, and a wide range of technically viable and workable solutions already exists in each of these sectors. Indeed, the broad scientific and policy-making community has been aware of the risks of climate change, and the solutions available, for well over two decades [4]. However, during that same period, on a global scale, the tangible effects of actions taken to reduce emissions of greenhouse gases (GHG) have been minimal. Not only has the world dramatically increased its consumption of all kinds of energy over the last thirty-five years, our energy mix has also remained essentially unchanged [5]. The proportion of overall energy requirements provided by fossil fuels has actually increased over this period, despite full and growing knowledge of the risks posed by climate change. On a global scale, we have, in effect, done little so far to combat it. Why? Climate change is a global issue, and does not respect national boundaries. Nor does it care about public opinion. We understand the science of the problem, at least in overall terms, if not in detail. We have the technology now to significantly reduce emissions. What has been lacking, so far, is the political, social, and economic will to deploy these existing technologies at scale. Action has been slow and insufficient because we do not want to pay the price of action. So we wait for the cost of abatement to drop, hoping that a new low-cost technology will be developed which can do the job. But there is a fundamental difference between the value of something and its price. Dealing with climate change may come at a relatively high price. But what matters are the benefits that result from that expenditure, not only to the emitters, but to society as a whole.

The Economics of Climate Change

Recently, there have been several attempts to quantify a global perspective on the economics of action and inaction on climate change. These studies have considered the costs of action to reduce emissions, and the economic costs that would result if climate change is allowed to take its course unhindered. A summary of the results of recent studies is provided in Table 1.

Table 1: Costs and benefits of acting to mitigate climate change

Study	Costs of Action	Benefits of Action	Comment
Stern (2006) [6]	+1 to −1% Global Product to 2030; approximately USD 250 bn/year over this period	Prevention of the loss of 5 to 20% of Global Product now and forever	Costs to the global economy of inaction are permanent, irreversible, and catastrophic
Garnaut (2008) [7]	−0.8% initial change in annual Australian Gross National Product (GNP), followed by −0.2 to −0.0% change in annual Australian GNP to 2050	0.0 to 0.2% increase in annual Australian GNP in second half of this century.	From 2050, mitigation adds to the growth rate of the Australian economy, as, at the margin, more new climate change damages are avoided than new mitigation costs added. By the end of the century, Australian GNP is higher than it would have been without mitigation.
IEA and OECD (2009) [8]	USD 4.1 tn over the next 20 years	Savings in fuel costs alone of over USD 7 tn; stabilisation of GHG concentrations in the atmosphere above 550 ppm CO_2e	Significant expenditure on R&D expected to bring overall cost of stabilisation down; CCS to play key role in overall mitigation globally

These studies largely focus on the benefits (as damage avoided) to the man-made economy of climate change mitigation. Associated benefits from the elimination of other types of air-emissions that would occur as part of GHG reductions (lower emissions of particulates, soot, NOx and SOx from power plants, for instance), and the value of benefits of biodiversity and ecosystem protection, for instance, are rarely considered, but could be significant.

Nevertheless, it is clear, from a macro-economic perspective, that the world is far better off with concerted action to achieve deep cuts in GHG emissions than it would be otherwise. However, our global economic system is not set up to either measure or reward firms (or individual countries) for acting in the common good. Firms of all kinds must seek to maximise profit so that they can remain in business and deliver shareholder returns. Society, and the governments that represent them, must therefore regulate the activities of the market to achieve desired social outcomes. Hence, widespread discussions are currently underway worldwide to put in place mechanisms which will effectively put a price on carbon emissions to the atmosphere.

Reducing emissions from electrical power generation is one of the most important steps than can be taken in an overall GHG mitigation effort. Electricity production contributes approximately 25% of the total of direct man-made GHG emissions today [9]. The widespread adoption of coal-fired power, especially in the rapidly developing economies of China and India, is predicted to significantly increase the overall emissions from this sector over the next twenty years. Given coal's abundance and low cost, it is unlikely that its use can be radically curtailed any time soon, no matter how quickly climate change affects the planet, or how rapidly the world responds.

On this basis, there is now significant agreement among policymakers in many countries that carbon capture and geo-sequestration (CCS) has a vital role to play in the overall efforts to reduce GHG emissions worldwide. In particular, our ability to retrofit existing coal-fired power plants, and to retro-fit other types of high-emission facilities with post-combustion capture, will be essential if we are to meet desired atmospheric stabilisation targets. The problem is that capture and geologic storage of CO_2 is generally considered to be expensive.

CCS IN PERSPECTIVE

CCS has been widely identified as a significant potential contributor to global strategies aimed at reducing emissions of GHG to the atmosphere. Much of the focus on CCS to date has been in the area of government funded research and development, both in terms of capture technology, and in studying the long term fate and mobility of CO_2 in various subsurface environments. In particular, CCS has been seen as a way to significantly reduce the GHG impacts of the widespread global use of coal for electrical power generation. On this basis, governments in several developed nations (such as the USA, Canada, Australia and some in Europe) continue to fund a range of demonstration projects designed to prove the technology, develop operational experience, and spearhead the drive to cost-efficiency. The technical feasibility of each of the individual components of CCS (capture, transport and geological sequestration) is well understood. In the gas industry, amine systems for removing entrained CO_2 in raw gas have been widely used for years. The basic technology is mature and robust. Equally, the transport of CO_2 via pipeline is well understood. Thousands of kilometres of CO_2 pipeline systems have been laid and operated, much of it associated with dedicated enhanced oil recovery operations. The geo-sequestration element of CCS is the least well-developed of the three components, but nevertheless the petroleum and waste management industries have decades of experience in injecting fluids of all types into geological formations for long-term storage. Nevertheless, there continues to be significant public opposition and concern about the risks associated with long term CO_2 leakage from storage sites [10]. It is perhaps in the combination of all of these elements into a fully-integrated project that the main challenges for CCS arise.

Globally, there are 62 active or planned commercial scale integrated CCS projects, comprising capture, transport and sequestration elements, sequestering over 1 Mtpa CO_2 [11]. Of these, however, only seven projects are currently in the operational stage; the remainder are in the evaluation, definition, or execution stages. To date, it has been in the petroleum industry that much of this full-scale operational application of CCS has occurred: six of the projects are at natural gas processing facilities and, of those, two are offshore. As will be discussed below, much of the reason for the leadership of the gas sector in CCS is that the marginal cost of applying CCS in this sector is generally significantly lower than in other sectors, particularly coal-fired power generation [12].

CCS ECONOMICS

The overall life-cycle environmental, social and economic sustainability of CCS is examined through considering three different applications: managing the CO_2 entrained in reservoir gas in the natural gas sector; retro-fit of CCS to stationary fixed coal-fired power generation; and reducing the GHG footprint of a liquefied natural gas plant (LNG). In these examples, the Environmental and Economic Sustainability Assessment (EESA) method is used, in which various options are considered by not only examining conventional financial costs of abatement, but also explicitly valuing the environmental and social externalities affected by each option [13]. While carbon emissions to atmosphere are clearly the major externality, other external costs and benefits also exist. In this analysis, a sustainable and economic solution is one which generates more benefit than cost, to all stakeholders, when all environmental, social and economic factors are considered across the full life cycle.

Managing CO_2 Entrained in Natural Gas

The natural gas industry currently leads the commercial scale application of integrated CCS (>1 Mtpa) worldwide. Two of the largest and most successful projects have been offshore, both in Norway, where a significant and long-standing carbon pricing mechanism (since 1991) has helped to drive development of CCS.

The Snøvit project, led by Statoil, has been in production since October 2007 and currently produces approximately seven billion cubic metres of gas per year from an offshore field. CO_2 is removed from the gas stream and piped about 150 km back to the field for injection through a dedicated well. Since April 2008, around 0.7 Mtpa of CO_2 has been safely injected and stored in the Tubåen sandstone (some 2,600 metres beneath the seabed). A monitoring program has been set-up to investigate the behaviour of CO_2 underground.

The Statoil Sleipner project, also in the North Sea off the coast of Norway produces natural gas with about 9% CO_2, which is too high for customer requirements. By capturing some of the CO_2 from the reservoir gas, the CO2 level is reduced to 2.5% to meet export and customer specifications. The Sleipner capture and storage gas processing facility, operational since 1996, is one of the global pioneers of CCS. It is the world's first fully operational offshore gas field with CO_2 injection. It is also the world's first CO_2 storage project in a geological formation 1000 metres below the sea floor. Approximately 1 Mtpa of CO_2 is separated from produced gas and injected into a saline aquifer above the hydrocarbon reservoir zones. Maximum injection is planned for 20 Mt, with 8 Mt injected to date.

Recent reviews have shown that the cost of CCS for CO_2 entrained in the raw reservoir gas is low compared to other applications. The Global Carbon Capture and Storage Institute found that on average, CCS increased the cost of production for stationary power production by between 39% (for IGCC) and 78% (for supercritical pulverized coal plants). However, in natural gas processing, CCS increased cost of production on average by only 1% (the lowest of all sectors). CO_2 capture is inherent in the design of gas processing facilities where the reservoir gas contains carbon dioxide, so the marginal cost of CCS is only in the transport and sequestration components, which typically represent less than 20% of the total cost of CCS. Pipelining costs for a single facility are typically in the range of USD 3 to 4/t CO2 (depending of course on the distances involved), and geo-sequestration costs are in the order of USD 3 to 8/t CO_2. Initial reservoir identification and characterization costs are typically in the range of USD 25 m to 150 m. The overall costs of CCS are likely to come down in future as CCS is more widely deployed worldwide.

For an offshore natural gas facility operating in Australia with 12% CO_2 in reservoir gas, CCS was examined as part of a wider GHG management strategy. The availability of a well characterized down-dip part of the producing reservoir for geo-sequestration added significantly to the overall technical feasibility of CCS. The average cost of CCS at this facility was estimated to be in the order of USD 15 to 25/t CO_2 over the project life-cycle. Other recent studies have suggested that CCS for CO2 in gas processing ranges from about USD 18/t CO_2 to as much as USD 40/t CO_2. Given that the Australian Carbon Pollution Reduction Scheme (abandoned in 2008) was predicted to generate an effective carbon price of as much as USD 40/t CO2e by 2010, and that the current government is discussing the imposition of a carbon tax in the near future, and considering that the current social cost of carbon is likely in the range of USD 50 to 100/t CO_2e right now, CCS for this application may already be economic (from society's perspective), and is very likely to be financially advantageous for project proponents within the near to medium term (in this discussion, economic refers to the full environmental, social and financial perspective, while financial refers only to the costs and benefits to the project proponent).

Coal-Fired Power Station Retrofit

An existing coal-fired power station in Australia was examined in a detailed engineering, economic and sustainability feasibility study to determine the practicality of applying CCS to dramatically reduce GHG emissions. The 25-year old 425 MW facility currently produces about 2 Mtpa of CO_2. Two capture scenarios were investigated: A 5500 t_{pd} system which would capture the balance of emissions at typical operating loads, and a second, larger system (8200 t_{pd}) designed to capture 100% of flue gas when the plant's boilers are operating at maximum continuous rate. The analysis considered all aspects of the retrofit, including plant layout and access, capture technology selection, transport of CO_2, and identification of suitable disposal sites, within a context of what can be achieved today, with existing technology, knowledge and resources.

Capture

A wide range of currently available capture technologies was considered.

As shown in Table 2, not all are applicable to post-combustion capture, and not all are commercially available. On this basis, and because of the limited process information available for most of the other technologies, the study was based on a monoethanolamine capture technology, which is both commercially available through a number of vendors, and is fully applicable to the large-scale retrofit being considered. Installation of the CO_2 capture system and compression at the plant will require an area of about 3,000 m² for the smaller option, and 4,500 m² for the larger capacity option.

Table 2: Capture technologies considered

Technology	Technology Type	Application to PCC	Commercial Status
Monoethanolamine	Chemical solvent	Yes	Commercial
Chilled Ammonia	Chemical solvent	Yes	Pilot
KS Solvents	Chemical solvent	Yes	Pilot
Aqueous Ammonia	Chemical solvent	Yes	Pilot
Methyl Diethanolamine	Chemical solvent	No	Commercial
Diethanolamine	Chemical solvent	No	Commercial
Selexol	Physical solvent	No	Commercial
Sulfinol	Mixed chemical-physical solvent	No	Commercial

Performance

Installation of the capture system has a significant effect on the performance of the plant. The monoethanolamine system puts a significant additional energy demand on the plant. Table 3 exhibits one of the ironies of CCS in this application: to capture CO_2, significantly more coal must be burned. Nevertheless, option 2 reduces overall annual emissions from 2 Mt to about 0.25 Mt.

Table 3: Capture performance

Option	Emissions to Atmosphere (Mtpa CO_2e)	Net Power Output at Boiler MCR (MW)	Emissions Captured (Mtpa CO_2e)
Base case (Current operations)	2.0	425	0
Option 1 (5,500 tpd)	0.8	325	2.0
Option 2 (8,200 tpd)	0.25	275	2.7

Transport and Storage

A permitted and available geo-sequestration site was assumed to exist approximately 500 km from the power station. Transport of CO_2 overland by pipeline was assumed. Pipeline is the established method for moving large volumes of CO_2 over long distances. Most of the current expertise in CO_2 transport lies in the petroleum industry, where CO_2 is widely used for enhanced oil recovery. Over 4,000 km of dedicated CO_2 pipelines are currently in operation in the USA [14].

Two scenarios were considered: one where a dedicated pipeline transports 3 Mtpa of CO_2 to the sequestration site, and another where several operators share a larger 12 Mtpa transport and sequestration system and share the associated economies of scale. Compression power of 40 MW is provided at the plant.

Cost Analysis

The total capital outlay for this CCS retrofit was estimated to range between about USD 0.5 b$_n$ and USD 1 b$_n$, paid out over a seven-year construction period. Table 4 shows the range of estimated unit costs per tonne of CO_2 avoided for CCS retrofit, in present value terms over an assumed operations life of 25 years, using a 10% discount rate. The least preferred scenario included dedicated infrastructure, poor sequestration reservoir performance, and high estimates for capture costs, while the preferred scenario involved shared infrastructure (with unit costs approximately one-third lower than dedicated infrastructure), low-end estimates of capture costs and optimal reservoir performance. Of these aggregated unit costs, about 80% of the cost was for capture, 10% for transport and 10% for sequestration. It is also important to note that the highest uncertainty in cost was associated with the sequestration component. The range of unit cost estimates reflects the commercial and engineering uncertainty inherent in delivering a complete CCS project at the present time.

Table 4: Financial cost summary, CCS retrofit (millions 2008 USD/tCO2e)

	Option 1	Option 2
Least-preferred scenario	120	145
Preferred scenario	71	90

External Costs and Benefits

This analysis, for a real facility, using technology available now, shows that under present policy positions, retrofitting existing coal-fired power stations is a not financially viable proposition for operators. The main benefit of employing

CCS is to create and environmental and social benefit associated with reducing carbon emissions to the atmosphere. The value of this benefit is expressed as the social cost of carbon, or the real value of the damage caused to society by each additional tonne of GHG emitted to the atmosphere. Using high estimates of the social cost of carbon, such as Stern's USD 85/t CO_2e, this CCS retrofit would be marginally beneficial using option 1 of the preferred scenario. If the social cost of carbon were to rise over time (as it will if global emissions are not curbed), then preferred scenario retrofit becomes increasingly beneficial from society's perspective.

However, until an effective price on carbon exists, operators have no financial incentive to deploy CCS. In this example, carbon price would have to reach at least USD 75/t CO_2e, or the costs of CCS would have to drop dramatically, before this operator could justify a CCS retrofit. A much more advantageous approach for this operator in the near term would be to examine other alternatives to removing carbon emissions from its overall portfolio, where this can be achieved at lower cost. This might include examining new build plants using more efficient super-critical designs, and pre-combustion options located closer to disposal sites.

While carbon emission reduction is the chief benefit of CCS, removal of other air pollutants such as oxides of nitrogen and sulphur, particulates, and even heavy metals, may also occur as a result of capturing and treating effluents. The valuation of these additional atmospheric benefits is discussed in more detail in the following example.

In addition, there is also a range of potential external costs associated with deep geological disposal of CO_2. Risks associated with geo-sequestration of CO_2 include gas leakage to surface in populated areas (7% to 10% CO_2 in air is sufficient to cause immediate danger to human life and health), acidification of groundwater supplies, and geological instability [15]. Any of these eventualities could generate significant external social and environmental costs. The likelihood and magnitude of these risks will vary considerably depending on the geological conditions of the reservoir, location, nearby population density, and the vulnerability of nearby aquifers [16]. Proper sequestration site selection, design and monitoring can significantly reduce the risk of leakage and the severity of impact should leakage occur [17]. All of these external costs would make CCS more costly from an overall environmental, social and economic perspective, over the long term. Only if the social cost of carbon is sufficiently reflected in an effective price for carbon, and if it rises significantly over time, will the operator, in this example, be able to justify deploying CCS.

Reducing the GHG Emissions from LNG Manufacture

LNG operations release GHG emissions at various stages of production, shipping, re-gasification, storage and distribution, including consumption. This example focuses on CCS applied to managing CO_2 emissions from a typical LNG plant. Given the predominance of cost associated with carbon capture, this discussion focuses on the capture element only, and assumes that CO_2 can be readily disposed of into suitable geological formations close to the facility. Offshore disposal into a depleted natural gas field would also reduce concerns over external costs associated with possible long term leakage of CO_2 from the reservoir.

Technical Overview of Options

In this analysis, seven CCS cases were examined and compared in terms of effectiveness and cost, and put into a larger context by including four other ways of reducing GHG emissions from the facility. CCS options involve various combinations of pre-versus post-combustion capture, central power station and direct drive, and retrofit versus new build installation. Table 5 lists each case along with a median estimate of the total capital and operational costs for the facility as equipped. For retrofit options, these figures include lost revenue from down-time, and assume a retrofit date of 2021. As shown, the capital expenditure associated with each case varies considerably. The most expensive of the cases represents a capital cost increase of about 60% of the total installed cost of the business-as-usual reference facility design.

Table 5: GHG management cases and cost data (million US $ 2010)

Case	Description	CAPEX	Annual OPEX
Ref.	Standard LNG facility currently in operation, business-as-usual, no CCS fitted	4,400	160
1	Carbon capture from the reservoir gas only + CCS	4,500	160
2	Retrofit post-combustion capture (liquefaction), aero-derivative gas turbine drive + CCS	5,700	160
3	Retrofit natural gas combined cycle (GTCC) central power generation, post-combustion capture, electric motor compressor drives in liquefaction trains + CCS	6,000	250
4	Option 3, but greenfield	5,500	160
5	Retrofit local pre-combustion capture, aero-derivative gas turbine drive + CCS	6,000	160
6	Retrofit central GTCC, pre-combustion capture, electric motor drives in liquefaction trains + CCS	6,500	230
7	Option 6, but greenfield	6,100	160
8	Local combined cycle power, best-in-class energy efficiency, additional gas turbine waste heat recovery + part steam turbine direct compression drives	4,600	170
9	Central combined cycle power, best-in-class, electric motor compressor drives in liquefaction trains	4,800	160
10	Surplus power generation, exporting to grid, and	6,000	165
11	Buy power from de-carbonised grid.	5,200	160

Figure 1 shows the annual CO_2 emissions of each case. Note that all of the CCS cases significantly reduce emissions and improve GHG emissions intensity performance. The reference plant is estimated to produce about 0.25 tCO_2/t LNG, whereas the CCS designs achieve in the range of 0.03–0.13 tCO2/t LNG. Syngas production and pre-combustion capture (Cases 5, 6 and 7) by steam methane reforming have the highest fuel gas use, highest CAPEX and a much poorer emission performance than the post-combustion capture cases (Cases 2, 3 and 4).

CO2 Emissions

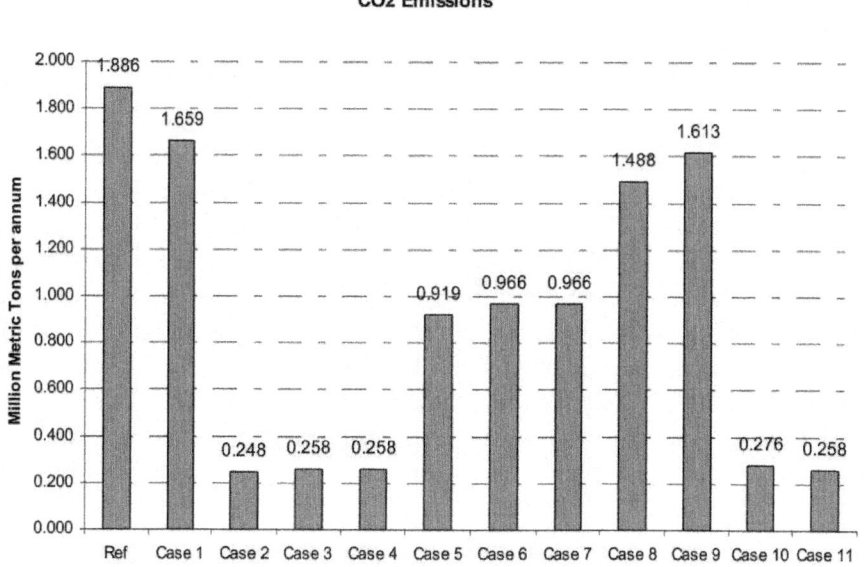

Figure 1: Annual CO_2 emissions of each case.

Life Cycle Environmental and Economic Sustainability Assessment (EESA)

Using the EESA methodology discussed above, the full life cycle environmental, social and economic sustainability of various CCS and non-CCS cases was evaluated, by considering all of the usual financial parameters (CAPEX, OPEX, energy costs), but also by including the value of various emissions to the atmosphere, notably CO_2, NOx and SOx. This type of analysis explicitly recognises the key issues associated with a project, and using the common metric of money, allows trade-offs between parameters to be examined across a wide range of possible future conditions. Cases which result in a net overall positive benefit to society as a whole (including the operator), within the limits of the study (in this case LNG production only, not including other parts of the LNG life-cycle) are deemed economic and sustainable. If a case produces less overall benefit to all stakeholders, over its life-cycle, than it costs, when all relevant environmental, social and financial aspects are considered, then it is uneconomic, and unsustainable (even if it is financially profitable to the operator).

Specific assumptions used in the analysis are listed in Table 6. For sensitivity analysis, high and low ranges are also provided for each key parameter. A base case value of USD 25/t CO_2e was selected For CO2, based on the long term average European Trading System price. The high estimate is based on the Stern Review, which set the social cost of carbon at approximately USD 85/t CO_2e. Nitrogen oxides (NO_x) and sulphur oxides (SO_x) emissions result in dis-benefits such as respiratory illness and acid rain. Markets for NO_x and SO_x emissions limit the volume of NO_x and SOx released and allocate the emissions in an economically efficient manner. The largest markets and auctions for these gases are in the USA. Recent EPA spot market prices for SOx and NOx emissions are USD 520/t and USD 640/t, respectively [18]. Clean Air Conservancy are offering offsets for SOx and NO_x at approximately USD 1653/t and USD 2646/t respectively [19]. Given the range of recognised values above, median values for SOx and NOx were chosen as USD 521/t and USD 637/t respectively. The analysis was conducted over a 40-year planning horizon, matching the estimated useful life of the facility. Retrofit was assumed to occur in 2021 in all cases. All values are in real 2010 dollars; inflation is not included.

Table 6: Assumptions for key parameters

Parameter	Low Value	Base Value	High Value
Discount rate	3.0%	3.5%	10%
Fuel gas price	USD 1/mmbtu	USD 3/mmbtu	USD 5/mmbtu
LNG price	-	USD 7.50/mmbtu	-
Power on-sale price (Opt 10)	USD 40/MWh	USD 50/MWh	USD 100/MWh
Power purchase price (Opt 11)	USD 50/MWh	USD 62.50/MWh	USD 125/MWh
Carbon cost	USD 0 /tCO₂e	USD 25/tCO₂e	USD 85/tCO₂e
SOx price	USD 0/t	USD 521/t	USD 1,860/t
NOx price	USD 0/t	USD 637/t	USD 2,360/t

Base Condition Results

The results of the base assessment are presented in Figure 2, with net present values (NPVs) for each option compared to the reference case. Where an option shows positive NPV, it performs that much better than the reference case, over the 40-year life cycle. The results cover the full economic analysis (inclusive of financial and social costs and benefits, and with transfer payments removed). This is a marginal assessment – the revenues generated by the sale of LNG are not included; only the differences associated with GHG management are considered. Options are listed in order of increasing CAPEX from left to right.

Under base conditions, only two of the six options were more economic than the reference case—Case 8 (best-in-class energy efficiency option; NPV + USD 268 million) and Case 1 (CCS of CO_2 from the reservoir gas stream: NPV +USD 5 million). Case 1 achieves carbon reduction benefits with greater value than the financial costs of achieving them. Case 8 also achieves significant carbon reduction benefits, though the majority of the benefits for this case are from fuel cost savings from more efficient plant design; the combination of these benefits far exceeds the financial costs of achieving them.

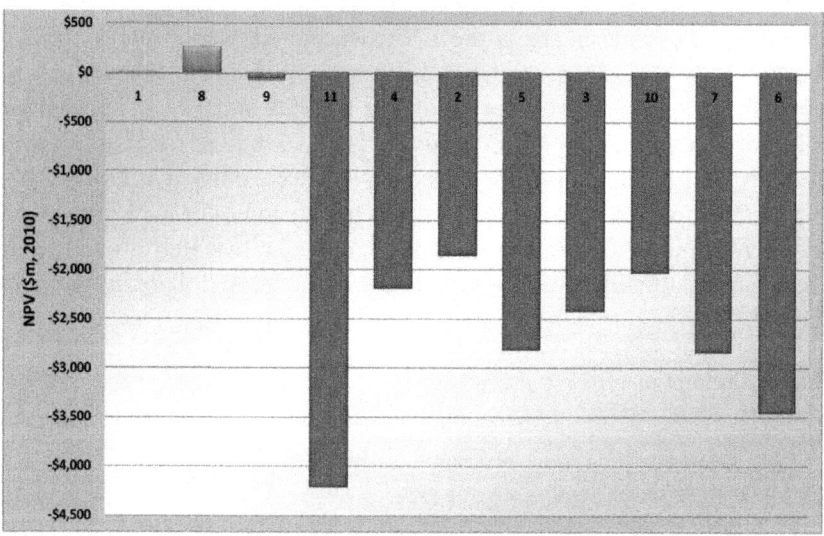

Figure 2: Base condition results—economic NPV compared to reference case (2010 USD m).

It should be stressed that the costs included for the CCS Cases 1–7 are for capture only—and do not include any of the sometimes significant costs for CO_2 transport and disposal. Equally, none of the potential external costs of sequestration are included (all of which would make CCS less attractive, from a full life-cycle environmental, social and economic perspective). Under base conditions, CCS is not economic, even using a low social discount rate and a reasonable median estimate for cost of carbon. In this case, the costs of implementation are simply too high to provide value for society—in fact, all

CCS cases, retro-fit and greenfield, are more than a billion dollars worse, in present value terms, than business-as-usual under base case conditions. This is before the financial and social costs of sequestration are included. The inference is that society should find other ways of reducing GHG emissions before CCS is deployed on LNG facilities, other than for reservoir CO_2 stripped from the raw natural gas.

Sensitivity Analysis

Any analysis of this type is inherently subject to uncertainty. Capital and operating costs are planning level estimates suitable for comparison purposes but subject to change. The valuation and estimation of external benefits and certain costs are subject to even larger variations, especially over a 40-year horizon. However, the key to the EESA is to reveal not absolutes in terms of dollars, but better and worse decisions overall, compared to the range of possible decisions that could be made.

From this perspective, sensitivity analysis is important because it allows the overall conclusions of the analysis to be tested across a wide range of parameter inputs. If a decision is favourable or economic over a wide range of parameter inputs, compared to other possible decisions, then despite the overall uncertainty in the actual dollar figures, the decision can be identified as superior among its competitors. This is particularly useful when examining the sustainability of options. By definition, sustainability is concerned with the future, which is inherently uncertain. By varying key input parameters over a wide but reasonable range, the implications of a range of possible futures can be examined. The ranges of values for key parameters for this assessment are presented in Table 6.

A fully interactive real time analysis tool, EcoNomics™ DELTΔ²™ [20], has been used to examine the effect of key parameters, across their full range, for each option. The NPV results are calculated for each value of each parameter, discretised across its full range, against every other possible combination of the other values, for each case. This in effect provides a database of every possible NPV result for each case, in which each result is considered to be equi-probable.

Table 7 shows the proportion of all conditions where a case is the most economic, or second most economic case, considering all possible combination of all values across their full range. Under approximately 92% of all possible combinations of conditions, Case 8 is the most economic, followed by Case 1 which is the most economic option under approximately 8% of all possible combinations of conditions. Case 1 is the second most economic option under over 73% of all possible combinations of conditions.

Table 7: Proportion of conditions where a case is the most economic

Option	Best case	2nd Best case
Option 8	91.6%	8.4%
Option 1	7.5%	73.0%
Option 10	1.0%	2.9%
Option 9	0.0%	15.7%

Fuel gas savings are the most significant contributor to the positive economics of Case 8, far exceeding the financial costs of achieving them. Associated benefits from GHG reductions are also significant. Increasing energy costs and a higher cost of carbon further increase the positive NPV achieved by this case. Case 1 is also economic under a wide range of possible future conditions, mostly driven by the GHG reductions achieved by this option compared to the reference case. It is the cheapest way to reduce GHG emissions. NOx and SOx values did not contribute significantly to the costs and/or benefits of each option.

Implications

For this facility, the analysis reveals that the environmentally, socially and economically optimum approach would be to implement Case 8 (best-in-class energy efficiency design across the facility) and Case 1 (CCS of CO_2 contained in the reservoir gas), either alone or in combination. The combined base condition economic NPV of these cases is USD 272 million better than business as usual. With rising energy costs and the increasing likelihood of rising carbon costs, each of the cases assessed delivers increasing overall economic benefits to society. This is an important finding with respect to longer-term capital investment decisions. Further insight can be had from examining narrower ranges of future possible conditions. In a "socially-minded" world, where carbon is assumed to be priced at greater than USD $50/tCO_2$e, and where the maximum discount rate considered is 6%, Case 8 is best under 95% of conditions. In a more "commercially-minded world", where carbon price is assumed to be always lower than USD $50/tCO_2$e, and discount rate is higher than 6%, Case 8 is still the most economic and sustainable under 79% of the remaining range of conditions over 40 years.

Option Value of Being Carbon Capture Ready

Making LNG plants carbon capture ready enables retrofitting at a lower cost at some point in the future, at the expense of a relatively small upfront investment [21–23]. The main requirements for being carbon capture ready are that sufficient space and access are provided for future deployment of capture equipment and systems. Figure 3 shows the difference in NPV over 40 years, at base values, between the CCS retrofit options on a capture ready plant and on a plant that is not capture ready (a plant where no provision for retrofit has been made), of similar overall design. Clearly, retrofitting a capture ready plant with CCS technology is more economic than on a carbon capture unready plant, no matter what year the retrofit is made. As shown above, Case 2 is the best of the CCS retrofit options.

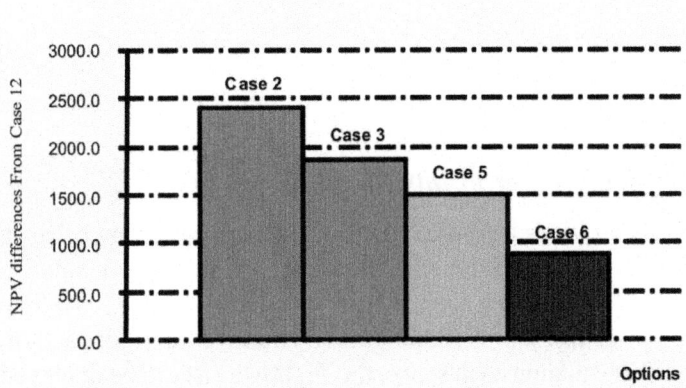

Figure 3: Base NPV differences between capture ready retrofit and capture unready plants.

The option value of being capture ready was also examined (the value of the capture ready pre-investment in implementing retrofit in a given year). Figure 4 shows the option value for being capture ready in implementing Case 2 (the best of the retrofit CCS options). As shown, the capture ready case value is highly dependent on carbon price and the timing of the retrofit. The longer one waits to undertake the retrofit, the less the option is worth, as one approaches the end of life of the facility (retrofitting in 2049 leaves only one year of carbon cost savings to recoup the investment).

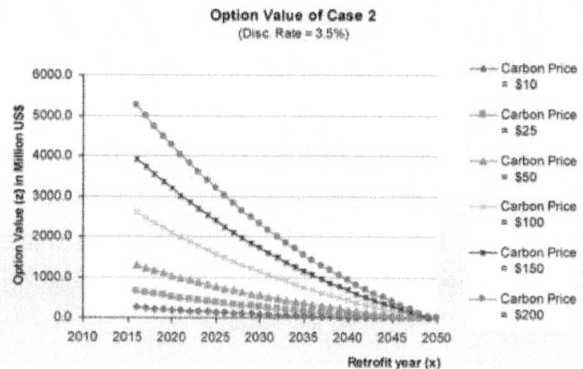

Figure 4: Option value of being capture ready for Case 2, base conditions.

Carbon Price Trigger Points for CCS

The analysis provides an indication of the threshold carbon price at which the decision to implement various carbon reduction measures would occur. Under base conditions, assuming social (3.5%) and commercial (10%) discount rates, the carbon price thresholds (in 2010 USD) that would trigger action are shown in Table 8. Note than in each sense, the first two cases to be deployed are Cases 8 and 1. Case 8 should be done now—it is already economically positive. The best of the CCS cases (Cases 4 and 2) are not economically or socially viable unless carbon prices/values are higher than USD $100/tCO_2$e. Also recall that the financial and external costs of transport and storage would have to be added to the figures shown, which would in all cases raise the threshold value for CCS implementation. In the context of current estimates of the social cost of carbon, CCS does not appear to be viable for managing emissions from facility liquefaction and compression.

Table 8: Carbon price threshold for capture

Case	Social ($i = 3.5\%$)	Commercial ($i = 10\%$)
8 – Energy efficient design	USD 0	USD 0
1 – CCS reservoir CO_2	USD 20	USD 25
9 – Centralised power, efficient design	USD 38	USD 120
10 – Surplus power generation	USD 85	USD 100
2 – Post-combustion CC retrofit	USD 100	USD 145
4 – Greenfield NGCC post-combustion retrofit	USD 116	USD 150

DISCUSSION

The examples presented above examine the wider environmental and economic implications of applying CCS in various applications. CCS is an expensive proposition, no matter how it is applied, both in terms of up-front capital costs, and in terms of the significant amounts of energy required to capture CO_2. Because of this, it is important that CCS performance be compared to other ways of reducing GHG emissions within a plant or portfolio of assets, if they exist. Strategically, this serves to put CCS into overall context as just one of many possible ways that the GHG intensity of an operation can be reduced.

In the case of existing coal-fired power stations, CCS may be one of the only options available for substantially reducing GHG emissions. In this case the question becomes which of a wide variety of CCS options available provides the best long term environmental, social and financial outcome. As the world moves to take action on climate change, and carbon prices rise in future, CCS will become increasingly attractive from a financial perspective. In other industries, such as LNG, a number of other alternatives are available for carbon emission reduction. Designing for energy efficiency and including maximum use of waste heat, may provide distinctly economic and sustainable alternatives. Under a wide range of future likely conditions, sequestering CO_2 from reservoir gas makes good economic sense and is a sustainable proposition today.

The analysis presented herein suggests that while all CCS options are not created equal (some are far more economic and sustainable than others), overall, CCS (other than capturing CO_2 directly from the raw reservoir gas) is unlikely to be economic or sustainable at carbon prices much below about USD 75/t CO2e, which is close to current estimates of the social cost of carbon. At Stern's USD 85/t CO_2e estimate, CCS is an economic choice right now for certain types of applications, as far as society is concerned. However, operators will have little incentive to deploy CCS unless the social value of carbon is wholly or partially reflected in a regulated price on carbon.

From a financial perspective, CCS costs need to come down significantly to make it viable in the near future for most applications, even as carbon prices are put in place around the world. Increased funding for research and development of new breakthroughs in carbon capture technology could substantially improve the viability of CCS in the coming years. Carbon sequestration also carries with it significant long-term cost and legislative and environmental risks which would also have to be factored into a complete analysis.

In some jurisdictions, there is the possibility that CCS will be required, by law, policy or regulation. If, for instance, a company self-regulates and sets a policy of zero carbon emissions, CCS may become one of the only ways to achieve this. It is also possible that carbon prices may rise to the kinds of levels which are needed for CCS to be justified from an overall environmental, social and economic cost-benefit point of view. The social cost of carbon continues to increase with each year of global inaction, as the concentrations of GHGs in the atmosphere steadily rise. As the social cost of carbon rises, CCS becomes an increasingly more economic proposition.

If we must retrofit, retrofitting a capture ready plant is a significantly more economic proposition than retrofitting a carbon capture ignorant plant. The cost savings of being carbon capture ready (with a relatively minimal pre-investment cost), compared to having to retrofit a plant that is not capture ready, are significant.

CONCLUSIONS

While there are many different ways for the power and energy industries to eliminate carbon emissions to atmosphere, they come at radically different costs, and produce different benefits to society. Worldwide, CCS is being intensely examined as a way of reducing emissions from fossil-fuel burning or producing operations. While full-scale deployment of CCS globally is still in its infancy, several successful long-term projects are underway (notably in the natural gas sector), and a suite of demonstrations projects have begun in various parts of the world, in various industry sectors. A wide variety of capture technologies are available, with widely differing capital costs, operational costs, energy requirements, and performance. Safe long-term geological sequestration requires detailed understanding of reservoir characteristics and appropriate long term institutional controls and monitoring. The external social and environmental costs of large scale releases from storage sites could be significant, if these eventuate, but thorough risk management in terms of site selection, monitoring and reservoir design have proven capable of significantly if not completely reducing the likelihood of such releases. The environmental and economic sustainability of CCS is a function of many factors, including the timing of deployment, the value placed on carbon, fuel and energy prices, the costs of disruptions to the business during retrofit, and the external costs of other associated air emissions and possible releases from storage sites. Analysis of three examples shows that all CCS is not created equal, and that industries should examine CCS in the context of other alternatives that exist to reduce emissions within their business, either at individual facilities or within a larger portfolio of assets. In addition, the wider external costs and benefits must be taken into consideration if the real value, and therefore the environmental and economic sustainability, of the action is to be understood.

REFERENCES

1.   IPCC Fourth Assessment Report: Climate Change 2007 (AR4): The Physical Science Basis; Solomon, S, Qin, D, Manning, M, Chen, Z, Marquis, M, Averyt, KB, Tignor, M, Miller, HL, Eds.; Cambridge University Press: Cambridge, UK, 2007.

2.   Pascala, S; Socolow, R. Stabilization wedges: Solving the climate problem for the next 50 years with current technologies. Science 2004, 305, 968–972. [Google Scholar]

3.   IPCC (2007a) Intergovernmental Panel on Climate Change Fourth Assessment Report. Mitigation of Climate Change; Cambridge University Press: Cambridge, UK, 2007; pp. 1–9.

4.   IPCC (2007) Intergovernmental Panel on Climate Change Fourth Assessment Report. The Physical Science Basis; Cambridge University Press: Cambridge, UK, 2007.

5.   International Energy Agency. World Energy Report; International Energy Agency: Paris, France, 2007. [Google Scholar]

6.   Stern, NH. The Economics of Climate Change: The Stern Review; Cambridge University Press: Cambridge, UK, 2007; pp. 2–9. [Google Scholar]

7.   Garnaut, R. Garnaut Climate Change Review—Draft Report; Commonwealth of Australia: Canberra, Australia, 2008. [Google Scholar]

8.   IEA/OECD. World Energy Outlook; International Energy Agency (IEA): Paris, France, 2009; pp. 2–4. [Google Scholar]

9.   National Academy of Sciences. Advancing the Science of Climate Change. Expert Consensus Report, America's Climate Choices; NAS: Washington, DC, USA, 2010. [Google Scholar]

10. Itaoka, K; Barton, A; Akan, MA. Path analysis for public survey data on social acceptance of CO2 capture and storage technology. Proceedings of the 8th International Conference on GHG Control Technology (GHGT-8), Trondheim, Norway, 19–22 June 2006.

11. Strategic Analysis of the Global Status of Carbon Capture and Storage: Report 5: Synthesis Report; McConnell, CH, Toohey, P, Thompson, M, Eds.; Global Carbon Capture and Storage Institute: Canberra, Australia, 2009; pp. 5–9.

12. Hardisty, PE. Analysing the role of decision-making for industry in the climate change era. J. Environ. Qual 2009, 20, 205–218. [Google Scholar]

13. Hardisty, PE. Environmental and Economic Sustainability; CRC Press: New York, NY, USA, 2010; pp. 5–12. [Google Scholar]

14. World Resources Institute. World Resources Institute CCS Community Engagement Guidelines: Workshop Survey; World Resources Institute: Washington, DC, USA, 2009. [Google Scholar]

15. De Best-Wladhober, MD; Daamen, A; Ramirez Ramirez, A; Faaji, C; Hendricks, C; de Visser, E. Informed public opinions on CCS in comparison to other mitigation options. Proceedings of the 9th International Conference on GHG Control Technologies, Washington, DC, USA, 16–20 November 2008.

16. Mathias, SA; Hardisty, PE; Trudell, MR; Zimmerman, RW. Screening and selection of sites for CO2 sequestration based on pressure buildup. Int. J. Greenh. Gas. Con 2009, 3, 577–585. [Google Scholar]

17. Intergovernmental Panel on Climate Change Special Report on CO2 Capture and Storage; IPCC, Ed.; Cambridge University Press: Cambridge, UK, 2005.

18. Allowance Markets Assessment: A Closer Look at the Two Biggest Price Changes in the Federal SO2 and NOx Allowance Markets; EPA: Washington, DC, USA, 2009.

19. Clean Air Conservancy, Homepage. Available online: http://www.cleanairconservancy.com (accessed on 22 September 2009).

20. Hardisty, PE; Sivapalan, M; Van Der Linden, S; Donohoo, S. WorleyParsons DELTA™ Toolset for Environmental, Social and Economic Life-Cycle Modelling; WorleyParsons: North Sydney, Australia, 2009. [Google Scholar]

21. Rohlfs, W; Madlener, R. Valuation of CCS-Ready Coal-Fired Power Plants: A Multi-Dimensional Real Options Approach; FCN Working Papers 7/2010; E.ON Energy Research Center, Future Energy Consumer Needs and Behavior (FCN): Aachen, Germany, 2010. [Google Scholar]

22. Rohlfs, W; Madlener, R. Cost Effectiveness of Carbon Capture-Ready Coal Power Plants with Delayed Retrofit; FCN Working Papers 8/2010; E.ON Energy Research Center, Future Energy Consumer Needs and Behavior (FCN): Aachen, Germany, 2010. [Google Scholar]

23. CO_2 Capture Ready Plants; Technical Study, Report No. 2007/4; International Energy Agency: Paris, France; May; 2007.

Chapter 3

PERFLUORINATED CHEMICALS IN SEDIMENTS, LICHENS, AND SEABIRDS FROM THE ANTARCTIC PENINSULA — ENVIRONMENTAL ASSESSMENT AND MANAGEMENT PERSPECTIVES

Juan José Alava[1, 2], Mandy R.R. McDougall[1], Mercy J. Borbor-Córdova[2], K. Paola Calle[2], Mónica Riofrio[3], Nastenka Calle[4], Michael G. Ikonomou[5] and Frank A.P.C. Gobas[1]

[1] School of Resource & Environmental Management, Faculty of Environment, Simon Fraser University, Burnaby, British Columbia, Canada

[2] Facultad de Ingeniería Marítima, Ciencias Biológicas, Oceánicas y Recursos Naturales, Escuela Superior Politécnica del Litoral (ESPOL), Guayaquil, Ecuador

[3] Instituto Antártico Ecuatoriano (INAE), Guayaquil, Ecuador

[4] Pacific Institute for Climate Solution (PICS), Simon Fraser University, Burnaby, British Columbia, Canada

[5] Institute of Ocean Sciences, Fisheries and Oceans Canada, Sidney, BC, Canada

ABSTRACT

Antarctica is one of the last frontiers of the planet to be investigated for the environmental transport and accumulation of persistent organic pollutants. Perfluorinated contaminants (PFCs) are a group of widely used anthropogenic substances, representing a significant risk to wildlife and humans due to their high biomagnification potential and toxicity risks, especially in food webs of the northern hemisphere and Arctic. Because the assessment of PFCs in the Antarctic continent is scarce, questions linger about the long-range transport and bioaccumulation capacity of PFCs in Antarctic food webs. To better understand the global environmental fate of PFCs, sediment, lichen (Usnea aurantiaco-atra), and seabird samples (southern giant petrel, Macronectes giganteus; gentoo penguin, Pygoscelis papua) were collected around the Antarctic Peninsula in 2009. PFC analytes were analyzed by LC/MS/MS, revealing the detection of PFHpA in seabirds' feather and fecal samples, and

PFHxS in lichens. PFBA and PFPeA were detected in 80% and 60% of the lichens, and PFTA in 60% of sediment samples. While oceanic currents and atmospheric transport of PFCs may explain the ubiquitous nature of these contaminants in the Antarctic Peninsula, military bases and research stations established there may also be contributing as secondary sources of PFCs in the Antarctic ecosystem.

INTRODUCTION

Past research shows that legacy persistent organic pollutants (POPs) such as dichlorodiphe-.nyltrichloroethane (DDT), polychlorinated biphenyls (PCBs), and hexachlorocyclohexanes.(HCHs) pose substantial problems related to environmental and ecosystem health on a global.scale [1–5.]. POPs can be transported over very long distances, biomagnify in food webs, and.cause adverse health effects in high trophic level species such as birds and marine mammals..Cold regions that are typically isolated from anthropogenic activity, such as the Arctic and the.Antarctic, are particularly vulnerable to POPs because of the global distillation phenomenon,.which causes many pollutants to concentrate in these regions [.6., .7.]. The Arctic Monitoring.Assessment Program (AMAP), in association with the United Nations Environment Pro-. gramme (UNEP) Stockholm Convention on Persistent Organic Pollutants, has played a key.role in documenting the fate, transport, and effects of these pollutants in the Arctic, and has.promoted global initiatives to monitor, manage, and control these substances [.6., .8.]. Despite.enhanced understanding of POP contamination in the Arctic, limited information exists on the.state of pollution in Antarctic food webs. Researchers have identified a lack of comparative. data between the polar regions of the world, where many efforts have been directed toward.understanding POP contamination in the high latitudes of the Northern Hemisphere such as.the Canadian Arctic and Greenland [8–12].. Ongoing research has identified emerging contaminants of concern, including perfluorinated.contaminants (PFCs), which are expected to pose significant risks to the environment and.wildlife, particularly in the Arctic and the Antarctic [13–15]. Although PFCs have been detected.in some Antarctic ecosystems and biota, the environmental transport and bioaccumulation.patterns of PFCs, mainly perfluoroalkyl acids (PFAAs) such as perfluorinated carboxylates. (PFCAs) and perfluorinated sulfonates (PFSAs), remain relatively unexplored within Antarc-.tica. PFCs are highly fluorinated anthropogenic compounds, often utilized as repelling agents,.with applications including coatings for paper or food packaging and textiles, industrial.surfactants, insecticides, and historically, aqueous film-forming foams [.16,17.]. Due to their.widespread use, PFCs are now considered environmentally ubiquitous substances, found in.all areas around the world. In response, numerous measures have been taken to reduce the.adverse impacts of PFCs on local and global scales [8]..PFCs are extremely persistent, can travel long distances (predominantly via ocean currents),.bioaccumulate in food webs, and achieve highest concentrations in marine mammals and.birds. PFCs are of particular ecological and toxicological concern due to their tendency to.biomagnify in food webs and cause adverse health effects, including reproductive damage,.immunotoxicity, and hepatotoxicity [.18]. Of further interest is the unique physicochemical. nature of PFCs. Whereas many legacy POPs are lipophilic and therefore accumulate in fatty.tissues, PFCs tend to accumulate primarily in protein-rich tissues, such as the liver. Two.PFAAs, perfluorooctane sulfonate (PFOS) and perfluorooctanoate (PFOA), represent the most.commonly investigated PFCs of significant risk to wildlife and humans due to their ubiquitous.nature, global fate and transport, high biomagnification potential, and toxicity risks, especially.in aquatic and marine food webs of the northern hemisphere and Arctic [18–21]. Phase out.programs designed to eliminate the production of PFOS were established for some regions in.the early 2000s, followed by the addition of PFOS to the list of restricted POPs under the.Stockholm Convention on Persistent Organic Pollutants in 2009 [.8.]. Despite these initiatives,. production of PFOS, PFOA, and several other PFAAs still take place around the world,.including several developing countries [22–25]. One of the priority actions under the Antarctic.Treaty is the assessment and monitoring of POPs, including PFCs, in Antarctica. Considering.that assessments of PFCs in the Antarctic are limited, questions linger about the long-range.environmental transport of these substances to the Southern Hemisphere, and the capacity of.these substances to bioaccumulate in Antarctic food webs..PFCs have been detected in various Antarctic environmental media and biota, typically in the.pg/g to ng/g range, though many samples return nondetectable levels or levels below the.minimum level of quantification [.9.]. Recent studies show that levels of many PFCs in Arctic.environments have been increasing, with concentrations of several PFCs equivalent to or.surpassing that of DDT, PCBs, PBDEs, and other organochlorine pesticides [.19.]. Similar.patterns are anticipated for PFCs in Antarctic environments as they are continuously delivered.from other geographic locations via long-range transport..Although some PFCs are already categorized as POPs, the majority of these substances are not.subject to global or local controls. To ensure that potential impacts of pollutants on Antarctic.wildlife are considered in the global environmental agenda and throughout negotiations on.commercial chemical production and use, it is important that a high-quality research program.is developed on the fate and effects of contaminants in Antarctic ecosystems and wildlife. As.part of an ongoing scientific initiative and collaboration between the Ecuadorian

Antarctic.Institute (INAE), Simon Fraser University (Canada), the Institute of Ocean Sciences (IOS,.Fisheries and Oceans Canada, DFO), and the Escuela Superior Politecnica del Litoral (Ecua-.dor), a study to investigate and monitor PFCs was initiated in Peninsula Antarctica around.the surrounding areas of the Ecuadorian Station "Pedro Vicente Maldonado" during the 2009.Ecuadorian–Antarctic expedition. In this chapter, we provide one of the primary findings on.PFCs in sediments and biotic matrices, including lichens as well as feces and feathers from the.southern giant petrel (.Macronectes giganteus.) and gentoo penguin (.Pygocelis papua.), and.evaluate the use of noninvasive techniques to monitor emerging organic contaminant of.concern in the Antarctic environment

MATERIALS AND METHODS

Study Area and sampling

The Ecuadorian Research Station "Pedro Vicente Maldonado" (Maldonado Station, hereafter).is located at Fort William Point, Greenwich Island (62°31'S; 59°46'W; Figure 1). The study area.encompassed the Barrientos Island (62°24'01"S; 59°43' 52"W), Dee Island (62°25'48.5" S;.59°47'69.6" W), Punta Ambato (62°26'33" S; 59°47'28.8" W), and the surroundings of the.Maldonado Station (62°27'59"S; 59°43'32.5"W), as illustrated in .Figure 1.. Sampling was.conducted using three tracks established by the Maldonado Station to access the coastline of..Fort Williams, which enclose two sampling zones: Ensenada Guayaquil and Bahia Chile. These.sectors are only used by technical and military personnel that work at the Station and visiting. scientists that come to the island for research purposes. Barrientos Island is used principally.as a tourist stopover for cruise ships where tourists land and walk around the island for bird-.watching. In Dee Island and Punta Ambato, sampling was deployed around the coastline. All.sampling was done during the Austral summer and seabird breeding seasons of 2009. The.collection of abiotic and biotic samples is described as follows.

Sediments.

Sediment samples were collected from three locations in the Antarctic Peninsula including Dee.Island (.n.= 1 site), Maldonado Station (.n.= 2 sites), and Punta Ambato (.n.= 2 sites) (.Figure 1)..Sediment samples were directly collected using 100 mL centrifugation tubes, stored at < 4°C.until transportation to the laboratory in Canada..

Seabirds

Gentoo penguins (.Pygoscelis papua.) and southern giant petrels (.Macronectes giganteus.), two.species of seabirds that inhabit the Antarctic Peninsula, were identified as potential bio-.indicators of PFCs contamination. The main reason for selecting seabirds is based on studies.showing that bird populations are most affected by contaminants, specifically POPs, among.wildlife species (see [.26.] for a review). Bird species have the greatest capacity to biomagnify.chemicals because of their highly energy-efficient metabolic system and also because of their.high trophic position within the food web. Bird populations are therefore often at high risk.from bioaccumulative substances, and can act as the "canary in the coal mine" for the larger.Antarctic ecosystem..In this context, we conducted a noninvasive sampling technique to minimize or completely.avoid the impacts of lethal or invasive sampling on the local bird populations. Sampling.focuses on the collection of shed/molted feathers and excreted fecal matter from nesting sites..Because of the very high affinity of PFCs for protein, feathers are good noninvasive sampling.media for PFCS, as they consist mainly of protein matter (i.e., keratin, a high molecular weight.protein). Feathers have also been used to successfully monitor mercury in seabird populations.such as brown skuas., Catharacta lonnbergi., chinstrap penguins, .Pygoscelis antarctica., and gentoo. penguins, .P..papua.), in our study area [.27.], as well as PFCs in the feathers of aquatic and marine.birds, including grey heron (.Ardea cinerea.) and herring gull.,.(.Larus argentatus.) from the.Northern Hemisphere [.28.]. Fecal matter is known to contain some of the highest concentration.of contaminants due to the gastrointestinal magnification that occurs in the intestinal tract of.consumer organisms. In addition, the contaminant concentrations in fecal matter are related.to compounds absorbed by the organism, such that they can provide a measure of accumulated.concentrations. The low capacity to migrate to the gaseous phase (i.e., air) and high octanol-.air partition coefficient (.K_{OA}.) of the analytes (Table 1) cause minimal losses of the contaminants.from feces or feathers to the air after feathers or fecal matter have been dropped. This means.that the concentrations of the chosen analytes can remain a measure of bird exposure levels.long after the feces have been excreted or feathers have been shed...Molted feathers were collected randomly in and around nests and colonies of petrels sur-.rounding the Maldonado's Station and stored in ziploc-type plastic bags (.n.= 5). Only one bag.of feather samples for gentoo penguin was collected from the Maldonado's Station. Fecal.matter samples from gentoo penguins were collected from nesting sites and colonies around. the Maldonado's Station (.n.= 9), Barriento Island (.n.= 7), and Dee Island (.n.= 3). All feces samples.were placed into 20 mL glass vials. Both feather and fecal samples were stored in coolers and.transported by airplane with dry ice (–20°C) until transportation to the lab in Canada...

Lichens

Lichen (.Usnea aurantiaco-atra.) samples (.n.= 5) were collected from rocky areas around the.surroundings of the Maldonado's Station and wrapped with clean, sterile aluminum foil and.stored in ziploc plastic bags until further transportation to the lab. The rationale to select lichens.is based on the premise that this biological matrix can be used as a potential monitor and.indicator of global atmospheric transport of some PFCs to the Antarctic Peninsula

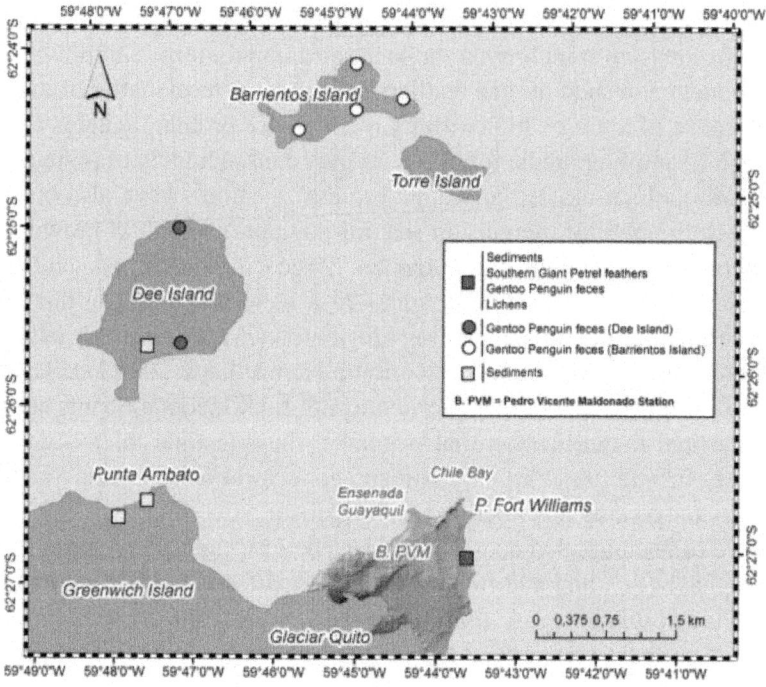

Figure 1: Geographical location of the study area and sampling sites in the Antarctic Peninsula

Table 1 summarizes the compiled physical–chemical properties for the various PFCs studied, including molecular weights (MW), log octanol–water partition coefficients (log KOW), log octanol–air partition coefficients (log KOA), and log D values. Because the physicochemical properties of PFCs are considerably different from that of many other legacy POPs (i.e., they can be ionized at environmentally relevant pH), it is important to recognize that relationships applicable to other POPs may be less relevant when applied to PFCs and other ionizable compounds. For instance, many organic compounds of concern, including numerous agricultural and pharmaceutical compounds, are lipophilic in nature, and will tend to accumulate in fatty tissues [3]. The octanol–water partition coefficient (KOW) has become a common property used to describe the tendency of a substance to partition into lipid, as the behavior of octanol and lipid are quite similar. Octanol thus serves as a suitable surrogate for lipid, particularly within predictive bioaccumulation models [29]. However, KOW describes the lipophilicity of neutral compounds, and is not necessarily applicable to ionizable organic compounds (IOCs) such as PFCs, where the measure of lipophilicity is pH-dependent [30]. Many PFCs are almost completely ionized at environmentally relevant pH [31]. A more applicable indicator for predicting the lipophilicity of ionizable substances is log D, where both the neutral and the ionic species of the compound are accounted for [30].

PFC ANALYSIS: EXTRACTION AND QUANTIFICATION

Sediment and biological samples were extracted and analyzed at the Institute of Ocean Sciences (IOS), Fisheries and Oceans Canada (DFO), Sidney, British Columbia, Canada. PFC concentrations were analyzed by liquid chromatography tandem mass spectrometry with double mass detectors (LC/MS/MS), as described elsewhere [19]. Analyte concentrations were determined with respect to the mass labeled quantification and internal standards using isotope dilution method. Fifteen PFCs were examined in this study (Table 1). High purity (>95%) analytical standards, including perfluorobutane sulfonic acid (PFBS), perfluorohexanesulfonic acid (PFHxS), PFOS, perfluorobutanoic acid (PFBA), perfluoropentanoic acid (PFPeA), perfluorohexanoic acid (PFHxA), perfluoroheptanoic acid (PFHpA), PFOA, perfluorononanoic acid (PFNA), perfluorodecanoic acid (PFDA), perfluoroundecanoic acid (PFUnA), perfluorododecanoic acid (PFDoA), perfluorotetradecanoic acid (PFTA), and perfluorooctanesulfoamide (PFOSA), were used. Mass-labeled internal standards included six PFCs ($^{13}C_2$ PFOA, $^{13}C_2$ PFDA, $^{13}C_2$ PFDoA, and $^{13}C_4$ PFOS, $^{13}C_4$-PFOA). Calibration curves were constructed from the analysis of calibration standard solutions (range 0.08–5.0 ng/mL).

Various calibration standards and standard additions were prepared and used as quality assurance/quality control (QA/QC). QA/QC measures included initial method validation work, consisting of analyte recovery experiments of native PFCs in clean sediments and biota. The method of detection limit (MDL) was set equal to the concentration of the method's level of quantification (MLOQ) for samples and subtracted from quantified concentrations of each analyte (Table 2). Only corrected data above the MLOQ are reported in this work. Concentrations of PFCs were expressed on a wet weight basis (ng/g ww). Extraction methods are briefly described as follows.

SEDIMENT

Sediment samples ($\approx$ 10 g wet weight) were added to 50 mL polypropylene centrifuge tubes and spiked with internal surrogate spiking solution (360 ng of $^{13}C_2$ PFOA, 120 ng of $^{13}C_2$ PFDA, 120 ng of $^{13}C_2$ PFDoA, and 120 ng of $^{13}C_4$ PFOS; Table 1). After 20 min, 10 mL of 0.1% acetic acid in MeOH was added, and samples were extracted on a shaker table for 16 h. After extraction and centrifugation, 1 mL was pipetted into 1.5 mL Ependorf vial containing 25 mg of activated carbon. Then the vial was subject to centrifugation for 30 min at 14,000 rpm; 300 μL of supernatant was taken and combined with 300 μL of water and 50 μL of 20 ppb of recovery standard and centrifuged again for 15 min at 14,000 rpm. Then, 300 μL of supernatant was used for LC/MS/MS analysis (i.e., injection volume=100 μL for LC/MS/MS).

FEATHERS

Approximately 0.74 g of feather was weighed, and then homogenized by adding first HNO_3 (e.g., 4 × 0.9 mL, 2 × 0.9 mL) with a series of vortexing steps until the whole particulates completely disappeared within 3 h. Samples were set up for digestion at room temperature (RT) for 12 h. Afterwards, 15 mL of 5 M NaOH prepared in water was added to samples and shaken on a shaker table for 5 min. The pH was measured to ensure the sample was acidic enough (i.e., pH~3–4) prior to direct injection in LC/MS/MS (large volume injection). After neutralization and extraction with 2.5 mL MeOH for a total volume 27.5 mL, ion suppression was found from recoveries; therefore, additional dilution (10×) was done until ion suppression was reduced (i.e., injection volume = 200 μL).

FECES

Penguin fecal matter (~0.65 g of feces) was weighed and homogenized with HNO_3 (e.g., 4 × 0.9 mL, 2 × 0.9 mL and vortexing). Samples were set at RT for digestion during 12 h. After digestion, samples were neutralized to pH equal to 3.2–4.2, brought up to 50 mL and centrifuged at 6000 rpm for 25 min; 100 μL of recovery standard (^{13}C4-PFOA) was added to an aliquot of 400 μL and injected into LC/MS/MS (i.e., injection volume= 200 μL).

LICHEN

Lichen (2 g) was extracted based on the methodology described in reference [32]. After extraction, 4 mL of solution was blown down to 2 mL, followed by collecting 1 mL aliquot and added into 1.5 mL Ependorf vial containing 25 mg of activated carbon. The vial was subject to centrifugation for 30 min at 14,000 rpm and 300 μL of supernatant was obtained and combined with 300 μL water and 50 μL of 20 ppb recovery standard and centrifuged for 15 min at 14,000 rpm. Then, 300 μL of supernatant was used for LC/MS/MS analysis (i.e., injection volume=100 μL).

Table 1: List of target perfluoroalkyl chemicals (PFCs) and radiolabeled surrogates monitored using LC/MS/MS

Chemical Name	Abbreviation	Formula	MW (g/mol)	Log* K_{OW}	Log* K_{OA}	Log** D
Target Analytes						
Perfluorobutanoic acid	PFBA	C_3F_7COO-	214.0	1.3	5.0	0.060
Perfluoropentanoic acid	PFPeA	C_4F_9COO-	264.0	2.1	5.3	0.54
Perfluorohexanoic acid	PFHxA	$C_5F_{11}COO-$	314.1	3.1	5.6	1.1
Perfluoroheptanoic acid	PFHpA	$C_6F_{13}COO-$	364.1	2.8	5.9	1.6
Perfluorooctanoic acid	PFOA	$C_7F_{15}COO-$	414.1	3.6	6.3	2.3
Perfluorononanoic acid	PFNA	$C_8F_{17}COO-$	464.1	4.5	6.6	2.9
Perfluorodecanoic acid	PFDA	$C_9F_{19}COO-$	514.1	5.4	6.8	3.5
Perfluoroundecanoic acid	PFUnA	$C_{10}F_{21}COO-$	564.1	6.4	7.1	4.2
Perfluorododecanoic acid	PFDoA	$C_{11}F_{23}COO-$	614.1	7.1	7.4	5.0
Perfluorotetradecanoic acid	PFTA	$C_{13}F_{27}COO-$	714.1	8.8	8	6.1
Perfluorobutane sulfonic acid	PFBS	$C_4F_9SO_3-$	300.1	N/A	N/A	-0.53
Perfluorohexane sulfonic acid	PFHxS	$C_6F_{13}SO_3-$	400.1	N/A	N/A	0.54
Perfluorooctane sulfonic acid	PFOS	$C_8F_{17}SO_3-$	500.1	4.3	7.8	1.7
Perfluorodecane sulfonic acid	PFDS	$C_{10}F_{21}SO_3-$	600.1	N/A	N/A	3.1
Perfluorooctane sulfonamide	PFOSA	$C_8F_{17}SO_2NH_2$	499.1	6.3	8.4	-
Mass Labeled Standards						
Perfluorooctanoic acid[a]	$^{13}C_2$-PFOA	-	-	-	-	-
Perfluorodecanoic acid[b]	$^{13}C_2$-PFDA	-	-	-	-	-
Perfluorododecanoic acid[c]	$^{13}C_2$-PFDoA	-	-	-	-	-
Perfluorooctane sulfonic acid[d]	$^{13}C_4$-PFOS	-	-	-	-	-
Perfluorooctanoic acid[e]	$^{13}C_4$-PFOA	-	-	-	-	-

MW: molecular weight

*Log K_{OW} and K_{OA} values of individual PFCs were compiled from published values calculated using SPARC general partitioning model [33].

**Log D values were calculated at pH = 7.5 and T = 21°C using SPARC.

^{a} used to quantify PFBA, PFPeA, PFHxA, PFHpA, PFOA, PFNA

^{b} used to quantify PFDA and PFUnA

^{c} used to quantify PFDoA and PFTA

^{d} used to quantify PFBS, PFHxS PFOS, PFDS, PFOSA

^{e} used to quantify recovery of mass labeled surrogates

Table 2: Method's limit of quantification (MLOQ) for PFC analytes measured by LC/MS/MS

| Analyte | Sample Type | | | |
| | Feather | Feces | Lichen | Sediment |
	MLOQ (ng/g) for 0.74 g sample	MLOQ (ng/g) for 2 g sample	MLOQ (ng/g) for 2 g sample	MLOQ (ng/g) for 10 g sample
PFBA	0.81	2.16	0.73	2.04
PFPeA	1.22	5.86	4.11	4.34
PFHxA	1.10	2.51	3.33	1.75
PFHpA	0.57	3.25	0.77	0.55
PFOA	1.14	3.08	3.34	2.34
PFNA	0.80	4.04	1.04	1.23
PFDA	0.82	2.80	0.51	0.40
PFUnA	2.78	4.01	0.78	0.17
PFDoA	1.79	7.10	0.22	0.26
PFTA	1.30	10.86	0.57	0.25
PFBS	0.03	0.92	0.09	0.12
PFHxS	0.10	0.45	0.19	0.25
PFOS	0.30	0.88	0.75	0.33
PFDS	2.44	1.42	0.12	0.13
PFOSA	1.43	N/A	0.76	0.32
FHUEA	0.64	2.96	0.19	0.25
FOUEA	1.06	1.55	0.08	0.27
FDUEA	4.02	4.73	0.31	0.25

RESULT AND DISCUSSION

Pfc Concentrations

Several PFC compounds showed concentrations above the MLOQ, as shown in Table 3. Perfluorotetradecanoic acid (PFTA), a chemical with a high *KOW* and high *KOA* that will persist for decades in humans, was measured in 60% of sediment samples, but undetected or below the MLOQ in lichens, feces, and feathers. Perfluoroheptanoic acid (PFHpA) was detected in all seabird feather samples (range = 1.60–2.85 ww ng/g; Table 3), and in 47% of penguin feces, ranging 0.37–22 ng/g ww. All lichen samples exhibited concentrations of perfluorohexanesulfonate (PFHxS), ranging 0.20–1.20 ng/g ww, while perfluorobutyric acid (PFBA), perfluoro-n-pentanoic acid (PFPeA), and PFHpA were measured in 80%, 60%, and 60% of lichen samples, respectively. PFOA and PFOS were not quantified in most samples (i.e., < MLOQ or ND; Table 3), except for the detection of PFOS in two penguin feces samples (2.8 and 3.14 ng/g ww), and PFOA in a single fecal sample (2.0 ng/g ww) and lichen (4.7 ng/g ww). The lack of PFOS and PFHxS detection in Antarctic seabird feathers contrasts with the levels of PFOS and PFHxS found in feathers of grey herons (PFOS: 247 ng/g dw; PFHxS: ≈ 20 ng/g dw) and herring gulls (PFOS: 79 ng/g dw; PFHxS: > 30 ng/g dw) from the Northern Hemisphere (Flanders, Belgium) [28]. However, the absence of PFOA in our feather samples is consistent with the lack of detection of this compound in bird feathers from the same region [28]. For comparison purposes, the PFOA concentration detected in a sample of gentoo penguin feces was 14 times lower than the PFOA concentration (28.2 ng/g ww) detected in a single herring gull liver sample from Belgium [28]. Despite samples from other parts of the world that indicate a continued increase or no change in PFOS levels following the 2002 phase-out [34–37], a fast decline in PFOS concentrations has been observed in wildlife over the past decade [38,39]. PFDA, PFUnA, PFDoA, and PFOSA were not detected (ND) or < MDL. Except for the compound PFHpA, lack of detection of most analytes in samples and small sample sizes preclude undertaking robust statistical analyses for multisite or/and inter-species comparisons.

Pfc Patterns

Figure 2 shows the composition of PFCs observed in biotic and abiotic samples. PFHpA was the only compound detected in feathers of both petrels and penguins, accounting for 100% of total PFCs, while PFTA was equal to 100% of PFCs in sediment samples (Figure 2). PFHpA was also found in feces and lichen samples making up 24.5% and 23% of total PFCs, respectively. PFDS contributed to 54% and 17% of PFCs in feces and lichens, contrasting with PFPeA and PFHxS, which accounted for 3.4% and 50%, and for 4% and 21% of PFCs in feces and lichen samples, respectively. PFNA accounted for 38% of the PFCs in feces. These patterns clearly show that both perfluorinated sulfonates (PFSAs) and carboxylates (PFCAs) exhibit different fractions in seabirds, reflecting the potential role of biotransformation in shaping the accumulation of these compounds.

Table 3: Quantification data of PFCs (ng/g ww) in feather, feces, lichen, and sediment samples collected in the Antarctic Peninsula. Data taken and modified from reference [63].

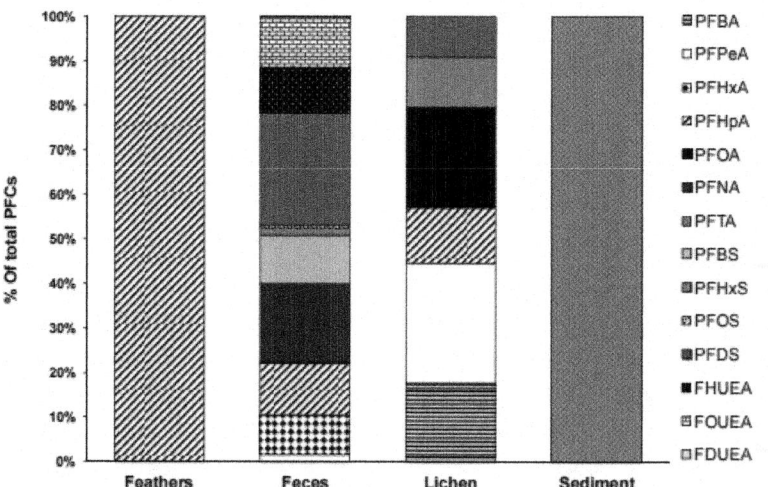

Figure 2: Composition pattern of PFC compounds detected in biotic (feathers and feces of seabirds, and lichen) and abiotic (sediment) samples from the Antarctic Peninsula. Abbreviations for PFC chemical names are defined in Table 1.

Bioaccumulation Of Pfcs

The biomagnification factor (BMF) [40] for PFHpA (i.e., BMF = C_B/C_D, where C_B is the PFHpA concentration detected in the predator, the giant petrel, and C_D is the PFHpA concentration observed in the diet/prey, gentoo penguin) was calculated using feather concentrations, as this was the only PFC compound readily detected in 100% of feathers samples. Hence, the concentrations of PFHpA in the petrel feathers (i.e., mean ±SD = 2.6 ± 0.60 ng/g ww; $n = 5$) and that of the penguin (1.60 ng/g ww; $n=1$; Table 3) were used as surrogates for concentrations in the tissues of the whole organism, assuming that the birds had been exposed to the compound for a sufficiently long time to allow the concentrations to reach steady state [40]. The criterion applied to indicate that PFHpA was biomagnified in petrels was a BMF > 1, such that a BMF greater than 1 indicates that the chemical is a bioaccumulative substance [41]. Here, we found that the BMF was close to 2 (i.e., 1.6), indicating that PFHpA biomagnifies in petrels. Although the concentrations of PFHpA in feces appear to be relatively higher than the concentrations found in lichen and feathers, comparisons of the PFHpA concentrations among biota samples show lack of significant differences (Welch's ANOVA, $p > 0.05$; Tukey–Kramer HSD (honest significant difference) test, $p > 0.05$), as shown in Figure 3.

To further illustrate the behavior of PFC concentrations in these samples, detected PFC compounds were plotted as a function of log D and log KOA, as shown in Figure 4. The majority of PFCs concentrations observed in biotic samples (i.e., feces and lichens) fall within log D values between 0 and 3, as seen in Figure 4A. While concentrations of PFCs tend to increase with increasing log D values from log D of 0 to log D of 3 in feces, PFC concentrations appear to decrease as the log D increases within the same range of log D values in lichens (Figure 4A). This observation may be an indication that both ionized and unionized forms of PFC compounds with low log D values (i.e., PFBA, PFPeA, PFHxA, PFHxS, PFHpA, PFOS, PFOA, PFNA, PFDS) are present in some organisms residing in this region and prone to potential transportation by oceanic currents (e.g., Antarctic Circulation Current) from either continental/ regional or local sources (i.e., international military bases and research stations) to the Antarctic Peninsula. Similarly, most PFCs concentrations observed in these samples, especially in lichens, fall within log KOA values of 5.0 and 6.5 (Figure 4B). Although concentrations for some PFC compounds show a tendency to decrease with increasing log KOA in feces (i.e., PFBS, PFHxS, PFPeA, PFHxA, PFHpA, PFOA, PFNA, PFOS), concentrations for similar PFCs seem to increase as log KOA increases in lichens (i.e., PFDS, PFHxS, PFBA, PFPeA, PFHpA, PFOA), as seen in Figure 4B. These trends may support the notion that low molecular weight compounds (e.g., 214-414 g/mol) with low log KOA are likely to be subject to long-range atmospheric transport and potentially reaching the region, where these compounds accumulate in biotic compartments, mainly in natural air samplers such as lichens and secondary in air-breathing organisms such as seabirds (petrels, penguins).

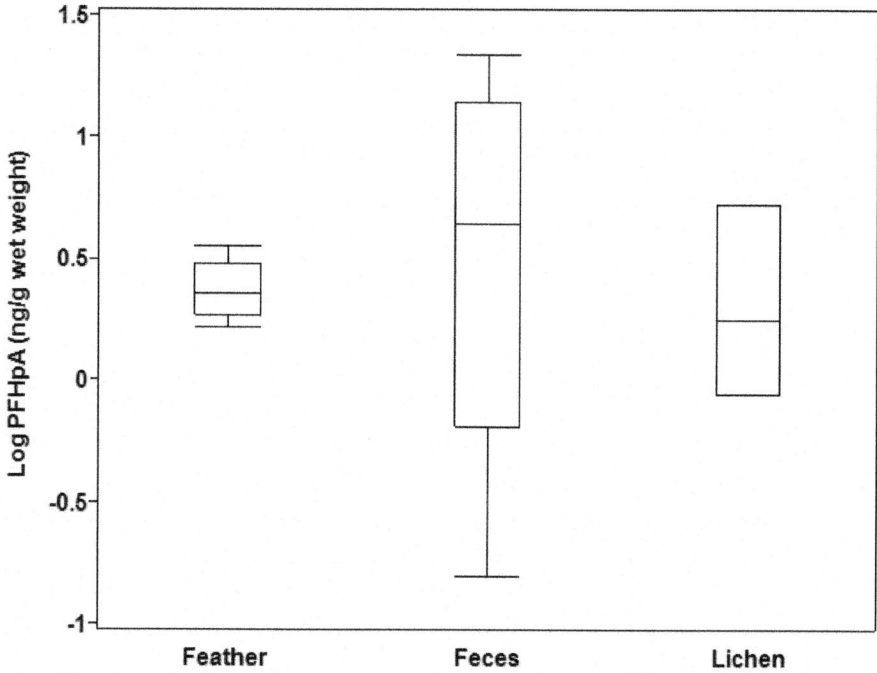

Figure 3: Box plots showing log transformed concentrations of PFHpA (ng/g ww) detected in lichen (n =3), and feather (n =6) and feces (n =9) of seabirds from the Antarctic Peninsula. The internal line across the box is the median; the ends of the box are the 25% and 75% quartiles; and the whisker bars are the minimum and maximum values. Because of unequal variances (i.e., heteroscedasticity; Bartlett test, $p <$ 0.005), a Welch's ANOVA, followed by a Tukey–Kramer HSD test, was used for the multicomparison, showing no significant differences in PFHpA concentrations among the biotic samples ($p > 0.05$).

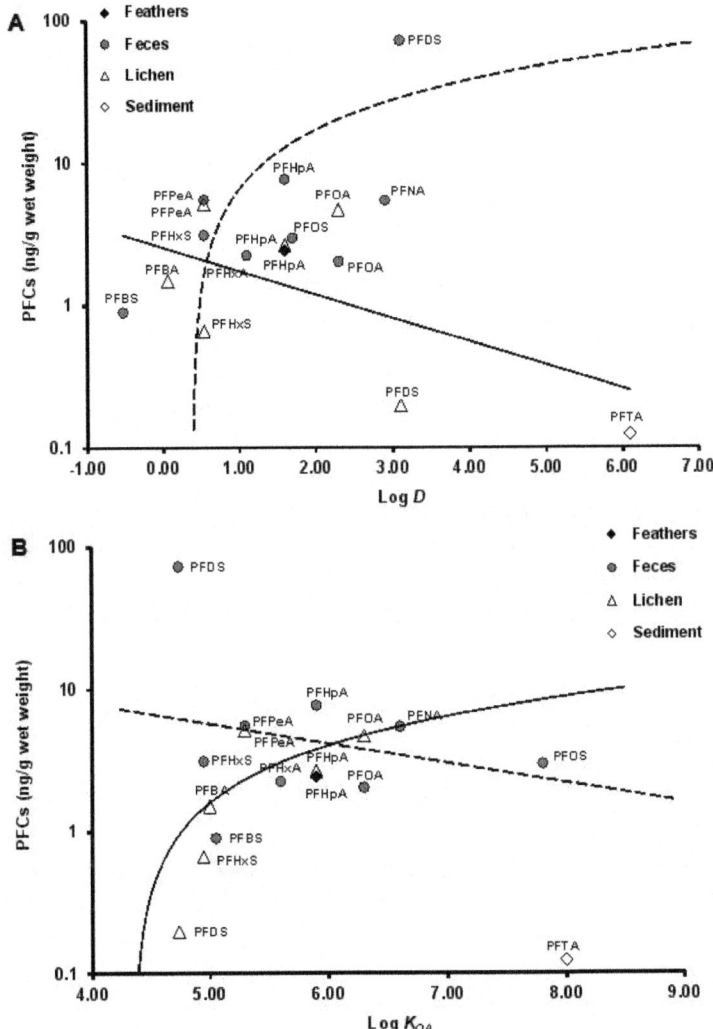

Figure 4: Concentrations of PFC compounds (logarithmic scale in ng/g wet weight) measured in sediments, lichens, and seabird feces and feathers from the Antarctic Peninsula as a function of log D (A) and log KOA (B). In Figure 4A, the solid line shows the behavior of detected PFC concentrations versus log D in lichen; the dashed line indicates the trend of detected PFC concentrations versus log D values in feces. In Figure 4B, the solid line indicates the behavior of detected PFC concentrations versus log KOA in lichen, while the dashed line shows detected PFC concentrations versus log KOA in feces. Abbreviations for PFC chemical names are defined in Table 1.

PFC HEALTH RISKS

Concentrations of PFCs detected in feathers and feces of the two seabird species studied here are well below the toxicological reference value (TRV) of PFOS (600 ng/g ww), calculated as an exposure threshold value for birds in nature, especially for apex avian predators [42]. This comparison indicates that gentoo penguins and petrels are not at risk by PFOS toxic effects.

TRANSPORT MECHANISMS, GLOBAL AND LOCAL SOURCES OF PFCS TO THE ANTARCTIC

There is still a degree of uncertainty surrounding the dominant pathway of PFC movement to the Antarctic, though researchers have highlighted two primary mechanisms generally accepted as the major modes of PFC transportation to the Antarctic: atmospheric and oceanic. Neutral, volatile precursor compounds, such as perfluorinated sulfonamide alcohols (FOSEs), perfluorooctane sulfonamides (FOSAs), and fluorotelomer alcohols (FTOHs), referred to as "flyers", are capable of being delivered to the Antarctic via fast, direct transport of contaminated wind, as opposed to cold trapping, as is common for many legacy POPs [12, 43–45]. Following deposition, these compounds are degraded via oxidation to form ionic PFCs, including PFSAs and PFCAs [12, 44–48]. Evidence supporting this mechanism of travel includes measurements of FTOHs from Europe to the Antarctica showing declining concentrations in the atmosphere with increased distance from sources in the Northern Hemisphere [12]. Given the far distances PFCs must travel to reach the Antarctic, in combination with short atmospheric residence times (ranging on average from 10 to 50 days), the level of effectiveness associated with atmospheric delivery of PFCs is relatively low. Additionally, the yield of ionic PFCs produced via oxidation of precursor compounds once transported to the Antarctic is often low [10, 11, 43–45, 47].

It is therefore expected that most PFCs are delivered to the Antarctic in their ionic, water-soluble state via the oceans [12, 14, 49]. Oceanic transport functions on a slower time scale for the Antarctic (in the order of decades, compared to days or weeks for atmospheric transport) because of the circulation patterns of the Southern Ocean, protecting the Antarctic from immediate fluxes in PFC concentrations as they are released elsewhere in the world. As time progresses, however, contamination from oceanic sources is anticipated to increase [10, 11, 45]. Slow oceanic transport is cited as the reason for increasing PFC concentrations in the Arctic since the 1950s [14]. Models designed for Arctic research show that if oceanic transport to the Arctic ceased, the quantity of PFCs and their precursors delivered to the Arctic via atmospheric transport could not account for the concentrations measured in water, and thus marine transport is considered to be more important than atmospheric transport [15, 43]. It is also important to note that atmospheric and oceanic transport may be difficult or impossible to discern. For instance, PFCs found in the ocean are made up of three inputs: direct emissions to water, atmospheric deposition into water, and precursor compounds into water followed by degradation to ionic PFCs [14].

Among the compounds found in this study, PFHpA, PFBA, and PFPeA are byproducts of stain/grease-proof coatings on food packaging, couches, and carpets, while PFHxS was used in fire-fighting foams and carpet treatments and phased out of consumer products along with PFOS and PFOA by the major manufacturer (3M Co.) in the early 2000s due to health risks. While long-range atmospheric and oceanic transport of PFCs may partially explain the ubiquitous nature of these contaminants in the Antarctic Peninsula, military bases and infrastructure of nations established there may also be contributing sources of PFCs in Antarctic ecosystems. Atmospheric long-range transport of PFAAs as marine aerosols and degradation of PFCA and PFSA precursors such as low molecular weight FTOHs and acrylates/acids (FTAs) or perfluoroalkyl sulfonamids (FASA) and sulfonamido ethanols (FASE), which are more volatile and released to the atmosphere during fluoropolymer production processes, can be considered as other major pathways [11, 14, 44, 50, 51] to reach and deposit on the Antarctic Peninsula.

Additional and potential sources of PFAAs in the Antarctic Peninsula include aqueous film-forming foams (AFFF) and emissions of a current use insecticide, sulfluramid (N-ethyl perfluorooctane sulfonamide), to control leaf-cutting and fire ants in South America [20, 52; J. Benskin, pers. comm., June 2012). AFFF formulations have consisted of perfluoroalkyl sulfonates (PFHxS, PFOS, PFDS) and more recently, fluorotelomer sulfonamide-based surfactants. While these latter materials can degrade down to short-chain perfluoroalkylcarboxylates (typically C4, C5, C6 PFCAs), sulfluramid can degrade to PFOS, FOSA, and PFCAs by abiotic and/or biological processes [53, 54]. Sulfluramid is manufactured in Brazil ($\approx$30 tons/year in 2007), and, in 2006, about 12 tons was exported to 13 other Central and South American countries [23, 55]. Because this insecticide is a semivolatile substance, it could be transported atmospherically to the Antarctic. Sulfluramid degradation products include PFOSA, PFOS, and potentially PFOA [52]. Despite high concentrations of PFOS, PFOA, and PFOSA measured off the Atlantic coast of South America (South Atlantic), increasing from Brazil to near Rio de la Plata (Argentina–Uruguay), attributed to the use of this substance [52], PFOS and PFOA were not detected in these Antarctic samples, with the exception of two feces samples and a lichen sample. This indicates that these two compounds have not yet fully reached the Antarctic Peninsula region, or local sources are not significant. The detection of several PFCA compounds in the present study is of particular importance as increasing trends of PFCA precursors (i.e., FTOHs) was observed in the Arctic with doubling times of 2.3–3.3 years from 2006 to 2012 [6].

IMPACT ASSESSMENT, ENVIRONMENTAL MANAGE-MENT, AND MONITORING IMPLICATIONS

The Ecuadorian Pedro Vicente Maldonado Scientific Station has been operated since 1988 shortly after Ecuador signed the Antarctic Treaty System (ATS) in 1987. In 1988, Ecuador became an associated member of the Scientific Committee for Antarctic Research (SCAR), and in November 1990 became a consultative member of the ATS [56, 57]. To accomplish this task, Ecuador fulfilled the Antarctic Treaty of "peaceful purposes" and "freedom of scientific investigation" [58]. The commitment to the protection of the Antarctic environment requires being in compliance with the Madrid Protocol, which since 1991 is the prime basis for environmental management of the Antarctic terrestrial and near-shore environments. At the Maldonado Station, the Antarctic environmental management program deploys and integrates a range of generic and international tools, including environmental impact assessments (EIAs), monitoring of pollutants in the marine environment, species and habitat protection, following the environmental principles of the Madrid Protocol, and the administrative and procedural mechanisms of the Committee for Environmental Protection (CEP) [58]. The Ecuadorian Antarctic Institute (INAE) has established good environmental practices, and trained their staff and visitors with a conduct code according to the Madrid Protocol. During the 2010–2011 period, an EIA was performed by the INAE [59] to establish the baseline conditions of the military base and research station, including its area of influence, developing the Environmental Management Plan for the activities taken place at the Maldonado Station.

Considering that local activities and maritime traffic can pollute the surroundings of the Maldonado Station and Antarctic Peninsula, results from impact assessments and monitoring of water quality and potential contaminants have revealed the presence of other anthropogenic pollutants such as hydrocarbons and pesticides in the marine environment [59]. For instance, analyses of total hydrocarbons (THCs) were performed in water samples at the Guayaquil Bay, where concentrations ranged from 0.3 mg/L in sites near the Maldonado Station to 0.85 mg/L at Chile Bay [59]. Pesticide concentrations at several sites of Chile Bay revealed the presence of the organochlorine insecticides, including lindane or γ-hexachlorocyclohexane (γ-HCH) (i.e., 0.335 mg/L) and β-hexachlorocyclohexane (β-HCH or beta-BHC) (i.e., 0.00072 mg/L), as well as traces of the organophosphate malathion and the herbicide atrazine [59]. The long-range atmospheric transport associated with direct deposition or precipitation of volatile organic chemical is now recognized as a major pathway by which pesticides can be transported and deposited in surface waters and ice of Antarctica thousands of kilometers far from their sources [60–62]. Although relatively low concentrations of some PFCs were observed in biota and sediments samples of the remote western Antarctic Peninsula environment and local sources associated with scientific stations and military bases appear to not be significant sources of PFCs, this study gives further evidence of background concentrations around the Antarctic. Results from this study are consistent with research showing that volatile PFCs are subject to atmospheric long-range transport to remote regions, contributing to the contamination of persistent PFCA and PFSA compounds in the Antarctic [11, 44].

Despite the limited sampling and the need for replication to confirm the findings of this study, the biological (lichens, feathers, feces) and abiotic (sediments) samples assessed in this work can be used as environmental matrices to track the fate of PFCs at various temporal and spatial scales in Antarctica. Of particular importance is the detection of several PFC compounds in seabird feces and lichens, which can be used as matrices for noninvasive sampling and long-term monitoring programs of PFCs in the Antarctic Peninsula. Long-term air monitoring and sampling of volatile PFCs in the Antarctica Peninsula is also recommended to elucidate atmospheric sources to the Maldonado's Station and surrounding environment.

ACKNOWLEDGEMENTS

We are in debt with the staff and military personnel of the Ecuadorian Antarctic Institute (Instituto Antártico Ecuatoriano, INAE), especially CPNV José Olmedo Morán, for the logistics and coordination for the 2009 Ecuadorian–Antarctic expedition. Special thanks to Dr. Patricia Ruano and Peter Olaya for their support to elaborate the map and to Dr. Jonathan Benskin for providing preliminary insights to this work.

REFERENCES

1.   Vallack HW, Bakker DJ, Brandt I, Brostrom-Lunden E, Brouwer A, Bull KR, Gough C, Guardans R, Holoubek I, Jansson B, Koch R, Kuylenstierna J, Lecloux A, Mackay D, McCutcheon P, Mocarelli P, Taalman RDF. Controlling persistent organic pollutants – what's next? Environ Toxicol

Pharm 1998;6:143–75.

2. Jones KC, de Voogt P. Persistent organic pollutants (POPs): state of the science. Environ Pollut 1999;100:209–21.

3. Mackay D, Fraser A. Bioaccumulation of persistent organic chemicals: mechanisms and models. Environ Pollut 2000;110:375–91.

4. Hansen KM, Christensen JH, Brandt J, Frohn LM, Geels C, Skjoth CA, Hedegaard GB. Overview of POPs in the environment: modeling their distribution and spread. Simulation and Assessment of Chemical Processes in a Multiphase Environment. NATO Science for Peace and Security Series C - Environmental Security. 2008; 255–72.

5. Tang H. Recent development in analysis of persistent organic pollutants under the Stockholm Convention. Trend Anal Chem 2013;45:48–66.

6. AMAP: Trends in Stockholm Convention persistent organic pollutants (POPs) in Arctic air, human media and biota. AMAP Technical Report to the Stockholm Convention. AMAP Technical Report No. 7 (2014); Oslo, Norway: Arctic Monitoring and Assessment Programme (AMAP). 2014;54p.

7. Wei S. Chen LQ, Taniyasu S, So MK, Murphy MB, Yamashita N, Yeung LWY, Lam PKS. Distribution of perfluorinated compounds in surface seawaters between Asia and Antarctica. Mar Pollut Bull 2007;54:1813––38.

8. United Nations Environment Program Secretariat of the Stockholm Convention on Persistent Organic Pollutants. The Nine New POPs: an introduction to the nine chemicals added to the Stockholm Convention by the Conference of the Parties at its fourth meeting. 2010. Available from: http://chm.pops.int/Programmes/NewPOPs/Publications/tabid/695/language/en-US/Default.aspx [Accessed: 2015-01-29]

9. Schiavone A, Corsolini S, Kannan K, Tao L, Trivelpiece W, Torres D Jr, Focardi S. Perfluorinated contaminants in fur seal pups and penguin eggs from South Shetland, Antarctica. Sci Total Environ 2009;407:3899–904.

10. Cai M, Yang H, Xie Z, Zhao Z, Wang F, Lu Z, Sturm R, Ebinghaus R. Per- and polyfluoroalkyl substances in snow, lake, surface runoff water and coastal seawater in Fildes Peninsula, King George Island, Antarctica. J Hazard Mater 2012;209–10:335–42.

11. Del Vento S, Halsall C, Gloia R, Jones K, Dachs J. Volatile per- and polyfluoroalkyl compounds in the remote atmosphere of the western Antarctic Peninsula: an indirect source of perfluoroalkyl acids to Antarctic waters? Atmos Pollut Res 2012;3:450–5.

12. Llorca M, Farre M, Tavano MS, Alonso B, Koremblit G, Barcelo D. Fate of a broad spectrum of perfluorinated compounds in soils and biota from Tierra del Fuego and Antarctica. Environ Pollut 2012;163:158–66.

13. Yamashita N, Kannan K, Taniyasu S, Horii Y, Petrick G, Gamo T. A global survey of perfluorinated acids in oceans. Mar Pollut Bull 2005;51:658–68.

14. Prevedouros K, Cousins IT, Buck RC, Korzeniowski SH. Sources, fate, and transport of perfluorocarboxylates. Environ Sci Technol 2006;40:32–44.

15. Zhao Z, Xie Z, Moller A, Sturm R, Tang J, Zhang G, Ebinghaus R. Distribution and long-range transport of polyfluoroalkyl substances in the Arctic, Atlantic Ocean and Antarctic coast. Environ Pollut 2012;170:71–7.

16. U.S. EPA: Fluorochemical Use, Distribution and Release Overview; Public Docket AR226-0550; 26 May 1999; St. Paul, MN: 3M Company. 1999.

17. Begley TH, White K, Honigfort P, Twaroski ML, Neches R, Walker RA. Perfluorochemicals: potential sources of and migration from food packaging. Food Addit Contam 2005;22:1023–31.

18. Lau C, Anitole K, Hodes C, Lai D, Pfahles-Hutchens A, Seed J. Perfluoroalkyl acids: a review of monitoring and toxicological findings. Toxicol Sci 2007;99:366–94.

19. Kelly BC, Ikonomou MG, Blair JD, Surridge B, Hoover D, Haviland L, Grace R, Gobas FAPC. Perfluoroalkyl contaminants in an Arctic marine food web: trophic magnification and wildlife exposure. Environ Sci Technol 2009;43:4037–43.

20. Armitage JM, Schenker U, Scheringer M, Martin JW, MacLeod M, Cousins IT. Modeling the global fate and transport of perfluorooctane sulfonate (PFOS) and precursor compounds in relation to temporal trends in wildlife exposure. Environ Sci Technol 2009;43:9274–80.

21. Giesy JP, Naile JE, Khim JS, Jones PD, Newsted JL. Aquatic toxicology of perfluorinated chemicals. In: Whitacre DM, ed. Rev Environ Contam Toxicol, Rev Environ Contam Toxicol 202. Springer Science + Business Media, LLC; 2010;p.1–52. DOI 10.1007/978-1-4419-1157-5_1.

22. Lim TC, Wang B, Huang J, Deng S, Yu G. Emission Inventory for PFOS in China: review of past methodologies and suggestions. Sci World J 2011;11:1963–80. DOI:10.1100/2011/868156.

23. Carloni D. Perfluorooctane sulfonate (PFOS) production and use: past

and current evidence. Vienna, Austria, 2009: United Nations Industrial Development Organization (UNIDO); 2009;56p. Available from: http://www.unido.org/fileadmin/user_media/Services/Environmental_ Management/Stockholm_Convention/POPs/DC_Perfluorooctane%20 Sulfonate%20Report.PDF [Accessed: 2015-01-29].

24. Wei Y. Obstacles in risk management of PFOS. In: Informal Workshop on Stakeholders' Information Needs on Chemicals in Articles/Products; 9–12 February 2009; Geneva, Switzerland: United Nations Environment Programme (UNEP); 2009. Available from: http://www.chem.unep.ch/ unepsaicm/cheminprod_dec08/default.htm [Accessed: 2015-01-29].

25. Han W. PFOS related actions in China. Workshop on Managing Perfluorinated Chemicals and Transitioning to Safer Alternatives; 12–13 February 2009; Geneva, Switzerland: United Nations Environment Programme (UNEP); 2009. Available from: http://www.chem.unep.ch/ unepsaicm/cheminprod_dec08/PFCWorkshop/default.htm [Accessed: 2015-01-29].

26. Rattner BA, Scheuhammer AM, Elliott JE. History of wildlife toxicology and the interpretation of contaminant concentrations in tissues. In: Beyer WN, Meador JP, eds. Environmental Contaminants in Biota: Interpreting Tissue Concentrations. 2nd ed. Boca Raton, FL: CRC Press; 2011. p.9–44.

27. Calle P, Alvarado O, Monserrate L, Cevallos, JM, Calle NL, Alava JJ. Mercury accumulation in sediments and seabird feathers from the Antarctic Peninsula. Mar Pollut Bull 2015;91(2):410–7. DOI:10.1016/j. marpolbul.2014.10.009.

28. Meyer J, Jaspers VLB, Eens M, de Coen W. The relationship between perfluorinated chemical levels in the feathers and livers of birds from different trophic levels. Sci Total Environ 2009;407:5894–900. DOI:10.1016/j.scitotenv.2009.07.032.

29. Fujita T, Iwassa J, Hansch CA. New substituent constant, π, derived from partition coefficients. J Am Chem Soc 1964;86(23):5175–80.

30. Kah M, Brown CD. Log D: lipophilicity for ionisable compounds. Chemosphere 2008;72(10):1401–8.

31. Armitage JM, Arnot JA, Wania F, Mackay D. Development and evaluation of a mechanistic bioconcentration model for ionogenic organic chemicals in fish. Environ Toxicol Chem 2013;32(1):115–28.

32. Ostertag SK, Tague BA, Humphries MM, Tittlemier SA, Chan HM. Estimated dietary exposure to fluorinated compounds from traditional foods among Inuit in Nunavut, Canada. Chemosphere 2009;75:1165–72.

33. Arp HP, Niederer C, Goss KU. Predicting the partitioning behavior of various highly fluorinated compounds. Environ Sci Technol 2006;40:7298–304.

34. Bossi R, Riget FF, Dietz R. Temporal and spatial trends of perfluorinated compounds in ringed seal (*Phoca hispida*) from Greenland. Environ Sci Technol 2005;39:7416–22.

35. Dietz R, Bossi R, Riget FF, Sonne C, Born EW. Increasing perfluoroalkyl contaminants in east Greenland polar bears (*Ursus maritimus*): a new toxic threat to the Arctic bears. Environ Sci Technol 2008;42:2701–7.

36. Chen CL, Lu YL, Zhang X, Geng J, Wang TY, Shi YJ, Hu WY, Li J. A review of spatial and temporal assessment of PFOS and PFOA contamination in China. Chem Ecol 2009;25:163–77.

37. Reiner JL, O'Connell SG, Moors AJ, Kucklick JR, Becker PR, Keller JM. Spatial and temporal trends of perfluorinated compounds in Beluga Whales (*Delphinapterus leucas*) from Alaska. Environ Sci Technol 2011;45:8129–36.

38. Butt, CM, Muir DCG, Stirling I, Kwan M, Mabury SA. Rapid response of arctic ringed seals to changes in perfluoroalkyl production. Environ Sci Technol 2007;41:42–9.

39. Hart K, Gill VA, Kannan K. Temporal trends (1992–2007) of perfluorinated chemicals in northern sea otters (*Enhydra lutris kenyoni*) from South-Central Alaska. Arch Environ Contam Toxicol 2009;56:607–14.

40. Gobas FAPC, Morrison HA. Bioconcentration and bioaccumulation in the aquatic environment. In: Boethling R, Mackay D, eds. Handbook of Property Estimation methods for chemicals: Environmental and Health Sciences. Boca Raton, FL, USA: CRC Press LLC; 2000. p.189–231.

41. Gobas FAPC, de Wolf W, Verbruggen E, Plotzke K, Burkhard L. Revisiting bioaccumulation criteria for POPs and PBT assessments. Integrat Environ Ass Manage 2009;5(4):624–37.

42. Newsted JL, Jones PD, Coady KK, Giesy JP. Avian toxicity reference values for perfluorooctane sulfonate. Environ Sci Technol 2005;39:9357–62.

43. Wania F. A global mass balance analysis of the source of perfluorocarboxylic acids in the Arctic Ocean. Environ Sci Technol 2007;41(13):4529–35.

44. Dreyer A, Weinberg I, Temme C, Ebinghaus R. Polyfluorinated compounds in the atmosphere of the Atlantic and Southern Oceans: evidence for a global distribution. Environ Sci Technol 2009;43:6507–14.

45. Bengston Nash S, Rintoul SR, Kawaguchi S, Staniland I, van den Hoff J,

Tierney M, Bossi R. Perfluorinated compounds in the Antarctic region: ocean circulation provides prolonged protection from distant sources. Environ Pollut 2010;158(9):2985–91.

46. Ellis DA, Martin J, De Silva AO, Mabury SA, Hurley MD, Sulbaek Andersen MP, Wallington TJ. Degradation of fluorotelomer alcohols: a likely atmospheric source of perfluorinated carboxylic acids. Environ Sci Technol 2004:38(12):3316–21.

47. Wallington TJ, Hurley MD, Xia J, Wuebbles DJ, Sillmann S, Ito A, Penner JE, Ellis DA, Martin J, Mabury SA, Nielson OJ, Andersen MPS. Formation of C7F15COOH (PFOA) and other perfluorocarboxylic acids during the atmospheric oxidation of 8:2 fluorotelomer alcohol. Environ Sci Technol 2006;40(3):924–30.

48. Schenker U, Scheringer M, MacLeod M, Martin JW, Cousins IT, Hungerbühler K. Contribution of volatile precursor substances to the flux of perfluorooctanoate to the Arctic. Environ Sci Technol 2008;42(10):3710–6.

49. Armitage JM, Cousins IT, Buck RC, Prevedouros K, Russell MH, MacLeod M, Korzeniowski SH. Modeling global-scale fate and transport of perfluorooctanoate emitted from direct sources. Environ Sci Technol 2006;40(22):864–72.

50. Paul AG, Jones KC, Sweetman AJ. A first global production, emission, and environmental inventory for perfluorooctane sulfonate. Environ Sci Technol 2009;43:386–92.

51. Lee H, D'eon J, Mabury SA. Biodegradation of polyfluoroalkyl phosphates as a source of perfluorinated acids to the environment. Environ Sci Technol 2010;44:3305–10.

52. Benskin JP, Muir CG, Scott BF, Spencer C, De Silva AO, Kylin H, Martin JW, Morris A, Lohmann R, Tomy G, Rosenberg B, Taniyasu S, Yamashita N. Perfluoroalkyl acids in the Atlantic and Canadian Arctic Oceans. Environ Sci Technol 2012;46(11):5815–23.

53. Tomy GT, Tittlemier SA, Palace VP, Budakowski WR, Braekevelt E, Brinkworth L, Friesen K. Biotransformation of N-ethyl perfluorooctanesulfonamide by rainbow trout (*Onchorhynchus mykiss*) liver microsomes. Environ Sci Technol 2004;38:758–2.

54. Martin JW, Ellis DA, Mabury SA, Hurley MD, Wallington TJ. Atmospheric chemistry of perfluoroalkanesulfona-mides: kinetic and product studies of the OH radical and Cl atom initiated oxidation of N-ethyl perfluorobutanesulfonamide. Environ Sci Technol 2006;40:864–72.

55. United Nations Environment Programme: Risk Management evaluation on perfluorooctane sulfonate. Report of the Persistent Organic Pollutants Review Committee on the work of its third meeting; 19−23 November, 2007. Geneva, Switzerland: United Nations Environment Programme (UNEP). 2007.

56. INAE (Instituto Antártico Ecuatoriano): Acerca del INAE [Internet]. 2015. Available from: http://www.inae.gob.ec/index.php/sobre-inae [Accessed: 2015-02-02].

57. COMINAP (Council of Managers of National Antarctic Programs): Instituto Antártico Ecuatoriano (INAE): About Ecuador's National Antarctic Program [Internet].2015. Available from: https://www.comnap. aq/Members/equador/SitePages/Home.aspx [Accessed: 2015-02-02].

58. [58]Hemmings AD. 'Environmental Management' as diplomatic method: the advancement of strategic national interest in Antarctica. In: Liggett D, Hemmings AD, eds. Exploring Antarctic Values. Proceedings of the the workshop Exploring Linkages between Environmental Management and Value Systems: The Case of Antarctica; 5 December 2011; Christchurch: University of Canterbury, Gateway Antarctica Special Publication Series 1301; 2013. p.70−89. Available from:http://antarctica-ssag.org/ wp-content/uploads/2013/05/SSAG-proceedings-2013.pdf. [Accessed: 2015-02-02].

59. Instituto Antártico Ecuatoriano: Estudio de Impacto Ambiental, Ex-Post. Estación Científica Ecuatoriana Pedro Vicente Maldonado; Verano Austral, Enero-Marzo 2010 y 2011. Isla Greenwich, Shetland del Sur, Antártida: Soluciones Ambientales Totales (SAMBITO). 2011. p. 117. Available from: http://es.slideshare.net/sambitoeco/eia-expost-pevima [Accessed: 2015-02-02].

60. Tanabe S, Hidaka H, Tatsukawa, R. PCBs and chlorinated hydrocarbon pesticides in Antarctic atmosphere and hydrosphere. Chemosphere 1983;12:277−88.

61. Larsson P, Järnmark C, Södergren A. PCBs and chlorinated pesticides in the atmosphere and aquatic organisms of Ross Island, Antarctica. Marine Pollut Bull 1992;25:281−7.

62. Sen Gupta R, Sarkar A, Kureishey TW. PCBs and organochlorine pesticides in krill, birds and water from Antarctica. Deep Sea Res Part II Top Stud Oceanog 1996;43:119−26.

63. Alava JJ, Ikonomou MG, Riofrío Briceño M, Gobas FAPC. Estudio de contaminantes orgánicos persistentes (COPs) en la Antártica:

Contaminantes Perfluoroalquilos (Perfluoroalkyl Contaminants, PFCs). Informe Final 2009. Ministerio de Defensa Nacional & Instituto Antártico Ecuatoriano (INAE); 15 Septiembre 2009. Guayaquil: INAE. 2009. p. 20.

Chapter 4

WHAT ARE THE ROLES OF NATIONAL AND INTERNATIONAL INSTITUTIONS TO OVERCOME BARRIERS IN DIFFUSING CLEAN ENERGY TECHNOLOGIES IN ASIA?: MATCHING BARRIERS IN TECHNOLOGY DIFFUSION WITH THE ROLES OF INSTITUTIONS

Masachika Suzuki[1]

[1]Faculty of Commerce, Kansai University, Suita-shi, Osaka, Japan

INTRODUCTION

While the international negotiation on climate change does not make much progress in designing the post-Kyoto scheme, technology innovation and transfer is becoming a central issue in the negotiation. In Cancun in 2010, the parties agreed to organize the Technology Executive Committee (TEC) and the Climate Technology Centre and Network (CTCN) (UNFCCC 2011). The developed countries have committed to provide $100 billion yearly to assist the developing countries in mitigation and adaptation through the Green Climate Fund (UNFCCC 2011).[1] - The scheme of the Fund is currently under discussion at the Transitional Committee for the design of the Green Climate Fund.

This paper consists of two parts. The first part of the paper attempts to show a broad landscape of barriers in technology diffusion in the developing countries by addressing two levels of barriers. The first level is about the barriers that are commonly observed among the developing countries (Section 2.1). The paper classifies these barriers into technological, financial and institutional barriers. The second level is about the barriers that are technology-specific (Section 2.2 and 2.3). Section 2.3 summaries the results of previous case studies that were conducted to uncover technology-specific barriers in diffusing clean energy technologies in Asia. These case studies include both technologies for industrial use such as wind, bio-energy and building energy efficiency and technologies for individual use such as LED (Light Emitting Diode) and Photovoltaic (PV) panels. It also contains technologies at the innovation stage such as Integrated Gasification Combined Cycle (IGCC) and Carbon Capture and Storage (CCS). Section 2.3 presents an analysis of the barriers through a comparison of the results of the case studies.

The second part of the paper explores roles of institutions to overcome identified barriers in diffusing clear energy technologies in Asia (Section 3). It addresses theoretical discussions on functions (or roles) of international and national institutions in technology innovation. It then attempts to match the barriers in technology diffusion identified in Section 2 with the functions of national and international institutions. The results of matching indicate that there are important roles of institutions both at the early and advanced stages of technological development to encourage R&D cooperation from the public site (early stage) and enhance the enabling environment and facilitate finance for the technologies (advanced stage).

STUDIES ON BARRIERS IN TECHNOLOGY DIFFUSION IN THE DEVELOPING COUNTRIES

Understanding barriers in technology diffusion lead to important lessons in designing policy instruments and institutions for diffusing clean energy technologies in the developing countries. With this understanding, researching about barriers has been part of the tasks under the UNFCCC as well as United Nations Environmental Program (UNEP) (UNFCCC 2011; UNEP Risø Centre on Energy, Climate and Sustainable Development 2011). Painuly indicates that there are several levels to explore and analyze such barriers. Painuly adds that the first level is a broad category of barriers and the lower levels include more detail and specific barriers (Painuly 2001). Section 2.1 illustrates barriers at the first level. Section 2.2 lists case studies that address barriers at a lower level that are more technology specific. Section 2.3 presents an analysis of the barriers through a comparison of the results of the case studies.

Barriers Commonly Observed Among The Developing Countries

The barriers at the first level are the barriers that are commonly observed among the developing countries. There are substantial amounts of research projects that have attempted to identify the barriers at this level including Painuly (2001), OECD/IEA (2001), Painuly and Fenhann (2002) andRaddy and Painuly (2004). Table 1 summaries key barriers identified through these and other research. The barriers are classified into technological, financial and institutional barriers[2] - [3] –

Table 1: Barriers (technological, financial and institutional) observed among the developing countries

Barriers	Barriers	Explanations	Source(s)
Technological	Limited capacity to assess, adopt, adapt and absorb technological options	• These technologies are primarily targeted at rural areas or poor customers, who have limited capacity to absorb these technologies. There is a general resistance to change, which is magnified due to lack of capacity to understand, adopt and adapt the technologies for greater benefit. The capacity constrains are not only linked to its use but in its production. There is limited manufacturing capacity and as a result not much innovation has taken place. Scale-up of manufacturing and therby reduction in the associated costs has not taken place. (Ravindranath and Balachandra pp.1010) • Technology not freely available in the market, technology developer not willing to transfer technology, problems in import of technology/ equipment due to restrictive policies/taxes etc. (Painuly pp.82)	(Ravindranath and Balachandra 2009) (Painuly 2001)
	Lack of knowledge of technology operation and management	• Lack of knowledge of technology operation and management as well as limited availability of spare parts and maintenance expertise (Doukas et al p.1139)	(Doukas et al 2009) (Luken and Rompaey 2008) (OECD/IEA 2001)
	Lack of skilled personnel/training facilities	• This can be a constraint for producers (Painuly p.80) • Lack of experts to train, lack of training facilities, inadequate efforts. (Painuly pp.83) • In China and much of South East Asia, there is a need for technically trained people and people with strong management skills. Where training of local workforce is provided, it should be recognized that Asians tend to learn more effectively by coping, rather than as individuals, when local language is used and with a practical "hands-on" approach. Also the issue of training in intellectual property rights is important. This is a long term issue but will be important for long term changes in attitudes to intellectual property rights in China. (Guerin pp.71)	(Painuly 2001) (Usha and Ravindranath 2002) (Jagadeesh 2000) (IPCC 2000) (Guerin 2001) (Worrell et al. 2001) (Flamos et al. 2008) (OECD/IEA 2001)
Technological	Lack of standard and codes and certification	• Product quality and product acceptability is affected. (Painuly pp.80) • Lack of institution/initiative to fix standards, lack of capacity, lack of facilities for testing/certification. (Painuly pp.83) • A degree of standardization would improve the penetration of photovoltaics (PVs), it would enable PVs to become more user friendly. (Oliver and Jackson pp.381) • Lack of standardization in system components resulting from the wide range in design features and technical standards, and absence of long-term policy instruments have resulted in manufacturing, servicing and maintenance difficulties of wind turbines. (Jagadeesh pp. 162)	(Painuly 2001) (Oliver and Jackson 1999) (IPCC 2000) (Joanna 2007) (Jagadeesh 2000) (OECD/IEA 2001) (Oltz and Beerepoot 2010)
Financial	Lack of access to financing	• High first costs and investments associated with mass manufacturing remain as barriers. Both the users and the manufactures have very low capital. This problem is further	(Ravindranath and Balachandra 2009) (Painuly 2001)

Barriers	Barriers	Explanations	Source(s)
		accentuated by the rigid lending procedures that limited access to financing even when financing is available on standard norms. (Ravindranath and Balachandra pp.1010)	(UNFCCC 2003) (Worell et al. 2001) (Jagadeesh 2000)
		• Capital costs may go up due to increased risk perception. Adverse effect on competition and efficiency. (Painuly pp.79)	(IPCC 2000)(Thorne 2008)
		• Small and medium scale enterprises (SMEs) above all lack the finances for cleaner technologies, but also contact with larger technology manufacturers and formal information channels. (UNFCCC 2003, p.12)	
		• Limited capital availability will lead to high hurdle rates for energy efficiency investments because capital is used for competing investment priorities...High inflation rates in developing countries and CEITs, lack of sufficent infrastructure increase the risks for domestic and foreign investors and limit the availability of capital (Worrell et al 2001, pp.6-7)	
		• International public finance is no longer going into energy (electricity) infrastructure, which is now seen as of interest to the private sector under the neo-liberal or privatization agenda (Thorne, p.3)	
Financial	Potential lack of commercial viability	• In general, technology imported from industrialized countries is more efficient but also more expensive than technology manufactured locally, and it therefore requires higher initial investment costs. This is of particular importance for the transfer of environmentally sound technologies. Furthermore, as a result of their typically early commercialization stage, environmentally sound technologies are often considered riskier than existing commercial technologies (Karakosta et al., p.1551)	(Karakosta et al, 2010)
	Lack of financial institutions to support renewable energy technologies, lack of instruments	• Adverse effect on competition and efficiency. (Painuly pp.79) • Under-developed capital markets, restricted entry to capital markets, instruments unfavorable regulations. (Painuly pp.83)	(Painuly 2001) (Jagadeesh 2000)
Institutional	Uncertain governmental policies	• Many of the renewable energy technologies in India are still in the development stage. There are no sufficient governmental regulations/ incentives to stimulate the adoption of renewable energy technologies by business and industries. They include: (a) lack of explicit national policy for renewable energy at end-use level; (b) incomplete transition to cost-based electric tariffs for most residential and some industrial customers; (c) poor availability of credit to the purchase of renewable energy technologies in the economy; and (d) lack of application of modern management skills in energy development agencies. (Reddy and Painuly pp.1436)	(Redd and Painuly 2004) (Painuly 2001) (Worell et al. 2001) (Schneider and Hoffman 2008) (Doukas et al. 2009) (Karakosta et al. 2010) (OECD/IEA 2001)

Barriers	Barriers	Explanations	Source(s)
		• It creates uncertainty and results in lack of confidence. May also increase cost of project. (Painuly pp.80)	
		• Uncertainty in policies, un-supportive policies, inadequately equipped governmental agency, red tape, lack of governmental faith in RETs, lack of policies to integrate renewable energy technologies products with the global market, inadequately equipped governmental agency to handle the product.(Painuly pp.84)	
		• National trade and investment policies may limit the inflow of foreign capital. This might be a barrier to technology transfer (Worrell et al. 2001, p.7)	
		• Uncertain ownership, lack of intellectual property-rights protection and unclear arbitration procedures. (OECD/IEA p.14)	
Institutional	Lack of infrastructure	• Problems related to availability of infrastructure such as roads, connectivity to grid, communications, other logistics. (Painuly pp.84) • The places where energy infrastructure has not yet been extended to are, by-and-large, areas where people are poor and unlikely to be able to cover the costs of infrastructure, nor would the users be able to consume sufficient service to make the investment financially feasible alone. Perversely, these are the development niches where many of the immature environmentally sound technologies may already provide least energy cost options. (Thorne pp.3-4)	(Painuly 2001) (Thorne 2008)
	Lack of information and awareness	• It increases uncertainty, and hence costs. (Painuly pp.79) • Lack/low level of awareness, inadequate information on product, technology, costs, benefits & potential of the renewable energy technologies, O&M costs, financing sources etc. Lack of agencies, or agencies ill equipped to provide information. Also, feedback mechanism may be missing or inadequate. Lack of knowledge/access to renewable energy technologies resource assessment data, implementation requirements. (Painuly pp.82) • It is generally believed that the adoption of renewable energy technologies are often not undertaken as a result of lack of information or knowledge on the part of the customer, or a lack of confidence in obtaining reliable information. Households and small firms and commercial establishments face difficulties in obtaining information on renewable energy technologies compared to the simplicity of buying conventional energy technologies. There is hardly any knowledge (software and/or hardware) about renewable energy technologies that is readily available and easily accessible for the consumers. Under these circumstances, information collection and processing consume time and resources which is difficult for small firms and individual households. (Reddy and Painuly pp.1435)	(Kathuria 2002) (IPCC 2000) (Painuly 2001) (Reddy and Painuly 2004) (UNFCCC 2003) (Worrell et al. 2001) (Flamos et al. 2008) (Karakosta et al. 2010) (Luken and Rompaey 2008) (OECD/IEA 2001)

Barriers	Barriers	Explanations	Source(s)
Institutional	Lack of consumer acceptance	• Adoption of renewable energy technologies are generally influenced by consumer perceptions of the quality and usefulness of these items when compared to conventional technologies. Renewable energy technologies are often perceived to be used with discomfort or sacrifice rather than as providing equivalent services with less energy and cost. Also, while purchasing a technology, consumers take the advice of their friends rather than obtaining information from the experts and take decisions which may not be economically rationale. (Reddy and Painuly pp.1436-1437) • Unknown product, aesthetic considerations, products lacks appeal, resistance to change, cultural reasons, high discount rates of consumers, inadequate information. (Painuly pp.84) • Many potential users of sustainable energy technologies have no or little experience with their application and the assistance provided in the development of such technologies is insufficient. Moreover, dissemination of EU experience sustainable energy technology implementation to other countries in the world has been limited (Flamos, p.5)	(Reddy and Painuly 2004) (Painuly 2001) (Flamos et al. 2008)

Technological barriers include not only limited access to the international technology market but also limited capacity to assess, adopt, adapt and absorb technological options (Ravindranath and Balachandra 2009; Painuly 2001). As the table indicates, lack of knowledge of technology operation and management as well as lack of skilled personnel/training facilities can be a major barrier for successful diffusion of clean energy technologies (Doukas et al. 2009; Luken and Rompaey 2008;Painuly 2001; Usha and Ravindranath 2002; Jagadeesh 2000; IPCC 2000; Guerin 2001; Worrell et al. 2001; Flamos et al. 2008; OECD and IEA 2001). Lack of standard and codes and certification can be a barrier too since product quality and product acceptability is affected (Painuly 2001).

A lack of financing is a major part of the financial barriers (Ravindranath and Balachandra 2009;Painuly 2001; UNFCCC 2003; Worell et al. 2001; Jagadeesh 2000; IPCC 2000; Thorne 2008).Ravindranath and Balachandra (2009) states that "high first costs and investments associated with mass manufacturing remain as barriers. Both the users and the manufactures have very low capital. This problem is further accentuated by the rigid lending procedures that limited access to financing even when financing is available on standard norms." At this point, Karakosta et al. (2010) further elaborates that "in general, technology imported from industrialized countries is more efficient but also more expensive than technology manufactured locally, and it therefore requires higher initial investment costs. This is of particular importance for the transfer of environmentally sound technologies." Lack of financial institutions to support renewable energy technologies as well as lack of financial instruments is also highlighted as part of the financial barriers (Painuly 2001; Jagadeesh 2000).

Institutional barriers include lack of explicit forms of institutions such as goals, policies, regulations and incentive programs as well as lack of implicit form of institutions such as information, awareness, social acceptance, and conditions of the surrounding environment. As for explicit forms of institutions, Painuly (2001) points out uncertainty in policies, unsupportive policies, inadequately equipped governmental agency, red tape, lack of governmental faith in renewable energy technologies, lack of policies to integrate renewable energy technologies products with the global market, inadequately equipped governmental agency to handle the product. Lack of infrastructure is another aspect of institutional barriers, pointed out by Painuly (2001), that is, problems related to availability of infrastructure such as roads, connectivity to grid, communications, and other logistics. As for implicit form of institutions, Painuly (2001) points out lack/low level of awareness, inadequate information on product, technology, costs, benefits and potential of the renewable energy technologies, O&M costs, financing sources. Flamos et al. (2008) addresses lack of customer acceptance as an institutional barrier. It points out that "many potential users of sustainable energy technologies have no or little experience with their application and the assistance provided in the development of such technologies is insufficient" (Flamos et al. 2008).

Section 2.1 addressed barriers that are commonly observed among the developing countries. Section 2.2 illustrates case studies addressing technology-specific barriers.

CASE STUDIES ADDRESSING TECHNOLOGY-SPECIFIC BARRIERS

There are a number of research initiatives that have attempted to identify barriers through the case study approach. The advantage of the case study approach is that it helps to uncover technology-specific barriers, while other studies looking at the developing countries or clean energy as a whole may overlook these barriers. Table 2 lists the case studies that are reviewed in this paper[4] - :

The Science and Technology Policy Research (SPRU) at University of Sussex and TERI in India jointly conducted a research project looking into barriers through several case studies in India including wind power, IGCC (Integrated Gasification Combined Cycle), LED (Light Emitting Diode), biomass, hybrid vehicles and photovoltaic (PV) panels (Case Study 1-6) (Ockwell, D., J. Watson et al. 2007; Ockwell, D., J. Watson et al. 2009). This is the most comprehensive research project thus far looking into barriers through the case study approach. The IIIEE at Lund University in Sweden conducted several case studies including Carbon Capture and Storage (CCS) and building energy efficiency (Case Study 7 and 8) (Dalhammar, C. et al. 2009). In addition, there are a number of case studies that are conducted on the individual basis (Case Study 9-13).

It is observed that many of these case studies are conducted in China and India. This is probably relating to the fact that these two countries have the largest potentials in diffusing clean energy technologies among the developing countries. Another point to note among these case studies is that two popular targets for a case study are wind power and bio-energy (including biomass/biogas). This is possibly due to the fact that these two technologies are at the stage where they are successfully implemented in some cases but there are still facing barriers to point out for further diffusion. On the other hand, Table 2 also indicates that there are a variety of research interests with respect to the targeted technologies for analysis. Some research interests are geared toward to the technologies at the innovation stage such as IGCC and CCS. Some research interests are directed to the products for individual use rather than industrial use such as hybrid vehicles, LEDs, and PV. The diversity in the targeted technologies for analysis may lead to interesting finding about barriers.

Table 2: List of case studies reviewed in this paper

	Research organization/individuals	Information on each case study		Sources
		Country	Technology	
Case study 1	SPRU (Science and Technology Policy Research) at University of Sussex and TERI in India	India	Wind power	Ockwell, D., J. Watson et al. (2009)
Case study 2	SPRU at University of Sussex and TERI in India	India	Integrated Gasification Combined Cycle (IGCC)	A: Ockwell, D., J. Watson et al. (2007) B: Ockwell, D., J. Watson et al. (2009)

	Research organization/individuals	Information on each case study		Sources
		Country	Technology	
Case study 3	SPRU at University of Sussex and TERI in India	India	LED (Light Emitting Diode)	Ockwell, D., J. Watson et al. (2007)
Case study 4	SPRU at University of Sussex and TERI in India	India	Biomass	Ockwell, D., J. Watson et al. (2007)
Case study 5	SPRU at University of Sussex and TERI in India	India	Hybrid vehicles	A: Ockwell, D., J. Watson et al. (2007) B: Ockwell, D., J. Watson et al. (2009)
Case study 6	SPRU at University of Sussex and TERI in India	India	Photovoltaic (PV) panels	Ockwell, D., J. Watson et al. (2009)
Case study 7	International Institute for Industrial Environmental Economics (IIIEE) at Lund University	Developing countries	Carbon Capture and Storage (CCS)	Dalhammar, C. et al. (2009)
Case study 8	IIIEE at Lund University	Developing countries	Building energy Efficiency	Dalhammar, C. et al. (2009)
Case study 9	United Nations Department of Economic and Social Affairs (DESA)	China	Wind power	United Nations, DESA
Case study 10	Lewis J.	India and China	Wind power	A:Lewis, J., (2007a) B:Lewis, J., (2007b)
Case study 11	Mizuno E. (on a publication by UNEP Risø Centre on Energy, Climate and Sustainable Development)	India	Wind power	Mizuno. (2011)
Case study 12	(Ravindranath and Rao on a publication by UNEP Risø Centre on Energy, Climate and Sustainable Development)	India	Bioenergy	Ravindranath and Rao (2011)
Case study 13	Suzuki, M., Okazaki B., and Jain K.	Thailand	Biogas	A: Suzuki, M., Okazaki B., and Jain K. (2010) B: Jain K., Okazaki B., Suzuki, M. (2011)

Comparative Study On Technology-Specific Barriers

Section 2.3 compares the results of the case studies identified in Section 2.2. Table 3 summarizes the results of the studies:

Table 3: Results of case studies

	Research organizations/ individuals	Information on case study			Barriers		
		Country	Technology	Technological barriers	Financial barriers	Institutional barriers
Case study 1	SPRU (Science and Technology Policy Research) at University of Sussex and TERI in India	India	Wind power			• IPR is the main issue. The transfer of technological know-how to Indian companies was restricted. (p.116) • The high cost of IPR acquisition. (p.118) • In the joint ventures and collaborative ventures, it had been noticed that the [Indian] companies had to depend on their European counterparts for all technical aspects and even operation and maintenance issues. (p.117)
						• It is very important to develop the indigenous capacity for technology development and manufacturing. Equally important would be to incentivize innovations from the viewpoint of national priority. (p.120)
Case study 2	SPRU (Science and Technology Policy Research) at University of Sussex and TERI in India	India	IGCC (Integrated Gasification Combined Cycle)	• Limited amount of testing of IGCC that has been done with Indian grade coal. All IGCC demonstration plants to date have been based on coals with different characteristics to Indian coal, especially ash content and ash fusion temperature.(A:p.58) • The long-term success of technology transfer in technologies such as gasification relies on building technological capacity within recipient countries. (A:p.58)	• The two key risks associated with IGCC are high capital costs and the lack of reliable operational history. The risks associated with high capital cost are amplified by the limited operational history and the new nature of this particular application of gasification. (A:p. 58)	• Premature to comment on IPR issues related to IGCC, since this technology is not considered to be commercial globally. (B:p.110)
Case study 3	SPRU (Science and Technology Policy Research) at University of Sussex and TERI in India	India	LED (Light Emitting Diode)	• Although the technical competency in India exists in the fields of material science, engineering, control electronics and other relevant fields, they have to be nurtured in the context of LED technology.(p.72) • Indigenous capacity is to be developed quickly	• No clear indication about the type of market that exists for LED. (p.69) • The leading players worldwide are not considering India as a potential region for investment as they do not see any market in India at present. (p. 72)	• It is a highly protected technology. As there are various processes involved in manufacturing LED chips, each process is patented and requires huge investment. At present the cost of investing in both chip manufacturing and resolving the IPR issues is substantially high compared to importing the chips. Therefore in India, the chips are imported

Research organizations/ individuals	Information on case study		Barriers		
	Country	Technology	Technological barriers	Financial barriers	Institutional barriers
			so that when technology is transferred it can be taken up. (p.74)	• Import of LED is much easier and cheaper than to manufacture it because of IPR issues. (p.69) • LED chip manufacturing requires several processes. Each process involves energy as well as capital-intensive equipment. The existing players in India are relatively smaller in size and are not ready/capable of investing huge amounts for LED chip manufacturing. (p.72)	primarily from China, Taiwan, Japan, the US and other countries. (p.72)
Case study 4	SPRU (Science and Technology Policy Research) at University of Sussex and TERI in India	India	Biomass	• The opportunity cost of power outages at briquetting plants. In many regions of India, electricity from the grid cuts out for hours at a time. (p.80) • The lack of accessibility to power presents problems. In India, where electricity connections are often unavailable in rural locales, the power requirement for briquetting machines could prove to be a major barrier to establishing plants in remote areas even if they are rich in	• Entrepreneurs and manufacturers alike identified working capital as a primary barrier to successful commercialization of briquettes. (p.79) • Banks are reluctant to finance agro residue projects. These products have traditionally been viewed as waste, with no collateral value. (p. 79) • Because of the low repayment record, briquetting has developed a poor	• As long as ram and die machines were selling and operating at an acceptable level, manufacturers were not willing to begin a new endeavor that carried with it some measure of uncertainty. (p.77) • The raw material situation is quite different in India, where sawdust is a commodity rather than a waste product and is in fact widely used, unprocessed, as a cooking fuel. (p.78) • The statistics about India's vast biomass resources and statements about the "virtually unlimited" supply of biomass in India can be

Research organizations/ individuals	Information on case study			Barriers		
	Country	Technology	Technological barriers	Financial barriers	Institutional barriers
			agricultural waste products. (p.80) • In the early days of biomass briquetting, Indian machines experienced more breakdowns and required more maintenance than anticipated. Indian entrepreneurs are experiencing high maintenance costs even with ram and die machines. (p.80)	reputation and been labeled as an irresponsible undertaking. Most stakeholders interviewed felt that subsidies are not the answer for the briquetting industry and that briquetting ventures will have to stand on their own. (p.80)	misleading....Competing uses for rice husk, coffee waste, bagasse, mustard stalks, and many other kinds of waste have caused the prices to rise dramatically. (p.79) • The lack of networking and information sharing among the manufacturers. (pp. 81-82)
Case study 5	SPRU (Science and Technology Policy Research) at University of Sussex and TERI in India	India	Hybrid vehicles	• It is as much a concern for governments in developed countries to encourage the development and uptake of this low carbon technology as it is for governments in developing countries. At present, however, all of the companies owning commercially viable hybrid technologies are based in developed countries. (A: p.89) • If foreign firms supplying hybrid technology maintain a high level of integration in their approach to transferring the technology this could make it more difficult for knowledge regarding the technology to diffuse		• Host country companies may be able to develop technological capacity through involvement in supplying parts for, or maintenance services for vehicles fitted with imported hybrid technology. Even so, there may be IPR issues associated with imitating patented hybrid drive trains. A better understanding of the extent to which IPRs might limit the development of new hybrid drive trains by developing country based manufacturers is an important issue that warrants further investigation.(A: p.95) • IPRs are dominated by a concentrated set of foreign companies rather than domestic players in India. Patents exist in a number of areas, including batteries, electric motors and power electronics, engines and system

	Research organizations/ individuals	Information on case study			Barriers	
		Country	Technology	Technological barriers	Financial barriers	Institutional barriers
				within the recipient country. (A: pp.94-95)		integration. In addition, patents exist for both products and processes. Thirdly, there is a general consensus by firms and other players (e.g. academic institutions) that they must work together to make advances in this area. (B: pp.84-85)
Case study 6	SPRU (Science and Technology Policy Research) at University of Sussex and TERI in India	India	Photovoltaic (PV) solar	• Mature production technology for silicon cells is available on the market without licenses since related patents have expired. (P.65) • Most Indian companies have focused on producing silicon solar modules, the fourth stage of the value chain. This is changing however, as an increasing number of Indian firms are planning on producing the entire PV value chain and are expanding into other areas, such as thin film technology. (P.65)		• Many informants also argue that recent PV industry development is largely driven by two additional relatively new national policies: 1. The Government of India's Semiconductor Policy Guidelines in September 2007, which is essentially a tax holiday until March 2010 and 2. Electricity Generation Based Incentives (GBI) providing a subsidy for grid connected PV power plants.(pp.74-75) • Regarding policies to support technological capacity, there are almost no policies in place to encourage collaboration at the national or international level.(p.76)
Case study 7	International Institute for Industrial Environmental Economics (IIIEE) at Lund University	Developing countries	Carbon Capture and Storage (CCS)	• An immediate conceptual difficulty with CCS is that it is to be made up of an integrated suite of technologies. Moreover, institutional components addressing the CCS chain will also be a crucial system component. As CCS is not		

Research organizations/ individuals	Information on case study		Barriers		
	Country	Technology	Technological barriers	Financial barriers	Institutional barriers
			market mature and does not have any commercial examples in operation, this report cannot address CCS system transfer. Rather, one example of an incipient technology transfer framework is noted here there are two transfer projects within its remit. (p.69)		
Case study 8	International Institute for Industrial Environmental Economics (IIIEE) at Lund University	Developing countries	Building energy efficiency	• A fragmented and complex construction process, with an inherent split incentives dilemma: Building markets prefer low initial costs, and get no benefits from life cycle energy savings, whereas users may be willing to pay a high upfront cost if significant economic benefits are possible during the use phase. (p. 92) • Uncertain energy savings from equipment due to the influence of users behavior. (p.92) • A lack of formal training and capacity building among construction workers makes it difficult to introduce new techniques and innovation in construction work. (p.93) • Lack of awareness of the potential and	• High initial costs for energy efficient and renewable energy equipment. This means that payback periods are long (up to 30 years) for many investments. (p.92) • The limited importance of energy expenditures as compared other household improvement or financial concerns. (p. 92)	• A lack of awareness and information of the opportunities, technologies and low cost of installing energy saving features. (p.92) • The lack of government interest in energy efficiency and renewable energy, and insufficient enforcement of existing policies also present barriers to energy saving in the building sector. • Poor enforcement of building codes and other mandatory standards, even among front-runner countries. (p.92) • Poor market surveillance and/or certification measures mean that low-quality products can enter the market and destroy consumer confidence in the technology. • Building codes tend to be less effective, due to insufficient implementation and enforcement, and corruption f or instance, in China the compliance rate is much higher

Research organizations/ individuals	Information on case study			Barriers		
	Country	Technology	Technological barriers	Financial barriers		Institutional barriers
			importance of energy efficiency measures, lack of financing, and lack of qualified personnel (p.92) • Mandatory energy audits and similar tools require training of auditors, however, there is often a lack of monitoring of quality of audits.(p.93) • Lack of evaluation and follow-up is a major concern.(p.93)			in large cities than in rural areas.(p.93) • Adaption to the local situation is crucial, not least for utility demand-side management (DSM) programs, and projects should be designed to fit the local situation.(p.93)
Case study 9	United Nations Department of Economic and Social Affairs (DESA)	China	Wind power			• Notably, the Chinese Government is considering the implementation of local IP requirements for wind power in an attempt to push international companies to transfer more technology. Such stipulations on IP requirements could be contested by international companies under the World Trade Organization or by simply limiting new FDI in this sector. (p.30)
Case study 10	Lewis J.	India and China	Wind power	• It took China and India less than 10 years to go from having companies with no wind turbine manufacturing experience to companies capable of manufacturing complete wind turbine systems, with almost all components produced locally. This was done		• Both China and India have excellent wind resources and aggressive, long-term government commitments to promote wind energy development...Some of the early support mechanisms in China and India, in particular, led to market instability as developers were faced with regulatory uncertainty, especially concerning pricing

Research organizations/ individuals	Information on case study		Barriers		
	Country	Technology	Technological barriers	Financial barriers	Institutional barriers
			within the constraints of national and international intellectual property law, and primarily through the acquisition of technology licenses or via the purchasing of smaller wind technology companies. While both companies pursued similar licensing arrangements to acquire basic technical knowledge, Goldwind's technology development model lacks Suzlon's network of strategically positioned global subsidiaries contributing to its base of industry knowledge and technical capacity. • Suzlon's growth model particularly highlights an increasingly popular model of innovation practices for transnational firms…Its expansive international innovation networks allow it to stay abreast of wind technology innovations around the world so that it can then incorporate into its own designs through its extensive research and development facilities. (8)		structures for wind power. In the early years of wind development in China and India, difficulties also resulted from a lack of good wind resource data, and a lack of information about technology performance stemming from little or no national certification and testing. • Policy reforms in the electric power sectors of both countries…has led to a series of regional renewable energy development targets in India, national targets in China, and additional financial support mechanisms for wind in particular. There are two key differences in the policy support mechanisms currently used in China and India: (1) China's recent reliance on local content requirements to encourage locally sourced wind turbines, which does not exist in India, and (2) India's use of a fixed tariff price for wind power, versus China's reliance on competitive bidding to set the price for most of its wind projects. (8)

	Research organizations/ individuals	Information on case study		Barriers		
		Country	Technology	Technological barriers	Financial barriers	Institutional barriers
Case study 11	Mizuno E. (on a publication by UNEP Risø Centre on Energy, Climate and Sustainable Development)	India	Wind power	• External factors such as the rapidly increasing high-tech characteristics of wind energy technology systems and the fast structural transformations of the industry at the frontier made it difficult for India to cope with the various changes. (p.46)		• A large market size and market certainty and continuity were lacking in India: even though many market demand characteristics were similar to those in the frontier market, without a sizable market and its own pulling power, technology upgrading through replicable technology transfer did not happen. The small market made all demands for technological improvement insignificant.(p.44) • India's experiences with wind technology have some important lessons for how to encourage private-sector replicable technology transfers from developed to developing countries. The small market size, the non- performance-oriented market mechanism, the policy inconsistency, the institutional problems of the power sector, the lack of technological capabilities to meet the increasingly higher quality requirements of wind energy technology and the persistent infrastructure deficiencies in India, along with tighter technology controls by technology providers and collaborators, all contributed to the increasing technology gaps in both product and capabilities with the frontier after the mid-1990s.(p.46)

	Research organizations/ individuals	Information on case study			Barriers		
		Country	Technology	Technological barriers	Financial barriers	Institutional barriers
Case study 12	Ravindranath and Rao (on a publication by UNEP Risø Centre on Energy, Climate and Sustainable Development)	India	Bio-energy (including biomass gasification, biomass combustion, biogas, efficient cook stoves)	• Gas cleaning systems are still not robust and hence high in terms of maintenance (p.136) • Poor understanding of managing moisture content (p.136) • Biomass drying techniques are not well established (p.136) • Lack of knowledge (p. 137) • Uncertainty and distrust in the source of information (p.137) • Inadequate training, capacity-building and user-education programs (p.137)	• Dual fuel systems do not seem economically feasible, and hence the focus is on producer gas. But 100% producer gas engines still are not very common, not readily available at all capacities (p.136) • The high initial costs of bio-energy technologies are perceived by many as a key barrier to the penetration of bio-energy technologies vis-à-vis conventional technologies. The principal capital cost of biomass power projects includes the costs of the gasifier, the engine generator, civil construction, biomass preparation unit, electricity distribution network and electrical and piping connections to the site of gasifier installation and need subsidization (p.138). • Mainstream financial institutions have been reluctant to take risks in lending due to a long history of poor	• The abundance of biomass was initially the push [by the government] needed to promote bio-energy technologies. There was therefore little or no interaction with rural communities in formulating the technologies. (p.135) • The institutional framework in India currently lacks a viable strategy to empower local communities. Community organizations and institutions are rarely involved in the planning, implementation and management of, say, the rural electrification program through biomass gasifiers. The failure of a large number of small village systems, such as biogas plants, and stand-alone gasifiers is to a large extent related to the fact that there is no coordinated local, institutional and government support. (p.137) • A critical problem has been overcoming issues arising out of bureaucracy...Many developers have mentioned the significant periods of delay in obtaining technical approvals.(p.137) • Climate change is not being seen an immediate threat or priority for rural communities. (p.137) • Social behavior and expectations.(p.137)

	Research organizations/ individuals	Information on case study			Barriers	
		Country	Technology	Technological barriers	Financial barriers	Institutional barriers
					recovery of loans in rural area.(p.138)	• Absence of an enabling environment. (p.137)
Case study 13	Suzuki, M., Okazaki B., and Jain K.	Thailand	Biogas	• There is no centralized information and orientation regarding biogas technologies and the equipments that are available . It is also very difficult to find data related to projects' performance and information about projects that have already been implemented. (A: p. 20) • There is a lack of awareness. There is also a lack of public support in terms of information, and little information regarding biogas is transferred. In addition to this, since the degree of education of the managers is low, the technology of anaerobic digesters and biogas production appears to the managers as being very complex issues. (A: p. 21) • The anaerobic digesters are complex and sensitive systems. Often, even the managers do not understand how it works. So, due to a low understanding of the new processes, managers rely heavily on the	• Most of the time, the focus of companies is to maximize the profit over a short period. Frequently the managers have little to no information about biogas or anaerobic digester systems and the subsequent technical implications and costs. (A: p.17) • Most technologies for wastewater systems and biogas came from developed countries (Parr et al., 2000). Proper transfer and adaptation to tropical climates requires investment and will result in costs being incurred (importation taxes, logistics, training, etc.). (A: p.20) • The tapioca and palm oil industries are traditional agro-industries, often managed by families with a basic application of management principles under a simple organizational	• The managers do not seek professional support when researching biogas technology due to financial reasons. On the other hand, often the managers do not know where to search for the information they need, since there are no standard guidelines or publicly available information about biogas performance and technologies. There is no support from the government and there are very few initiatives in R&D in regions where biogas is prominent. (A: p.18) • The starch and palm oil industries are traditional agro-industries, normally run by families in an informal manner and structure. In addition, many companies have an incorrect perception of the reality of the market. In these circumstances, a long term strategy or the development of a business plan is not realistic, nor is it a common practice for these industries. (A: pp.19-20)

Research organizations/ individuals	Information on case study		Barriers		
	Country	Technology	Technological barriers	Financial barriers	Institutional barriers
			technology provider. In order to remain focused on the core production process, or to save costs, often the managers do not provide adequate or appropriate training for the operators on the new wastewater/ biogas processes and systems. (A: p.22)	structure. In addition, biogas production is not considered as important as the core business. Thus, on many occasions the operators are not motivated to perform due to a lack of a company performance reward policy or due to a different remuneration compared to his coworkers in the core production business. (A: p.21)	

Barriers For Technologies For Industrial Use: Wind, Bio-Energy, And Energy Efficient Building

Starting from wind power, the results of Case Study 1 and 11 suggest that there are institutional and technological barriers for diffusion in India and China. According to Case Study 1, the cost of IPR acquisition is a major barrier in India. Case Study 1 points out that "the [Indian] companies had to depend on their European counterparts for all technical aspects and even operation and maintenance issues." Case Study 11 addresses a similar view that technologically, the wind power in India still hinges upon the external development of the industry. It states that "external factors such as the rapidly increasing high-tech characteristics of wind energy technology systems and the fast structural transformations of the industry at the frontier made it difficult for India to cope with the various changes." On the other hand, Case Study 10 provides a positive evaluation on the development of local wind power production in India and China. It observes that "it took China and India less than 10 years to go from having companies with no wind turbine manufacturing experience to companies capable of manufacturing complete wind turbine systems, with almost all components produced locally." The results of these case studies on wind in India and China indicate that although there is a great level of success in producing indigenous local power technologies, there are still technological as well as institutional barriers for further diffusion in these countries.

Bio-energy is similar with wind power with respect to its successful implementation in the developing countries. On the other hand, the results of the case studies on bio-energy suggest that it faces different types of barriers for further diffusion. According to Case Study 12, implementations of bio-energy projects in India have met both technological and institutional barriers in the operational phase such as poor understanding of managing moisture content, lack of knowledge, uncertainty and distrust in the source of information and inadequate training, capacity-building and user education programs. The case study on biogas power generation in Thailand comes to a similar conclusion (Case Study 13). It recognizes the "no centralized information and orientation regarding biogas technologies and the equipments" as well as the lack of understanding and awareness as the major barriers for successful implementation of the technologies. The results of these case studies suggest capacity building and knowledge development play an important role in the successful implementation of bio-energy technologies.

The case study on building energy efficiency also suggests that the technological barriers such as lack of knowledge and awareness as well as the institutional barriers such as lack of information on available technologies are major barriers in this case too (Case Study 8). The results of Case Study 8 highlights, as the technological barriers, uncertain energy savings from equipment due to the influence of users behavior, a lack of formal training and capacity building among construction workers, lack of awareness of the potential and importance of energy efficiency measures, lack of financing, and lack of qualified personnel. In the case of building energy efficiency, lack of institutional support is another area of institutional barrier. It points out the lack of government interest in energy efficiency and renewable energy, and insufficient enforcement of existing policies, poor enforcement of building codes and other mandatory standards as major institutional barriers.

Barriers For Technologies For Individual Use: Hybrid Vehicles, Leds, And Pv

Other than wind power, there are studies that identify IPRs as a major barrier for technological diffusion. The case study on hybrid vehicles in India is one of them. It indicates that IPRs are the major barrier in this case as well since "IPRs are dominated by a concentrated set of foreign companies" (Case Study 5). It states "all of the companies owning commercially viable hybrid technologies are based in developed countries." The results of the case study on LED also suggest that IPRs are the key barrier for the diffusion of LED (Case Study 3). They case study demonstrates that "it is a highly protected technology. As there are various processes involved in manufacturing LED chips, each process is patented and requires huge investment. At present the cost of investing in both chip manufacturing and resolving the IPR issues is substantially high compared to importing the chips." In this regard, there may be important lessons to learn from the previously mentioned case on wind power for producing local technologies despite the existence of IPRs-related barriers. In the case of LED, however, the results of the study indicate there is a separate key barrier for the diffusion of the technology in India. The case study identifies the size of the market as a major financial barrier for technology diffusion in India. It states that there is "no clear indication about the type of market that exists for LED." Furthermore, it stresses that "the leading players worldwide are not considering India as a potential region for investment as they do not see any market in India at present."

Interestingly, in contrast to hybrid vehicles and LEDs, the results of the case study on PV in India suggest that IPRs are not an essential barrier for the diffusion of the technology in India (Case Study 6). It maintains that mature production technology for silicon cells is available on the market without licenses since related patents have expired. Moreover, an increasing number of Indian firms are planning on producing the entire PV value chain and are expanding into other areas, such as thin film technology.

Barriers For Technologies At The Innovation Stage: Igcc And Ccs

The results of the case studies on IGCC and CCS indicate that technological barriers are dominant for technologies at the innovation stage (Case Study 2 and 7). Financial and institutional barriers are not relevant for the technologies at the innovation stage. As for CCS, Case Study 7 states "As CCS is not market mature and does not have any commercial examples in operation, this report cannot address CCS system transfer." As for IGCC, Case Study 2 states "It might be premature to comment on IPR issues related to IGCC, since this technology is not considered to be commercial globally".

Thus far, Section 2.3 discussed technology-specific barriers. Another barrier, which this paper could not address this time, are country-specific barriers. It is recognized that in order to design proper policy instruments and institutions, understanding of barriers that are specific to a certain country or region is equally important. With this regard, Case study 10 is an exception among the selected case studies in highlighting several differences between India and China as to how these two countries overcome barriers to diffuse wind power technologies. It demonstrates that "there are two key differences in the policy support mechanisms currently used in China and India; 1) China's recent reliance on local content requirements to encourage locally sourced wind turbines, which does not exist in India; and 2) India's use of a fixed tariff price for wind power, versus China's reliance on competitive bidding to set the price for most of its wind projects." In addition, it discusses key differences on corporate strategies between two Chinese and Indian wind turbine manufacturing firms. This type of comparative studies are much needed in order for us to have better understanding of barriers in the diffusion of clean energy technologies.

Roles of institutions to overcome identified barriers in diffusing clear energy technologies in Asia

Section 2 presented the barriers commonly observed in the developing countries as well as the technology-specific barriers. Section 3 explores roles of institutions to overcome these barriers in diffusing clear energy technologies in Asia. Section 3.1 addresses theoretical discussions on the functions of international and national institutions in technology innovation. Section 3.2 attempts to match the barriers in technology diffusion identified in Section 2 with the functions of national and international institutions.

THEORETICAL DISCUSSIONS ON THE FUNCTIONS OF INTERNATIONAL AND NATIONAL INSTITUTIONS IN TECHNOLOGY DIFFUSION

There are theoretical explorations about the roles of institutions in changing a system in the area of innovation economics and innovation theory. For Joseph Schumpeter, who is the patron of innovation economics, an evolving institution is an important factor for economic growth. Inspired by Schumpeter, scholars in innovation theory attempt to define functions or roles of institutions in changing a system. Borrás, for example, defines that they are 1) competence-building and generation of incentives including production of knowledge, diffusion of knowledge, financial innovation, alignment of actors, guidance of innovators; 2) generation of incentives and reduction of uncertainty including appropriation of knowledge, reduction of technological diversity; and 3) establishment of limits and reduction of uncertainty including reduction of risk and control of knowledge usage (Borrás 2004). Another example is a study by Suurs and Hekkert. According to Suurs and Hekkert, there are seven functions of institutions including 1) entrepreneurial activities; 2) knowledge development; 3) knowledge diffusion; 4) guidance of the search; 5) market formation; 6) resource mobilization; and 7) legitimization (Suurs and Hekkert 2009).

There are also research initiatives that attempt to understand the roles of institutions in diffusing clean energy technologies both at the national and international level, although the focus of research is geared toward the national level rather than the international level. At the international level, a study conducted by de Coninck et al. is an example of such research (de Coninck et al. 2008). This study classifies technology-oriented agreements (TOAs) addressing climate change into four broad categories including 1) knowledge sharing and coordination; 2) research, development and demonstration (RD&D); 3) technology transfer; and 4) technology deployment mandates, standards, and incentives (de Coninck et al. 2008). According to a more recent study by Benioff et al., there are three roles of international institutions for innovation and transfer of clean energy technologies including research, development, and demonstration (RD&D) cooperation, enhancement of enabling environment, and financing facilitation and support (Benioff et al. 2010).

It is important to note here that the roles of institutions differ along the technological development of clean energy technologies. At the early stages of technological development, institutional support for the empowerment of research groups is needed to demonstrate and deploy technologies (Suzuki 2012). As the case studies on CCS and IGCC indicated in Section 2, the technologies at the innovation stage require strong R&D efforts to remove technological barriers in order to move forward to the next stage. At the innovation stage, the empowerment of network between international and local research groups is needed to enhance the R&D efforts, especially with a stronger initiative from the public side (Benioff et al. 2010; Morey et al. 2011; UNFCCC 2009).

At the advanced stages of technological development, institutional support as well as policy arrangement for the involvement of the actors in the private sector such as project developers, equity investors, manufactures, and commercial banks is essential in technology diffusion (GtripleC 2010;Carmody et al. 2007). Providing economic incentives for the private sector are an important measure to improve investment conditions and encourage its participations. Therefore, clean energy and carbon finance vehicles may be also effective to introduce technologies at the advanced stage. For example, the economic policy instruments such as CDM may take an instrumental role. If they are designed well, the schemes under discussion for the post-Kyoto regime such as the bilateral carbon crediting mechanism and the sectoral or program-based crediting mechanism can be also a good policy candidate for technology diffusion. At the national level, an introduction of a feed-in-tariff program has received greater attentions among the developing countries, while other economic instruments such as subsidy, emissions trading, and renewable energy certificate scheme can be also recognized as possible policy options.

The investment schemes such as co-investments and loans or risk guarantees may help to reduce risk associated with investment from the private sector (Suzuki 2012). In addition, such an arrangement for building a partnership between the private and the public (Public-Private Partnership: PPP) may leverage the interests of the private sector in developing technologies that would not be attracted to clean energy technologies otherwise.

MATCHING THE BARRIERS IN TECHNOLOGY DIFFU-SION WITH THE FUNCTIONS OF NATIONAL AND IN-TERNATIONAL INSTITUTIONS

Section 2.3 illustrated technology-specific barriers among different technologies. Section 3.2 attempts to match those barriers with the functions of national and international institutions that were identified in Section 3.1.

The case studies on wind as well as on hybrid vehicles and LED indicated that difficulties associated with IPRs are major barriers in technology diffusions. Indeed, IPRs are complex issues and providing opportunities to learn about the issues can be an important institutional arrangement as the first step.Ockwell, D., J. Watson et al. (2009), on the case of wind in India, states that "there was a need to create awareness among the industry players who do not have deeper understanding of implications of IPR rules and regulations, including those in the context of WTO regime." Preparing patent pools for licensing inventions is often discussed as a necessary arrangement in diffusing clean energy technologies but it requires careful institutional design not to remove incentives for the private sector and discourage its innovational efforts. At the international level, the World Intellectual Property Organization (WIPO) can facilitate such venues for the private sector in the developing countries to learn about IPRs-related issues.

The case study on LED identified the size of the market as a major barrier. This case, together with the case on building energy efficiency, also pointed out high capital cost as a major barrier. In order to overcome these barriers, the roles of institutions in facilitating and supporting finance are important. On LED, Ockwell, D., J. Watson et al. (2007) states that "as government is already promoting PV integrated energy efficient lighting systems for rural lighting applications, incentives could be provided for LED based PV integrated systems." As for the case on biomass, low priority in finance is recognized as a major barrier. In this case, knowledge sharing and coordination is the key in overcoming the barrier in technology diffusion. At this point, Ockwell, D., J. Watson et al. (2007)demonstrates that "all the briquetting machine manufacturers felt that there is practically no collaboration or communication among them. The lack of networking and information sharing among the manufacturers is one of the greatest constraints to diffusion of technological developments in the sector. Hence projects aimed at promoting knowledge sharing among the manufacturers and users of biomass briquettes will be very useful for the sector".

The case studies on bio-energy, biomass, and building energy efficiency all emphasized that lack of the enabling environment is the key barrier in technology diffusion. The case study on bio-energy in India highlighted "poor understanding of managing moisture content, lack of knowledge, uncertainty and distrust in the source of information and inadequate training, capacity-building and user education program" as a major hindrance. The case study on biomass in Thailand pointed out a lack of formal training and capacity building among construction workers, lack of awareness of the potential and importance of energy efficiency measures, lack of financing, and lack of qualified personnel. In order to overcome these barriers associated with a lack of the enabling environment, the case study on bio-energy in India suggested promoting collaboration between industry and academia, for field demonstrations, and promoting feedback and communication between developers and implementers (Ravindranath and Rao 2011). It stated that "the development of training schemes could provide a route to alleviating this skill shortage. It is important to ensure that all staff involved in training and development have been adequately trained themselves. Use of R&D institutions in training could be beneficial" (Ravindranath and Rao 2011).

As for the technologies at the early stage of technological development, the cooperation in R&D between the pubic and the private sectors as well as the cooperation between local and overseas actors are inevitable in order to overcome technological barriers. As emphasized earlier, the strong initiatives from the public side are needed since it is difficult to expect the private sector to play an important role if the business model is not yet visible. The case study on CCS indicated that "given current policy and market conditions, carbon markets appear marginal or inadequate for CCS applications such as industrial-scale demonstration plants to be economically viable without (potentially significant) additional support" (Dalhammar, C. et al. 2009). The case study on IGCC concluded that "one possible approach to overcoming the risks of high capital costs is for government to share the funding of demonstration activities with industry... Financial support from developed to developing countries would be needed to provide for incremental costs and technology transfer fees, through international financing mechanism" (Ockwell, D., J. Watson et al. 2007; Ockwell, D., J. Watson et al. 2009). Table 4 illustrates both identified barriers and roles of institutions to overcome the identified barriers

Table 4: Identified barriers and roles of institutions to overcome the identified barriers

	Early stage	Advanced stage
Barriers	• **Technological barriers:** Case Study 2 (IGCC), 7 (CCS) • High capital cost: Case Study 2 (IGCC)	• **IPRs:** Case Study 1 (wind), 9 (wind), 11 (wind), 5 (hybrid vehicles), and 3 (LED) • Market size: Case study 3 (LED) • High capital cost: Case study 3 (LED), 8 (building energy efficiency) • Low priority in finance: Case Study 4 (biomass) • Lack of enabling environment: Case Study 8 (building energy efficiency), 12 (bio-energy),13 (biogas) • Lack of policy support: Case Study 6 (PV), 8 (building energy efficiency)
Roles institutions	In theory... • R&D cooperation • Financing facilitation and support ("resource mobilization" and "market formation") • Entrepreneurial activities	In theory... • Knowledge sharing and coordination (including "guidance of the search") • Enhancement of enabling environment (including "legitimization") • Financing facilitation and support (including "market formation" and "resource mobilization")
	Identified roles	**Identified roles**
	R&D cooperation • Public-supported centers for technology innovation and transfer. • Strengthening bilateral and multilateral network for R&D. Financing facilitation and support • Technology funding mechanisms for the developing country participants in R&D. • Global clean technology venture capital fund. Entrepreneurial activities • Clean energy incubator incentives.	Knowledge sharing and coordination/enhancement of enabling environment • Patent pools for licensing inventions. • Various capacity building programs covering a whole supply-chain. • Business matching venues among various business actors such as project developers, manufacturers and investors (local and international). Financing facilitation and support • Various clean energy finance and carbon finance vehicles including CDM, bilateral crediting scheme, co-benefit approach at the int'l level, feed-in-tariff, subsidy at the national level. • Co-investments, loans or risk guarantees. • Public-Private Partnerships (PPPs).

CONCLUSION

This paper consisted of two parts. The first part of the paper attempted to show a broad landscape of barriers in technology diffusion in the developing countries by addressing two levels of barriers: generic barriers and technology-specific barriers (Section 1 and 2). Section 2.3 summarized the results of previous case studies that were conducted to uncover technology-specific barriers in diffusing clean energy technologies in Asia.

The second part of the paper explored roles of institutions to overcome the identified barriers in diffusing clear energy technologies in Asia (Section 3). It attempted to match the barriers in technology diffusion identified in Section 2 with functions of national and international institutions. The results of matching indicated that there are several different roles of institutions including the role to encourage R&D cooperation from the public site for the technologies at the early stages of technological development and the role to enhance the enabling environment and facilitate finance for the technologies at the advanced stages of technological development.

It is recognized that the existing institutions both at the national and international levels have already been working to overcome barriers in diffusing clean energy technologies. For example, at the national level, the governments in the developing countries are conducting various capacity building programs to enhance knowledge of the private sector about clean energy technologies. At the international level, the financial institutions such as the World Bank and Asian Development Bank are facilitating financial support to encourage diffusion of clean energy technologies. At the innovation stage, there are both bilateral (such as the Global CCS Institute for building a network between Australia and the developing countries) and multilateral (such as the Asia-Pacific Partnership on Clean Development and Climate concluded in April 2011) network to encourage technology innovation. Further research is needed to investigate whether these existing institutions are playing a role in overcoming the barriers that were illustrated in this paper.

ACKNOWLEDGEMENTS

This work was supported by the Global Environment Research Fund of the Ministry of the Environment, Japan (S-6-3). The author wishes to thank for the support.

NOTES

[1]   The text of the COP document states that [The Conference of the Parties] recognizes that developed country Parties commit, in the context of meaningful mitigation actions and transparency on implementation, to a goal of mobilizing jointly USD 100 billion per year by 2020 to address the needs of developing countries (paragraph 98); agrees that, in accordance with paragraph 1(e) of the Bali Action Plan, funds provided to developing country Parties may come from a wide variety of sources, public and private, bilateral and multilateral, including alternative sources (paragraph 99); and decides that a significant share of new multilateral funding for adaptation should flow through the Green

Climate Fund (paragraph 100).

[2]   It is not possible to clearly distinguish barriers into the three classifications. Many barriers relate to more than two classifications. Under the circumstances, the paper attempts to fit each barrier into the most appropriate classification.

[3]   Table 1 includes some technology-specific barriers as well as country/region-specific barriers. It is also noted that the table contains selected major barriers only.

[4]   This paper looks into key case studies in Asia only, although there are case studies being conducted in other parts including South America and Africa.

REFERENCES

1.   Benioff, R., de Coninck, H., et al. (2010) Strengthening Clean Energy Technology Cooperation under the UNFCCC: Steps toward Implementation, National Renewable Energy Laboratory.

2.   Borrás, S. (2004) System of innovation theory and the European Union, Science and Public Policy, Volume 31, Number 6, 425-433.

3.   Carmody, J. et al. (2007) Investing in Clean Energy and Low Carbon Alternatives in Asia, Asian Development Bank.

4.   Dalhammar, C., P. Peck, N. Tojo, L. Mundaca, and L. Neij (2009) Advancing Technology Transfer for Climate Change Mitigation: Considerations for Technology Orientated Agreements Promoting Energy Efficiency and Carbon Capture and Storage (CCS), IIIEE Report.

5.   de Coninck, H. et al. (2008) International technology-oriented agreements to address climate change, Energy Policy 36(1).

6.   Doukas, H., C. Karakosta and J. Psarras (2009) RES technology transfer within the new climate regime: a "helicopter" view under the CDM, Renewable and Sustainable Energy Reviews, 13: 1138-1143.

7.   Flamos, A., W. Van der Gaast, H. Doukas, and G. Deng (2008) EU and Asian countries policies and programmes for the diffusion of sustainable energy technologies, Asia Europe Journal, 6(2): 261-276.

8.   GtripleC (2010) Engaging Private Sector Capital at Scale in Financing Low Carbon Infrastructure in Developing Countries, Asian Development Bank.

9.   Guerin, T. F. (2001) Transferring environmental technologies to China recent developments and constraints, Technol Forecast Soc Change, 67: 55-75.

10. IPCC (2001) Methodological and Technological Issues in Technology Transfer, New York: Cambridge University Press.

11. Jagadeesh, A. (2000) Wind energy development in Tamil Nabu and Andhra Pradesh, India Institutional dynamics and barriers-A case study, Energy Policy 28:157-168.

12. Jain, K., Okazaki. B., and Suzuki, M. (2011) Challenges and Barriers in Technology Transfer and Performance of Biogas Plants in Southeast Asia. An Analysis of Tapioca and Palm Oil Industries Associated with CDM Business in Thailand, Working paper for IMRE Alumni Conference 2011, Institute of Technology Bandung, Indonesia.

13. Karakosta, C., H. Doukas and J. Psarras (2010) Technology transfer through climate change: setting a sustainable energy pattern, Renewable and Sustainable Energy Reviews, 14: 1546-1557.

14. Lewis, J., (2007) A Comparison of Wind Power Industry Development Strategies in Spain, India and China, Prepared for the Center for Resource Solutions.

15. Lewis, J., (2007) Technology acquisition and innovation in the developing world: wind turbine Development in China and India. Studies in Comparative International Development, 42.3-4.

16. Luken, R. and F. Van Rompaey (2008) Drivers for and barriers to environmentally sound technology adoption by manufacturing plants in nine developing countries, Journal of Cleaner Production, 16(S1): 67-77.

17. Morey, J. et al. (2011) Moving Climate Innovation into the 21st Century: Emerging Lessons from Other Sectors and Options for a New Climate Innovation Initiative, Clean Energy Group.

18. Ockwell, D., J. Watson, G. MacKerron, P. Pal, F. Yamin, N. Vasudevan, and P. Mohanty (2007) UK–India Collaboration to Identify Barriers to the Transfer of Low Carbon Energy Technology: Final Report.

19. Ockwell, D., J. Watson, G. MacKerron, P. Pal, F. Yamin, N. Vasudevan, and P. Mohanty (2009) UK-India Collaborative Study on the Transfer of Low Carbon Technology: Phase II Final Report.

20. OECD/IEA (2001) Technology without Borders: Case Studies of Successful Technology Transfer, OECD/IEA, Paris.

21. Painuly, J. (2001) Barriers to renewable energy penetration; a framework for analysis, Renewable Energy, 24(1), 73–89.

22. Painuly, J. and J. V. Fenhann (2002) Implementation of Renewable Energy Technologies - Opportunities and Barriers, UNEP Collaborating Centre on Energy and Environment, Risø National Laboratory, Roskilde.

23. Ravindranath, N. H. and Blachndra (2009) Sustainable bioenergy for India: Technical, economic and policy analysis, Renewable Energy, 34, pp.1003-1013.

24. Reddy, S. and J. P. Painuly (2004) Diffusion of renewable energy technologies-barriers and stakeholders' perspectives, Renewable Energy, 29, pp.1431-1447.

25. Schneider, M., A. Holzer, and V. H. Hoffmann (2008) Understanding the CDM's Contribution to Technology Transfer, Energy Policy, 36: 2930-2938.

26. Suurs, R., Hekkert, M. (2009) Cumulative causation in the formation of a technological innovation system: The case of biofuels in the Netherlands, Technological Forecasting & Social Change, 76: 1003 - 1020.

27. Suzuki, M. (2012) Addressing a Portfolio of Effective Policy Measures and Financial Mechanisms to Encourage Technology Innovation and Transfer of Clean Energy Technologies in the Asia-Pacific Region, International Society for Ecological Economics, Rio de Janeiro, June 18th 2012.

28. Suzuki, M., Okazaki B., and Jain K. (2010) Identifying Barriers for the Implementation and the Operation of Biogas Power Generation Projects in Southeast Asia: An Analysis of Clean Development Projects in Thailand, Economics and Management Series Working Paper, EMS-2010-20, International University of Japan.

29. Thorne, S. (2008) Towards a Framework of Clean Energy Technology Receptivity, Energy Policy, 36: 2831-3838.

30. UNEP Risø Centre on Energy, Climate and Sustainable Development (2011) Diffusion of Renewable Energy Technologies: Case Studies of Enabling Frameworks in Developing Countries, UNEP Collaborating Centre on Energy and Environment, Risø National Laboratory, Roskilde.

31. UNFCCC (2003) Enabling Environments for Technology Transfer, Technical Paper, FCCC/TP/2003/2.

32. UNFCCC (2009) Advance Report on Recommendations on Future Financing Options for Enhancing the Development, Deployment, Diffusion and Transfer of Technologies under the Convention, Note by the Chair of the Expert Group on Technology Transfer, FCCC/SB/2009/INF.2.

33. UNFCCC (2011) Report of the Conference of the Parties on its sixteenth session, held in Cancun from 29 November to 10 December 2010, Addendum, Part Two: Action taken by the Conference of the Parties at its Sixteenth Session, FCCC/CP/2010/7/Add.1.

34. Usha R.K. and Ravindranath N.H. (2002) Policies to overcome barriers to the spread of bioenergy technologies in India, Energy for Sustainable Development 2002; 6(3): 59–73.

35. Worrell, E., R. van Berkel, Z. Fengqi, C. Menke, R. Scaeffer, and R. O. Williams (2001) Technology Transfer of Energy Efficient Technologies in Industry: A Review of Trends and Policy Issues, Energy Policy, 29: 29-43.

Chapter 5

PUNCTUATIONS AND DISPLACEMENTS IN POLICY DISCOURSE: THE CLIMATE CHANGE ISSUE IN GERMANY 2007-2010

Volker Schneider[1] and Jana K. Ollmann[1]

[1] Department of Politics and Public Administration, University of Konstanz, Germany

INTRODUCTION

Climate change or "global warming" is a major issue within the debate about the sustainability of social and natural systems. In this context it has become the prime example for policy problems that are characterized by long time horizons, large uncertainty and high ambiguity [1]. In such a policy context, problem definitions get vague and unstable, preferences become unclear, and the potential of social conflict is high [2]. Under these circumstances, policy making heavily relies on public discourse in which issues and interest conflicts are collectively debated and shared "definitions of the situation" are constructed. For this reason, the issue of climate change has attracted high attention among policy researchers interested in discourse analysis since the early 1990s.

While many empirical studies focus on the rise and decline of discourse activities, some critics have questioned the relevance of climate change discourses at all. For instance, even scholars of cultural studies such as [3] call for a *"return from the world of discourses and systems back to the actions and strategies with which social beings try to manage their existence"*. Such a perspective implies that policy problems are seen as objectively given and self-evident, without any need to be collectively defined and represented. In an epistemological perspective, this is a naïve version of realism [4]. According to our perspective, however, public discourse is an essential part of policy-making, besides the interests, preferences and strategies of all involved actors and the institutional constraints in which policies are decided and implemented.

Policy-controversies and debates are not just "surface phenomena" of political processes but are rather an integral part of power structures and exchange relations in policy-making. The analysis of public debates and policy discourses – in a qualitative or quantitative manner – can therefore be seen as an important component of policy analysis [5].

In this paper we will apply a specific form of quantitative discourse analysis to the debate on global warming and related policy decisions. Since qualitative discourse analysis runs short in terms of transparency, comparability and replicability, we use various methods of quantitative structural analysis to specify the role of actors and their interrelations within the policy discourses on climate change [6]. Recent methodological developments, namely the combination of category-based, computer-assisted, qualitative content analysis and social network analysis [7-9] provide new possibilities to analyze discourse coalitions, actor constellations, conflict structures, and their dynamics at the level of discourses and policy debates.

The specific goal of our paper is to trace and interpret the evolution of German public discourse on climate change in terms of punctuated equilibrium theory (PE theory), which is a distinctive version of evolution theory in the natural and social sciences. It rejects gradualist assumptions and emphasizes discontinuities in processes at all levels which have been triggered by great and singular events [10]. When applied to social developments, PE theory explains policy change as a result of major shifts in the public perception of a policy issue, which in turn is triggered by focal, and often "external" events [11]. These processes are intermediated by negative and positive feedback mechanisms that accelerate or slow down developments.

Our study will assess core propositions of PE theory with respect to the impact of the financial crisis on the German climate discourse between 2007 and 2010. Germany has been widely acknowledged to be a front-runner in climate policy on the European and global level. A commonly accepted explanation is that intense public participation and strong public consensus based on "ecological modernization" have contributed to this success. Even though this consensus has dominated the German discourse for over two decades, some scholars [12, 13] have issued concerns that it might prove to be unstable. Since its peak in 2007, public attention to the issue of climate change has been declining. Our data show that this down-swing seems to have been strongly amplified due to the financial and economic crisis in 2008 and 2009. In the context of this massive downturn, actors changed their discursive behavior, impacting actor positions and frame constellations.

Our paper proceeds in three steps. In the next section we will give a short outline of various theoretical perspectives in the analysis of policy discourse, emphasizing punctuated equilibrium theory. Our third section proposes a formal and quantitative approach to structural analysis of discourse configurations that are linked to actor networks. In the fourth section we will apply this approach to policy discourse in the domain of global warming in Germany under the influence of the recent economic crisis. In the conclusions we summarize our findings and raise some question for further analysis.

The complexity of policy discourse

In the study of public responses to social and environmental problems two opposing perspectives have dominated the academic debate: the objectivist (or naïve realist) and the social constructivist approaches. Objectivists define social problems as objectively given, self-evident, without any measurement problem. From this perspective, changes in the atmosphere and their consequences can be determined in an objective and definitive way. They also assume that rational and well informed actors can develop an optimal adaptation strategy [14]. Yet from a constructivist point of view, a social problem *"exists primarily in terms of how it is defined and conceived in society"* [15]. Thus, climate change only turns into a social problem when individuals or groups conceive it to be a threat to nature and society. The individual as well as the collective perception of risk are thereby influenced by social, cultural and political contexts [14, 16].

Over the last twenty years environmental issues have inspired discourse analysis within different sub-disciplines of social sciences, e.g. communication science, science and technology studies, as well as policy science. These studies share the conviction that the constructivist perspective is especially fertile with respect to issues that are characterized by long time frames, large uncertainty and high ambiguity. Climate change matches all these characteristics:

Long time frames. Significant changes within the atmosphere emerge "creepingly" over long periods that do not correspond to the time horizon of everyday life experience. Society thus depends on scientific research to detect, anticipate and communicate these risks. In that way, scientific facts only attract public interest and political concern if they can be linked to social threats and possible solutions [1, 14].

Uncertainty. While human influence on climate change is widely accepted in contemporary science, uncertainties remain about its future development and consequences [17, 18]. Scientific forecasts vary along modeling techniques and measurement methods [14]. Uncertainty complicates risk assessment and communication. Under these conditions, objective cost-benefit analysis of precautionary measures turns into a "mission impossible".

Ambiguity. This property can be defined as a *"state of having many ways of thinking about the same circumstances or phenomena"* [19]. While uncertainty may be reduced by further information, additional information does not reduce ambiguity. Even if there is a complete spread of scientific information, different people will have different perceptions of the problem. For instance, climate change can be understood as a risk to biodiversity, human health, economic development, social equity or political stability. These different problem definitions may not be reconcilable, and hence create vagueness, confusion and conflict [2].

Some social scientists concerned with climate change see their research on a "pragmatic middle ground" between objectivism and constructivism, denying neither that threats are objectively given nor that public perception is subject to significant variation. Especially risk communication researchers are concerned with how objective expert information can be effectively communicated to the public [20]. However, these approaches fail to acknowledge the role of actors and their particular interests to influence public perception – from risks assessment to the reporting and public interpretation of these risks [14]. The constructivist perspective highlights that actor relations and cultural contexts in science, culture and the public sphere are more relevant for the debate on climate change than is the quality of information [14]. Thus, public discourse has to be linked to actor constellations in the policy process. It is this relational dimension that differentiates our method of discourse analysis from traditional forms of discourse analysis within other sub-disciplines of social sciences.

Policy research has traditionally regarded policy making as a linear problem-solving process of a simple "conceive-decide-implement" sequence, starting with problems that are defined in an objectivist perspective [21]. However, the growing complexity of policy problems nurtured skepticism about the rationality of such processes. Policy analysts now increasingly acknowledge that distinct value orientations, specific information processing capacities, and subjective lines of argumentation and interpretation are influencing the policy process. In this perspective, public discourses have to be seen as essential components of policy making [6, 22].

With respect to discourses on climate change, a number of studies have examined the rise and decline of issue attention in public arenas as well as the evolution of political agendas in this policy domain. Studies focused on changes of the public perception of climate change as a social problem as well as the role of different social sub-systems such as science, politics and the media during the successive stages of the issue's career [12, 23, 24]. Some scholars tried to map problem perceptions and conflict lines to explore the possibilities of policy consensus [25, 26]. Malone [27] used a network approach to map similarities between "families" of arguments. Analyzing narrative structures within environmental discourse, Hajer [28] examined how actors build discourse coalitions around story lines that integrate situational factors, general problem interpretations and policy interests within a coherent narrative. Fisher et al. [29], using a methodology similar to ours, analyze discussions about climate change within US Congress and display how consensus around the issue emerged during its 110th session (January 3, 2007–January 3, 2009).

While all these studies have emphasized the need for communication and mediation in public debates, only some of them have conceptualized discourse as an integral part of the policy process. In addition, some of the studies display quite serious methodological deficiencies. For instance, interpretative "process tracing" and "case studies" often raise problems with respect to transparency, replicability and comparability [5]. Qualitative approaches inherently concentrate attention only to relatively few actors and relations, without taking into account the vast plurality and heterogeneity of actors, the multiplicity of linkages, and the complexity of discourse configurations.

For some time, there have been certain theories in policy analysis in which emphasize discourse elements such as ideas or beliefs. One example is the Advocacy Coalition Framework, which describes the policy process as a struggle between different coalitions that share similar belief systems and tries to establish these beliefs as the dominant policy interpretation within a policy subsystem [30]. In the perspective of the Multiple Stream Approach, policy entrepreneurs use discursive tactics to link policy problems to their preferred policy solutions [2].

In the present paper we use Punctuated Equilibrium Theory (PE theory), which emphasizes the role of policy venues, policy images, and the impact of large singular events: policy actors try to alter the institutional arena within which a given issue is negotiated (venue) to promote their values and policy beliefs (policy image). Actors attempt to transform the overall "issue culture" by persuading undecided participants or mobilizing hitherto uninvolved actors. In this content external and/or internal focal events can have a deep impact on policy development. Public attention to specific issues may suddenly rise or shift towards other issues, thereby attracting new policy actors and restructuring policy discourse. However, it depends on policy feedbacks whether this intrusion creates a serious challenge to the dominant policy image or the deep-rooted actor constellations. Positive feedbacks (e.g. bandwagon effects, social learning or political entrepreneurship) enhance policy change whereas negative feedbacks (e.g. access barriers or coalition building to sustain the present policy image) reinforce existing images or constellations [11, 31-33].

In general, most analyses based on PE theory track policies over long periods to identify patterns of policy stasis and abrupt punctuations. Our analysis will concentrate on rather short intervals, starting one year before the financial crisis as a "punctuating event" and ending in the first quarter of 2010.

This study conceptualizes discourses as communication processes permitting *"policy issues and conflicts to be collectively understood and defined, (...), meanings to be shared and reconstructed, and arguments to be set forth, debated, and eventually institutionalized"* [34]. Such a view of policy discourse implicitly uses a network perspective of policy making in which decisions and programs are not merely structured by formal institutions and few governmental actors but rather by complex informal relations between multiple and heterogeneous policy actors [35]. Policy systems are functionally differentiated into various sub-systems evolving around specific policy issues, and are composed of actors who regularly seek to influence policy processes that are guided by beliefs and interests.

It is useful to distinguish among two types of policy discourse: *sub-system specific* and more *general public discourse*. Discourses take place in different forums or arenas in which individual or collective actors present their issue interpretation while the audience is observing and evaluating. Actors contribute to discourses in order to persuade others and the audience to adopt their issue perspective [36, 37]. According to PE theory, a high degree of consensus within sub-system specific discourses favors policy making in terms of routine procedures, and in most cases policies evolve in an incremental manner [32]. During the normal course of policy making, actors tend to communicate predominantly within established policy circles [38]. If one or several sub-system members disagree with the dominant problem perception, they try to change the venue of discussion, i.e. they push the issue to the public arena where a broader and more heterogeneous audience can be addressed.

Based on different communication technologies, there are various kinds of public discourse arenas of which this study considers the mass media to have the largest impact on the policy debate. Although principally everybody can participate in the mass media forum (at least as a member of the audience), editors and journalists enjoy privileged positions since they exert some control with respect to who can say what, when and how. Thus, public arenas can be biased by power coalitions in which media actors play an important role as well. In contrast to discourses at the sub-system level, issues are discussed controversially in public discourses. Heterogeneous actors contribute to different problem definitions, and dominant or consensual policy images are established only by way of tedious debates.

Another facet of discourse arenas is their limited *carrying capacity*: only few problems can be addressed at once [39]. While the respective subsystems specialize on a given issue, in most cases a bunch of issues compete for attention.

Their competitiveness depends on *novelty* and *dramatic value*. According to Downs [40], public attention follows a cyclical pattern of rise and decline. Such issue cycles have been extensively discussed in the literature, wherein two points have been emphasized: firstly, major external events catalyze issue attention because they create a sense of dramatic crisis that cannot be sustained in their absence [41, 42]. Secondly, claims-making activities alone cannot explain that one issue attracts more public attention than another [43], but they play an important role in connecting a specific event to the definition of a policy problem [44].

This study distinguishes four stages of collective problem redefinition within public discourse [45]:

Entry and exit. Individual and corporate actors (just like the discourse arena) have limited carrying capacities. Because of limitations in time, budget and personnel, they can only process few issues at a time [39]. When there is extensive media coverage of an issue, some actors that were not interested in the issue prior to a media hype now become engaged in public discourse either because they realize the problem›s importance or because they use it as an occasion for self-promotion or other policy strategies.

Framing. Problem definitions depend on framing, which is *"a way of selecting, organizing, interpreting, and making sense of a complex reality to provide guideposts for knowing, analyzing, persuading and acting"* [46]. Frames enable actors to get some understanding of complex situations and facilitate communication and action with regard to a perceived problem. The way an issue is framed impacts on whether people notice a problem, how they understand it and what viable solution they take into consideration. The framing of an issue is not necessarily constant – neither the individual nor the collective way of framing. Actors aim to get their frame recognized as the authoritative version of "reality" [36].

Salience. This concept describes how much a frame dominates the discourse [47]. A frame has a low salience if it is used rarely by few actors whereas its salience is high if it is used repeatedly by many actors. When attention to an issue rises or declines, shifts in actor constellations also generate changes in frame constellations. New actors contribute to new frames while old frames vanish when their supportive actors leave the discourse arena [45].

Proposal and debate. Changes in the collective framing of an issue also changes influence debates on policy measures. Based on the multiple stream approach, proponents of PE theory expect that political actors are sometimes more interested in making sure that "their" policy solutions are adopted than in what problem these solutions address [45, 48]. During phases of collective problem redefinition policy entrepreneurs promote their policy ideas as solutions for the problem under discussion. As these ideas do not derive rationally from problem perceptions, they are nevertheless expected to be compatible with different problem interpretations.

An extended application of this approach to policy discourse would suggest that we have data on various discourse arenas and policy venues. Within the constraints of this study we had to concentrate on the discourse at the mass media level. In this respect, our theory-based expectations are that the financial and economic crisis was a punctuating event with regard to actor and discourse dynamics. During and after the crisis we expect significant change in the actor constellation and in the structure of discourse. Both will be measured and described with some methods of social network analysis.

DISCOURSES AS NETWORKS

In this study we use a formal and quantitative approach to discourse analysis. As exposed in the previous section, discourses consist of sets of individuals and organizational actors, groups of actors, and sets of concepts such as frames or positions. All of these refer to issues under discussion and emergent relations in terms of communication. Concepts do not float freely in the air or "hover above society" but are instead attached to concrete actors that use them within discourses to persuade others of their own problem interpretation. Discourse coalitions emerge among actors that are connected by similar issue positions and policy frames. Specific frames and problem definitions must not be mutually exclusive but differ with respect to their reconcilability [25]. This study assumes that the same actor is able to consider a problem from different perspectives and to use different frames within a given discourse. An actor might do so out of conviction or with strategic motives. In any case, two concepts that are used by the same actor in the same way (in the case of positions, the actor supports both or opposes both positions) can be assumed to be reconcilable to a certain extent.

Discourse network analysis formalizes these multiple relations by means of graph theory: A graph G consists of nodes from the set of actors $A=\{a_1, a_2, a_3...a_m\}$ and/or from the set of concepts $C=\{c_1, c_2, c_3...c_n\}$ and edges from the set of interrelations between nodes $E=\{e_1, e_2, e_3...e_p\}$ (Figure 1). Based on these formal concepts several types of graphs can be created:

- an actor network
- a concept network (based on positions or frames)
- an affiliation network linking actors and concepts
- an actor group network aggregating actors into groups

- a positions-frames network aggregating positions into frames.

These networks can be analyzed by conventional tools of social network analysis. This study is interested in the standing of actors as well as in the salience of concepts. Standing designates an actor›s visibility in terms of how much he/she contributes to the public discourse. An actor›s standing depends not only on his/her commitment but also on whether he/she succeeds in positioning his/her problem interpretation within the media arena. In terms of network analysis, standing designates the actor›s centrality within the discourse. Salience designates how much a concept is incorporated in a collective problem definition, how often it is used and how central it is within the affiliation network. Issue coalitions, groups of actors that share similar policy ideas are subgroups within the actor network in terms of network analysis. The reconcilability of issues is reflected by their interconnectedness within the issue network.

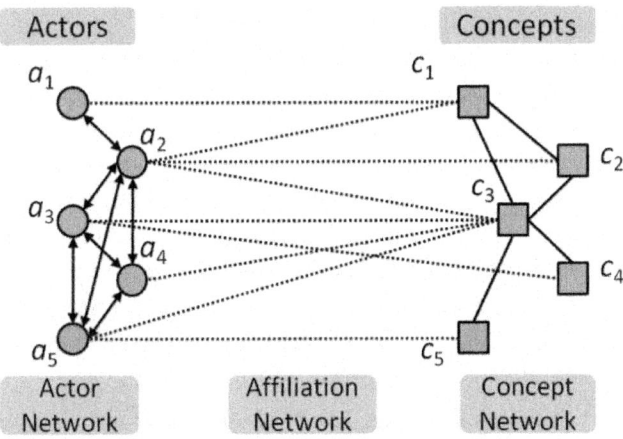

Figure 1: Discourses as Networks (Source: (6))

The sets of actors, concepts and edges change over time as actors enter or leave the discourse and change their respective problem interpretations. When one actor leaves the discourse, this reduces not only the set of actors but also the set of edges

a)   within the actor network by those edges that previously connected this actor to other actors,

b)   within the affiliation network by those edges that connected this actor to concepts, and

c)   within the concept network by those edges that connected the different concepts which had been used by this actor.

When an actor leaves the discourse, that used to apply many different frames and to comment several policy measures, significant structural changes can be observed in all networks

DISCOURSE NETWORKS ON GLOBAL WARMING IN GERMANY

This study applies discourse network analysis to the German discourse on climate change, assessing key propositions of PE theory. The following section portrays the German case and explains data collection, analysis and interpretation of results.

CLIMATE POLICY IN GERMANY: THE BACKGROUND

Previous studies concerned with climate change policy have discussed Germany as an extreme case because of its outstanding achievements in this policy area [49]. They found that public discourse has played an important role in this development. The dominant perception of climate change and climate policy has been considerably stable: climate change is perceived as a problem that requires state intervention, whereby climate protection also bears economic opportunities. This perception has become known as "ecologic modernization" paradigm. However, there are indications that the economic crisis might have threatened the "German consensus" and changed the dominant perception of the climate issue as suggested by PE theory. The financial crisis provides the opportunity to conduct such a natural experiment.

The issue of climate change entered the sphere of German public discourse and high politics for the first time during the mid-1980s [13]. In 1986, a press release of the German Physical Society (DPG) and its subgroup, the Working Group on Energy (AKE), depicted climate change as an "impending catastrophe" requiring immediate political action and initiated an extensive coverage of the climate change topic within the mass media. Initially, the political sphere remained skeptical, doubting the scientific soundness of these warnings. However, it could not ignore increasing public concern and call for action. After the Chernobyl catastrophe in April 1986, Chancellor Helmut Kohl swiftly established the Ministry of Environment (June 5, 1986). Shortly after that, in March 1987, he declared the climate issue to be one of the world's most pressing environmental problems [13]. From this point on, Germany has emerged as a forerunner in domestic climate protection and as a pacemaker at the European as well as at the global level. A large range of policies for the reduction of greenhouse gas (GHG) emissions has been passed during the last two decades. Energy efficiency has been raised in all economic sectors [13]. By 2008 Germany had already reached its target of a 21% reduction by 2010 compared with 1990, and the recent government pursues a reduction target of 40% until 2020 [50].

Scholars have largely attributed this success to the broad public participation and the consensual style of German policy making. This style has been enhanced by the integration of the Green movement into political institutions during the late 1980s, by federalism and by the German electoral system of proportional representation [13, 49]. Scholars and policy makers have pointed out that public support for the German government's initiatives is strongly based on the public perception that policy interventions in the field of environmental policy do not weaken but strengthen economic growth. This perception has its seeds in experiences made during the 1980s when demanding and costly measures with respect to another environmental issue, air pollution, did not hamper economic growth, but instead enhanced employment, technological innovation and the modernization of industries. The public perception of "ecological modernization" as a win-win-strategy in solving environmental problems has proven to be very stable, despite of an attention decline with respect to climate issues during the 1990s. However, while the government and proponents of a strong global climate change policy have provided the public with considerable information about net benefits for the country as a whole, they have kept quiet about redistributional effects of current and planned domestic programs and international commitments [13]. They issue the concern that reliance on "ecological modernization", combined with some kind of "distributional opaqueness", might turn out to be a drawback to German consensus. Furthermore, the drive for consensus might backfire as soon as doubts enter the discourse with respect to the reliability of scientific findings on global warming [12]. This could threaten the legitimacy of political decisions based on scientific knowledge.

The issue of climate change passed through the issue-attention cycle for the first time in the second half of the 20th century [23]. Though it has never completely vanished from the public agenda, attention to climate change was relatively low during the second half of the 1990s. A new attention-cycle started at the beginning of the 21st century and reached its peak in 2007 when the IPCC published its Fourth Assessment Report. Since 2007, attention to the issue of climate change has been falling again. The financial crisis seems to have intensified this down-swing since it had more dramatic value and a higher degree of novelty than the issue of climate change [24]. Thus, when trade markets crashed in September 2008, this event drew media attention away from the climate problem, as predicted by the arena model of Hilgartner and Bosk [39]. The application of discourse network analysis on the German climate discourse within this study allows taking a closer look at actor constellations and frame configurations.

DATA SELECTION, CODING, AND NETWORK ANALYSIS

This study is based on newspaper articles published in the Frankfurter Allgemeine Zeitung (FAZ) and the Süddeutsche Zeitung (SZ) within the first quarter of the years 2007 to the first quarter of 2010 which treated climate change as a main topic. Both newspapers were chosen as data sources due to their prestigious status and high circulation rates (about 2 million copies each). Both are regarded as important reference media by other journalists and are read most frequently by the members of the German Parliament (Deutscher Bundestag). Hence, they can be assumed to have an influence on the society as a whole as well as on decision makers. Furthermore, both newspapers cover the main political spectrum of German politics. The FAZ has a rather conservative profile, while the SZ is considered to be more social-liberal.

The articles were selected from the online archives of both papers, including the complete news coverage for all days of appearance and all news sections. Within a two-step selection process "Klimaschutz*" (climate protection), "Klimawandel*" (climate change) and "Globale* Erwärmung*" (global warming) were identified as the most valid and effective choice of key words [24]. Articles that contained at least one of these keywords in the headline and/ or lead paragraph were copied to the JAVA based software *Discourse Network Analyzer* (DNA) programmed by Philip Leifeld [8]. The data set was manually reviewed and articles that contained the keywords but were not really about meteorological climate change (e.g. "Klimawandel*" in the sense of working atmosphere) were excluded. Opinion columns were excluded as well because of low intracoding-reliability.

Statements were edited within DNA. The unit of analysis was a statement, a part of the text where an actor expresses his beliefs or solution concepts for a policy problem [8]. In this study we look at two kinds of concepts: frames and positions. A first step of coding considered only frames. Tags were assigned to each statement that coded the individual speaker, the organization that he or she was affiliated with and the frame that he or she used. Thereby an actor was defined as an identifiable speaker that is not only mentioned in the article, but is given the opportunity to express his opinion by means of direct or indirect quote. Only those statements were coded that could clearly be attributed to a specific actor – an individual person or an organization. If an actor gave his opinion with regard to a specific policy measure within a statement (i.e.

rejection or support for a specific measure), the statement was edited a second time. This time, positions were coded instead of frames. A dummy variable indicating agreement or disagreement with regard to a position was recorded.

This study uses a typology of *frames* which was inductively developed on the basis of a random sample of 10% of all articles sampled for the first quarter of 2008 and 2009. The coding is based on methods and procedures developed by Gerhards and Schäfer [51]. In this way, different arguments of actors are grouped into interpretative patterns that are subject to several strategies of reduction. These in turn are assigned to several categories, following the idea that arguments and actors can be grouped according to the different rationalities of societal sub-systems. Applied to our subject, actors can use political, economic, scientific, ethical, ecological, and policy arguments. Our study thus assumes that with respect to viable policy responses it is important to differentiate whether political responsibility is attributed to the local, national, European or international level. Accordingly, political arguments are grouped into these four sub-categories (Table 1).

The next step of analysis refers to *positions* which are specific policy measures that an actor opposes or supports. The list of positions was inductively extended whenever an actor issued a policy measure not yet on the list. If a policy instrument was suggested several times but each time with respect to another sector (e.g. emission limits for the car industry or energy producers), these measures were categorized respectively.

From this data, several networks were generated with the help of *DNA* [8] and *UCINET 6* [52]. Analysis related to centrality positions and their visualizations were conducted with *visone,* a JAVA based software for the visualization and analysis of social networks [53-54].

Table 1: Frames

Frame	Description
Cultural frames	
1 Individual lifestyle	Statements about practices of individual and community living, consumption patterns, private insurances covering for damages resulting from impacts of climate change, etc.
2 Popular culture	References to information campaigns aiming to raise public awareness of the issue of climate change, books, films, etc.
Ecological/meteorological frame	
References to ecological and meteorological impacts of climate change that are already observable, e.g. rising sea levels, melting ice, heat waves, issues of biodiversity etc.	
Economic frames	
1 Microeconomic considerations	Statements on business aspects of climate change, e.g. economic costs imposed on companies by climate change mitigation policies or business opportunities for companies arising from green technologies
2 Macroeconomic considerations	Considerations regarding national location attractiveness, competition between German and foreign companies, creation of jobs, or economic growth
Ethical and social frames	
1 Sharing responsibility between industrialized and developing world	Discussion on how much commitments industrialized countries can demand from developing countries or on whether they have to compensate poor countries for increased climate risks and damages
2 Moral feeling of responsibility to mitigate climate change	Moral feeling of obligation to mitigate climate change, e.g. in the sense of intergenerational responsibility
3 Financial burden imposed on population	Discussion on who should bear the cost of climate change mitigation measures – i.e. the state, major polluters or the population – and what cost the population can be expected to pay for climate change mitigation
4 Social impacts of climate change	Considerations regarding social impacts such as migration and civil commotions
Politics and policy frames	
Debates on (potential) climate change mitigation or adaptation measures and on responsibilities of different actors in the policy arena	
1 Local level	Local governments take action/are called into account
2 National level	National governments take action/are called into account
3 European level	European institutions take action/are called into account
4 International level	International government actors take action/are called into account
Scientific frames	
1 Causes of climate change	Ideas or beliefs about the geophysical causes of climate change (e.g. the role of human-produced greenhouse gases)
2 Consequences of climate change	Predictions on the ecological consequences of climate change, e.g. changes in Atlantic circulation
3 Effects of climate change mitigation measures	Discussions on the potential effectiveness of mitigation measures and on whether anthropogenic climate change can still be maintained at a non-critical level at all

Frame		Description
4	Technology and applied science	Statements on new technologies (developed by scientists of research institutes and private companies) and applied science that may be employed to mitigate or adapt to climate change
5	Validity of scientific data and methods	Discussion on the proceedings of scientific research and the soundness of scientific pronouncements

DISCOURSE NETWORK ANALYSIS: FINDINGS AND IN-TERPRETATIONS

Our analysis is based on a structural content analysis of 774 articles and 1459 statements. Table 2 gives an overview on how many articles were published on the issue of climate change during the first quarters of the years 2007 to 2010 within the FAZ and SZ. It also displays the dimensions of the discourse networks within the respective quarters, the number of statements, organizations and positions. These numbers may also be influenced and biased by global policy developments, since the first quarter of 2007 was marked by the release of the IPCC Fourth Assessment Report, while possible important events like the UN Climate Conferences Copenhagen took place during quarters of the other years that are not included in the data. Table 2 shows that the discourse networks strongly vary between the different years with respect to actor participation and conceptual affiliation. This has to be kept in mind when we interpret the following findings.

Table 2: Media Coverage of the Climate Issue (FAZ and SZ), First Quarters 2007-2010

	Articles	Statements	Actors	Position Categories	Positions	Positions/ Statements
2007 Q1	380	774	194	40	268	0,35
2008 Q1	187	303	110	20	59	0,19
2009 Q1	112	206	87	13	39	0,19
2010 Q1	95	176	78	14	44	0,25
Sum	774	1459	469	87	410	0,28

In Figure 2 the evolution of media attention is depicted in the context of the main economic indicators. It shows that the economic downturn started in September 2008 with a plunge at the stock exchange and reached the real economy in 2009. As recovery was quick, at least in Germany the crisis was over in the beginning of 2010.

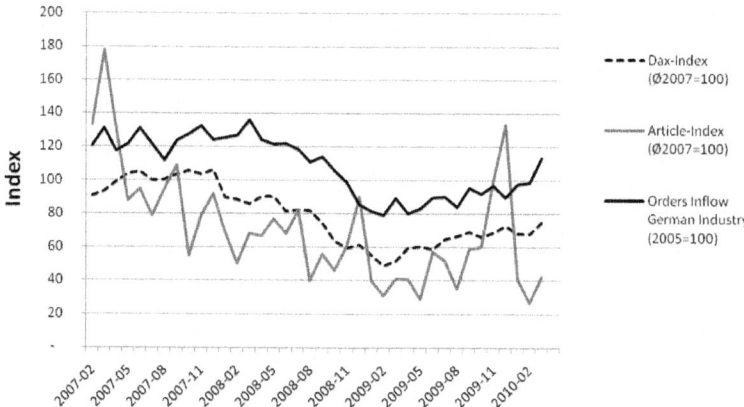

Figure 2: Articles on Climate Change and Economic Indicators

CHANGING ACTOR CONSTELLATIONS AND FRAME CONFIGURATIONS

A first step of analysis relates to possible changes in actor constellations due to the economic crisis. In this respect we are interested, firstly, in the overall actor dynamics in the field of discourse, and secondly in the relative standing of the various actors and significant changes in these positions.Figures 3 and 4, and table 3 give an overview on the dynamics. Figure 3 depicts data on entry, exit, and discourse continuation during the four years. The overall picture suggests a dynamic and pluralist policy arena in which many new actors are entering and constellations are changing.

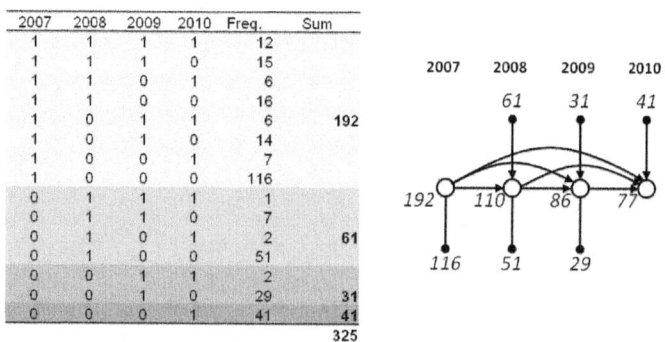

Figure 3: Participation Profiles

Table 3: Top 25-Actor's Statements

Name	# of statements				Participation Profiles				Normalized statements			
	2007	2008	2009	2010	2007	2008	2009	2010	2007	2008	2009	2010
Federal Ministry for the Environment (BMU)	76	18	4	10	1	1	1	1	9,82	5,94	1,95	5,81
Federal Environmental Agency	6	6	3	9	1	1	1	1	0,78	1,98	1,46	5,23
International Panel of Climate Change (IPCC)	22	4	6	5	1	1	1	1	2,84	1,32	2,93	2,91
Potsdam Institute for Climate Impact Research (PIK)	10	7	2	5	1	1	1	1	1,29	2,31	0,98	2,91
US Government	17	3	7	3	1	1	1	1	2,20	0,99	3,41	1,74
United Nations (UN)	7	1	2	3	1	1	1	1	0,90	0,33	0,98	1,74
Greenpeace	15	2	1	2	1	1	1	1	1,94	0,66	0,49	1,16
Small or medium-sized businesses	9	1	1	2	1	1	1	1	1,16	0,33	0,49	1,16
European Commission, DG Environment	29	4	10	1	1	1	1	1	3,75	1,32	4,88	0,58
Christian Democratic Party (CDU)	24	1	4	1	1	1	1	1	3,10	0,33	1,95	0,58
Green Party (Buendnis 90/Die Gruenen)	17	6	6	1	1	1	1	1	2,20	1,98	2,93	0,58
Oeko-Institut	1	2	2	1	1	1	1	1	0,13	0,66	0,98	0,58
Munich Re	7	0	6	8	1	0	1	1	0,90	-	2,93	4,65
Federal Ministry of Finance (BMF)	4	0	1	6	1	0	1	1	0,52	-	0,49	3,49
Chinese Government	6	0	1	5	1	0	1	1	0,78	-	0,49	2,91
Federation of German Consumer Organizations (vzbv)	3	3	0	5	1	1	0	1	0,39	0,99	-	2,91
UN Framework Convention on Climate Change (UNFCCC)	9	0	2	4	1	0	1	1	1,16	-	0,98	2,33
World Wildlife Fund (WWF)	9	5	0	3	1	1	0	1	1,16	1,65	-	1,74
US Democrats	2	0	1	3	1	0	1	1	0,26	-	0,49	1,74
US Environmental Protection Agency (EPA)	0	1	2	3	0	1	1	1	-	0,33	0,98	1,74
French Government	7	1	0	2	1	1	0	1	0,90	0,33	-	1,16
European Parliament (EP)	6	2	0	1	1	1	0	1	0,78	0,66	-	0,58
Danish Government	2	0	2	1	1	0	1	1	0,26	-	0,98	0,58
EP, Progressive Alliance of Socialists and Democrats (S&D)	2	1	0	1	1	1	0	1	0,26	0,33	-	0,58
German Chancellor	1	1	0	1	1	1	0	1	0,13	0,33	-	0,58

Table 3 lists the standings of the 25 top policy actors that participated in at least three years up to spring 2010. In order to control for variation in discursive activities, we normalized their figures with respect to the yearly total numbers and depicted them as percentages. Figure 4 correlates the four columns of the table (activity profiles) and shows interesting results. While the correlation between the actors' standings between 2007 and 2008 is rather high, the correlations dropped to .47 and .39 in the following years during the economic crisis. The pre-crisis actor configurations differ greatly from within- and post-crisis constellations.

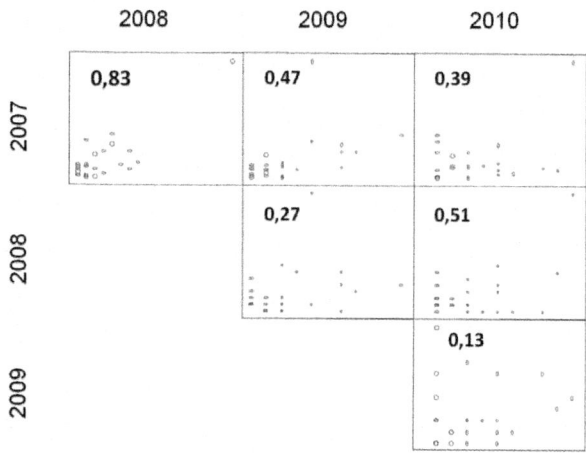

Figure 4: Correlation of 25 Top-Actor's Participation Profiles

A further key question is how the economic crisis affected frame configurations. If Maslow's hierarchy of needs also applies to policy discourse we can assume that economic frames gain importance and possibly crowed out non-economic frames. Figures 5-7 can give a partial answer to that question. They show which organization is utilizing certain frames within the discourse on climate change during the respective quarters. The thickness of links between actors and frames corresponds to the frequency that an actor uses the respective frame. The size and arrangement of frames indicate indegree centrality of frames which equals the relative frequency that a frame is cited by all actors. Thus, the frame with the biggest node area and the most central position within the circular arrangement is used the most often within the respective time period. Between 2007 and 2009, shifts in the frame constellation can be observed from year to year. Each year, another frame occupies the most central position within the discourse. Thereby, the first quarter of 2009 differs from the other periods of observation in two respects: Firstly, while a politics and policy frame dominates the discourse in all other quarters, the macroeconomic frame is the most central one in the first quarter of 2009. Secondly, while the discourse is evolving around few frames in 2007, 2008 and 2010, it is characterized by a much more heterogeneous frames distribution in 2009. The first observation seems to support the proposition that the financial crisis had a direct impact on the individual perception of climate change in so far that it highlights economic aspects of the problem. But, taking into consideration the low degree of overall centralization, the domination of the macroeconomic frame in 2009 is not very strong. It is only short-lived. In addition, this frame is mainly connected to actors from the business sector or foreign political actors, but less to domestic political actors who largely use the national policy frame in 2009. In all other years, the macroeconomic frame is used by a more heterogeneous set of actors. Over all time periods the macroeconomic frame has the most stable position and is always among the three most central frames. Its shift to the center in 2009 is a result of the overall fragmentation of the discourse. In the other years, as the thickness of links shows, governmental actors push forward the respective central frame. In 2007, the Federal Ministry for Environment, Natural Conservation and Nuclear Safety (BMU) promotes the national policy and politics frame, in 2008, the DG Environment of the European Commission pushes forward the European policy and politics frame and in 2010, different federal ministries promote the two politics and policy frames at the center. Such a strong political commitment is lacking in 2009.

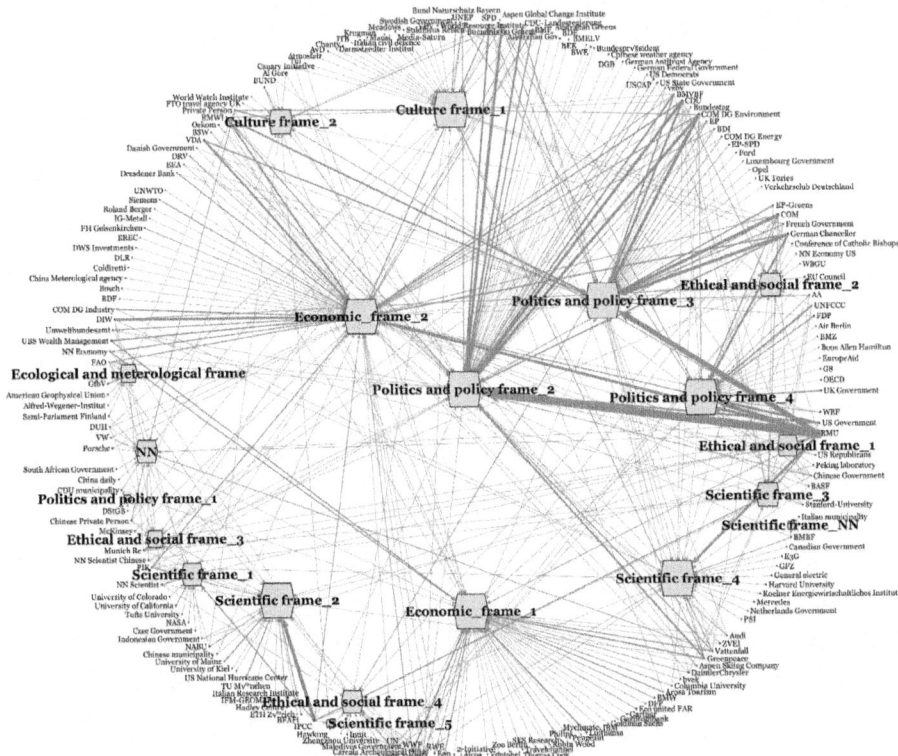

Figure 5: Affiliation Network Organizations-Frames 2007

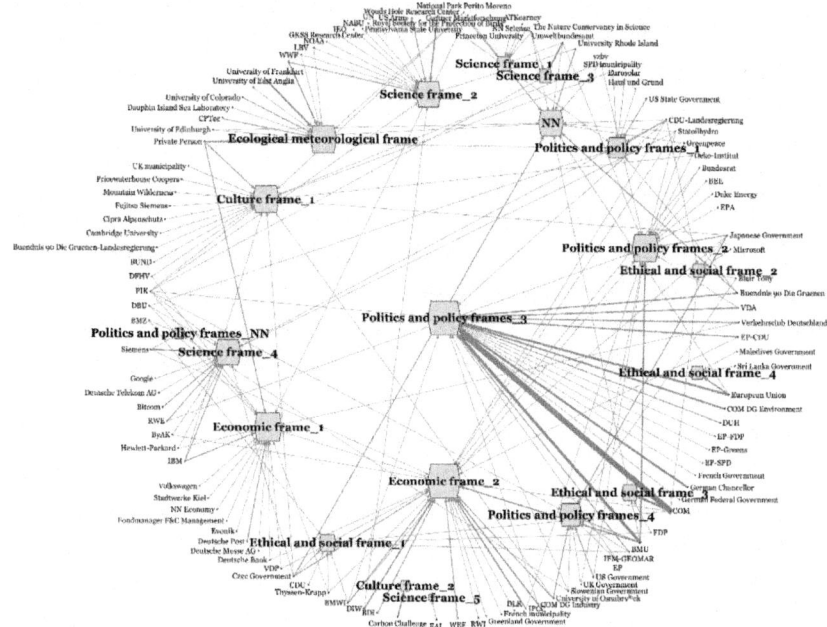

Figure 6: Affiliation Network Organizations-Frames 2008

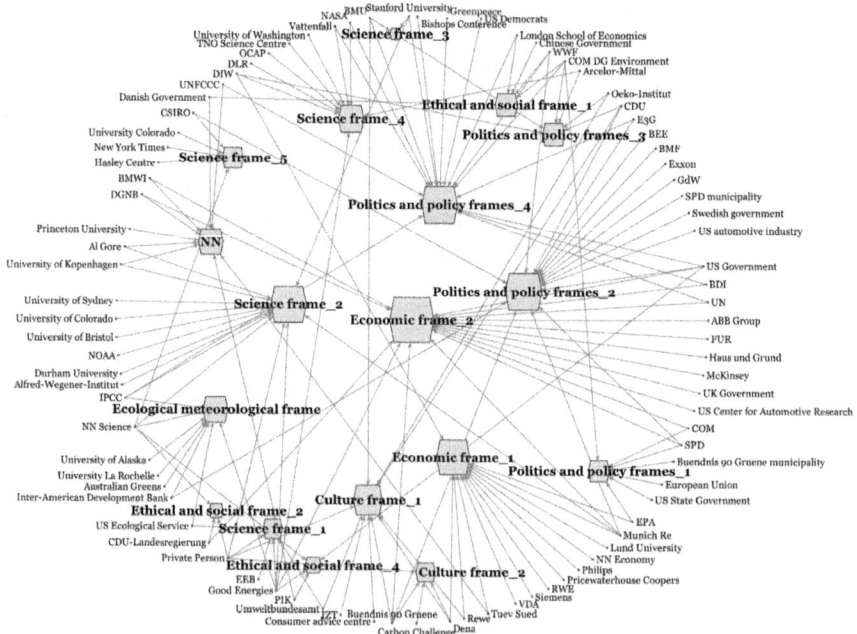

Figure 7: Affiliation Network Organizations-Frames 2009

A further step is the depiction of various organizations and their policy positions. Figures 8 and 9 show which actors have a position towards policy measures and whether they support (light links) or oppose (dark, dashed links) the relative measure or are undecided (dark continuous line). The size of the nodes reflects *indegree centrality* of the respective policy proposition which equals the relative frequency that a proposition is commented.

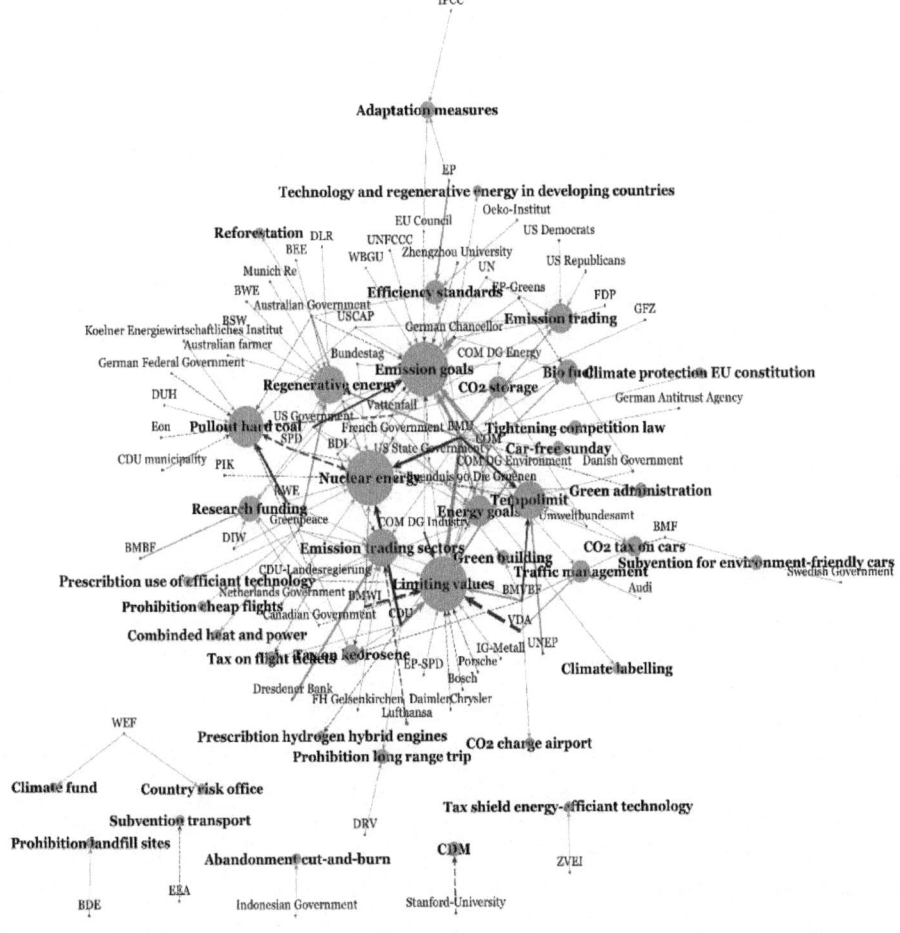

Figure 8: Affiliation Network Organizations-Positions 2007

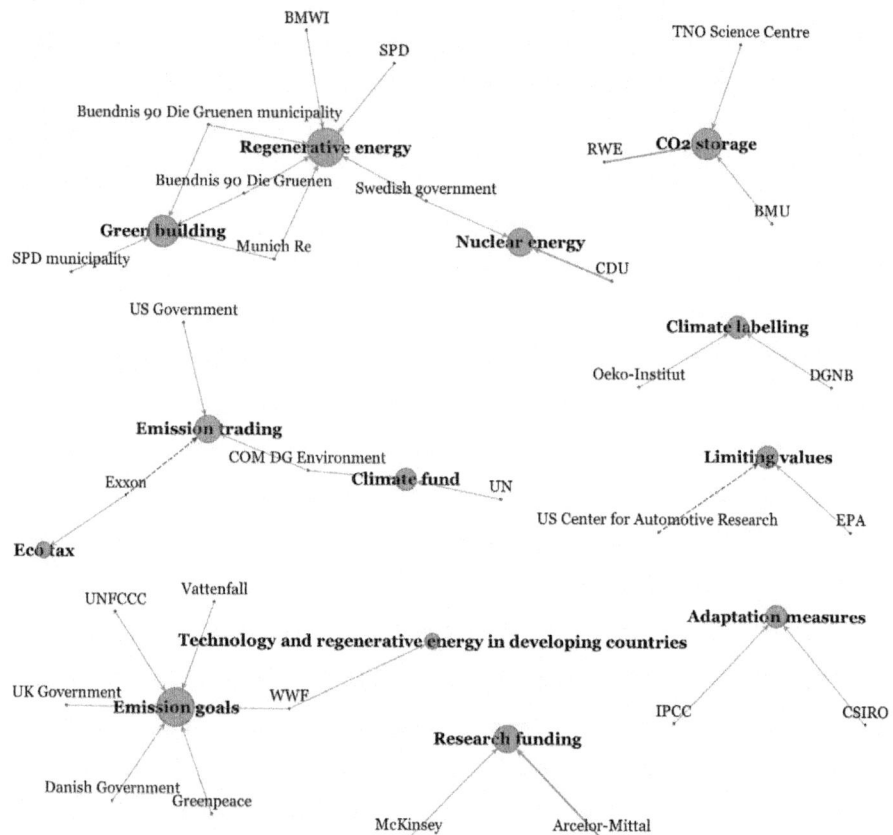

Figure 9: Affiliation Network Organizations-Positions 2009

Only in 2007 and 2008 there is a controversial debate on policy measures. Again, few governmental actors – especially the Federal Ministry of Environment (BMU) – hereby take a central position. The networks are decomposed into several components in 2009 and 2010. We can see that there is a consensus on the promotion of regenerative energy. It might be that - as political conflicts on how to tackle the financial crisis intensified - actors became less inclined to settle political conflicts in the area of climate policy.

With respect to frame analysis, Figure 10 shows the co-occurrence of frames in 2008. The width and darkness of links visualizes the strength of interconnection in terms of how many organizations use both interconnected frames.

Comparing the connections between frames within the different periods of observation, it can be stated that the macroeconomic frame is the frame, which is best connected to other frames, especially in 2008. The connection between the macroeconomic frame and the national politics and policy frame is especially strong in all years. Apart from that, the macroeconomic frame is always strongly connected to the relative dominating frame, which could explain the success of the "ecological modernization" paradigm. The financial crisis does not reduce the interconnectedness of the macroeconomic frame. The actors who hold up this interconnection are the insurance company Munich Re, the green party (Bündnis 90/Die Grünen) and the social democratic party (SPD). It seems that advocates of strong climate protection in the light of the financial crisis adopt economic frames to link the climate issue to the crisis and mobilize against the decline of attention to climate change.

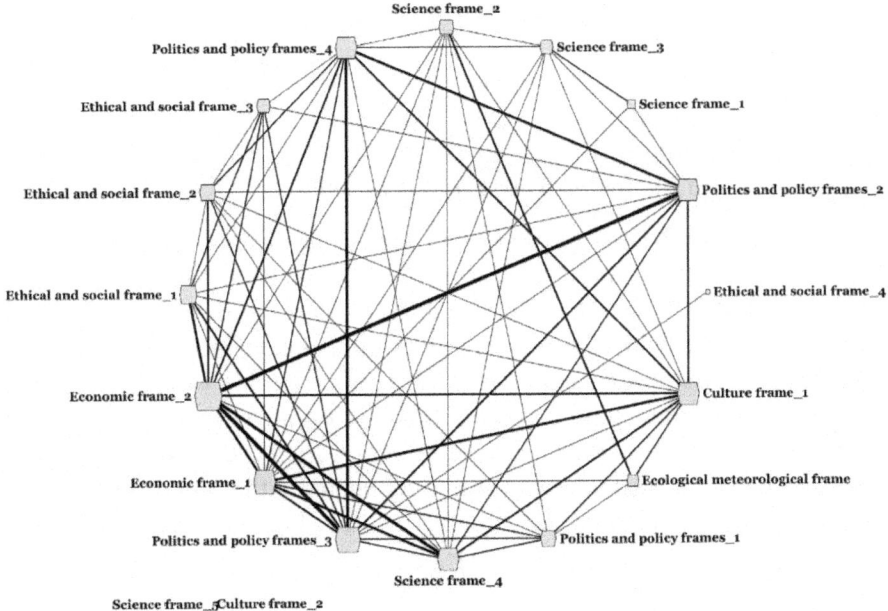

Figure 10: Co-occurrence Network Frames 2008

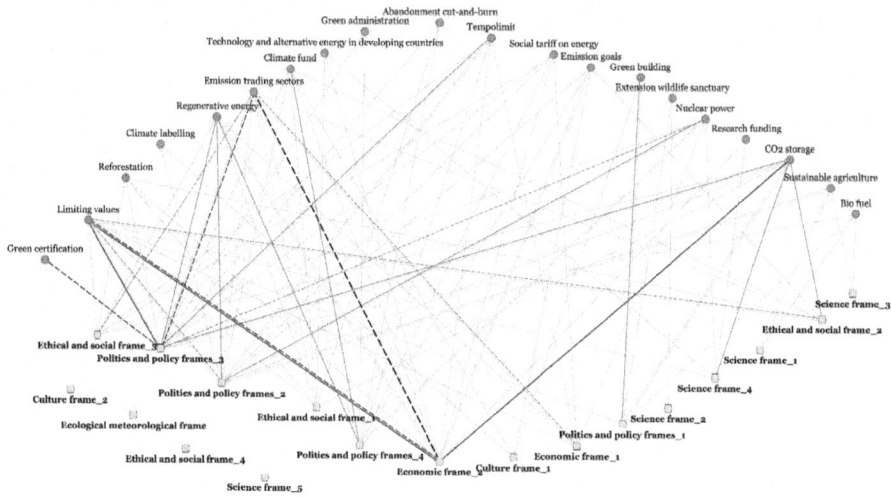

Figure 11: Affiliation-Network Frames-Positions 2008

A further analytical step refers to the frame-position network. In this respect, Figure 11 shows the co-occurrences of and conflicts between frames and positions in 2008. The width and darkness of links correspond to the number of actors that use the frame and the position that the relative link connects in the same way (continuous link) or in opposing ways (dashed link). The network is quite dense; it is not possible to infer the position from the frame or the other way round. The same observation can be made during the other periods of observation. This finding supports the proposition that the linkage between problems and solutions at the collective level is not straight forward [45].

CONCLUSION

While previous studies have often examined climate discourses from a perspective of functional social sub-systems and have conceptualized discourses primarily as a means to overcome difficulties of risk communication, this study conceptualized discourses as an essential part of modern policy making which is characterized by high interdependence and connectedness between societal subsystems. This perspective shifts the focus from differences in communication to the mobilization of actors within an integrative policy process. Public action is not primarily constrained by difficulties of communication on different problem perceptions and policy preferences, but by the limited capacity of actors to process many problems at once. Furthermore, different actors pursue different individual interests and strategically use public discourses to influence the structure of participation within a policy-subsystem. Punctuated Equilibrium Theory suggests that a big focal event may attract high public attention and thereby provides an opportunity to change the public perception of a policy issue and to restructure the policy arena.

Our paper has shown that the financial crisis amplified the decrease of public attention to the issue of climate change in Germany. Analyzing the media discourse on climate change between 2007 and 2010 by means of network analysis we show that actors are strongly involved in cross-sectoral communication and that specific policy positions cannot be directly derived from perceptions of climate change. This may facilitate co-operation in managing global warming across societal subsystems. Our analysis also demonstrates that a sincere debate on different policy measures is only possible when specific governmental actors claim political responsibility. In the aftermath of the financial crisis, when issue attention towards climate change declined, political commitment weakened as well, and the discourse became more fragmented. This fragmentation of the public discourse may impede policy innovation and hinder the management of climate change, especially if it is reflected in the subsystem specific discourse. However, the principle of "economic modernization" associated with the success of past German climate policy seems to have sustained after the financial crisis.

So far little is known about the relation of general public discourses and sub-system and policy specific discourses. Further research is necessary to gain a better understanding of how policy making is shaped by discourses at various levels and subsystems. Therefore, discourse network analysis proves to be a promising tool to grasp the complexity of discourse dynamics which are influenced by structural constraints as well as by strategic actor behavior. Other than qualitative frame analyses, it links actor and frame constellations in a specific and transparent way. This allows for case sensitive modeling but also for replicability and comparability.

ACKNOWLEDGEMENTS

The coding scheme for the data set used for this paper is based on preparatory work conducted by Katrin Vogt who accomplished her master's degree at the Department of Politics and Public Administration at University of Konstanz in 2009. We therefore address our special thanks to her.

REFERENCS

1.   Schneider V., Malang T., Leifeld P. Coping with creeping catastrophes: National political systems and the challenge of slow-moving policy

problems. In: Siebenhüner B., Arnold M., Eisenack K., Jacob K. (eds). Long-term Governance of Social-Ecological Change, London: Routledge, forthcoming.

2.   Zahariadis N. The Multiple Streams Framework: Structure, Limitations, Prospects. In: Sabatier PA. (ed.) Theories of the Policy Process. Boulder: Westview Press; 2007. p. 97-115.

3.   Heidbrink L., Leggewie C., Welzer H. Von der Natur- zur sozialen Katastrophe. Wo bleibt der Beitrag der Kulturwissenschaften zur Klima-Debatte? Ein Aufruf. Die Zeit. 1 November 2007.

4.   Bunge M. Finding Philosophy in Social Science. New Haven: Yale University Press; 1996.

5.   Schneider V., Janning F. Politikfeldanalyse. Akteure, Diskurse und Netzwerke in der öffentlichen Politik. Wiesbaden: VS Verlag; 2006.

6.   Janning F., Leifeld P., Malang T., Schneider V. Diskursnetzwerkanalyse. Überlegungen zur Theoriebildung und Methodik. In: Schneider V., Janning F., Leifeld P., Malang T. (eds.) Politiknetzwerke Modelle, Anwendungen und Visualisierungen. Wiesbaden: VS-Verlag; 2009. p. 59-92.

7.   Leifeld P. Political Discourse Networks - The missing link in the study of policy-oriented discourse. Paper presented at ECPR Joint Sessions in Münster, 2010. http://www.philipleifeld.de/cms/upload/Downloads/ leifeld_ecpr_paper.pdf (accessed 10 October 2012).

8.   Leifeld P. Die Untersuchung von Diskursnetzwerken mit dem Discourse Network Analyzer (DNA). In: Schneider V., Janning F., Leifeld P., Malang T. (eds.) Politiknetzwerke Modelle, Anwendungen und Visualisierungen. Wiesbaden: VS-Verlag; 2010. p. 391-404.

9.   Leifeld P., Haunss S. Political discourse networks and the conflict over software patents in Europe. In: European Journal of Political Research. 2012; 51(3):382–409.

10. Gould SJ., Eldredge N. Punctuated equilibria: the tempo and mode of evolution reconsidered. Paleobiology. 1977;3(2) 115-51.

11. Baumgartner FR., Jones BD. Agendas and Instability in American Politics. Chicago, London: University of Chicago Press; 1993.

12. Weingart P., Engels A., Pansegrau P. Risks of communication: discourses on climate change in science, politics and the mass media. Public Understanding of Science. 2000;9(2000) 261-83.

13. Weidner H., Metz L. German climate change policy. A success story with some flaws. The Journal of Environment & Development. 2007;17(4)

356-78.

14. Stehr N., Storch H. von The social construct of climate and climate change. Climate Research. 1995;5:99-105.

15. Blumer H. Social Problems as Collective Behaviour. Social Problems. 1971;18(3) 298-306.

16. Spector M., Kitsusue JI. Social Problems: A Re-formulation. Social Problems. 1973;21 145- 59.

17. Lindzen RS. Is the Global Warming Alarm Founded on Fact? . In: Zedillo E. (ed.) Global Warming – Looking Beyond Kyoto. Washington D.C.: Brookings Institutions Press; 2008. p. 21-33.

18. Rahmstorf S. Anthropogenic Climate Change: Revisiting the Facts. In: Zedillo E. (ed.) Global Warming – Looking Beyond Kyoto. Washington D.C.: Brookings Institutions Press; 2008. p. 34-54.

19. Feldmann MS. Order without Design: Information Production and Policy Making. Stanford: Stanford University Press; 2003.

20. Wiedemann PM. Strategien der Risio-Kommunikation und ihre Probleme. In: Jungermann H., Rohrmann B., Wiedermann PM. (eds.) Risikokontroversen, Konzepte, Konflikte, Kommunikation. Berlin: Springer-Verlag; 1991.

21. Hajer MA. A frame in the fields: policy making and the reinvention of politicy. In: Hajer MA., Wagenaar H. (eds.) Deliberative Policy Analysis Understanding Governance in the Network Society. Cambridge: Cambridge University Press; 2003. p. 88-110.

22. Fischer F. Reframing Public Policy. Discursive Politics and Deliberative Practices. Oxford, New York: Oxford University Press; 2003.

23. Trumbo C. Constructing climate change: claims and frames in US news coverage of an environmental issue. Public Understanding of Science. 1996;5 269-83.

24. Vogt K. A Kink in the Natural Life Cycle of Climate Change? The Effect of the Economic Crisis on the Discourse about Climate Change. University of Konstanz; 2009.

25. Dryzek JS. The Politics of the Earth: Environmental Discourses. Oxford: Oxford University Press; 1997.

26. Addams H., Proops J., editors. Social Discourse and Environmental Policy. An application of Q methodology. Chaltenham, Northampton: Edward Elgar; 2000.

27. Malone EL. Debating Climate Change. Pathways through argument to agreement. London, Sterling: Earthscan; 2009.

28. Hajer MA. The Politics of Environmental Discourse. Ecological Modernization and Policy Process. Oxford: Oxford University Press; 1995.

29. Fisher DR., Leifeld P., Iwaki Y. Mapping the ideological networks of American climate politics. Climatic Change Online First; 2012. http://dx.doi.org/10.1007/s10584-012-0512-7 (accessed 20 July 2012).

30. Sabatier PA., Weible C. The Advocacy Coalition: Innovations and Clarifications. In: Sabatier PA (ed.) Theories of the Policy Process. 2 ed. Boulder: Westview Press; 2007.

31. Baumgartner FR. Punctuated Equilibrium Theory and Environmental Policy. In: Repetto R, editor. Punctuated Equilibrium and the dynamics of US Environmental Policy. New Haven, London: Yale University Press; 2006. p. 24-46.

32. Ture JL., Jones BD., Baumgartner FR. Punctuated Equilibrium Theory – Explaining Stability and Change in American Policymaking. In: Sabatier PA. (ed.) Theories of the Policy Process. Boulder: Westview Press; 1999. p. 97-115.

33. Repetto R. Introduction. In: Repetto R. (ed.) Punctuated Equilibrium and the dynamics of US Environmental Policy. New Haven, London: Yale University Press; 2006. p. 1-23.

34. Dayton BP. Policy Frames, Policy Making ans the Global Climate Change Discourse. In: Addams H., Proops J. (eds.) Social Discourse and Environmental Policy An application of Q methodology. Chaltenham, Northampton: Edward Elgar; 2000. p. 71-99.

35. Kenis P., Schneider V. Policy Networks and Policy Analysis: Scrutinizing a New Analytical Toolbox. In: Marin B., Mayntz R. (eds.) Policy Networks: Empirical Evidence and Theoretical Considerations. Boulder: Westview Press; 1991. p. 25-59.

36. Gerhards J., Lindgens M. Diskursanalyse im Zeit- und Ländervergleich – Methodenbericht über eine systematische Inhaltsanalyse zur Erfassung des öffentlichen Diskurses über Abtreibung in den USA und der Bundesrepublik in der Zeit von 1970 bis 1994. Discussion Paper FS II Wissenschaftszentrum Berlin; 1995. p. 90-101.

37. Schwab-Trapp M. Diskurs als soziologisches Konzept – Bausteine für eine soziologisch orientierte Diskursanalyse. In: Keller R., Hirseland A., Schneider W., Viehöfer W. (eds.) Handbuch Sozialwissenschaftliche Diskursanalyse: Band 1: Theorien und Methoden. Wiesbaden: VS Verlag; 2006. p. 263-86.

38. Leifeld P., Schneider V. Information Exchange in Policy Networks.

American Journal of Political Science. 2012;56(3) 731-44.

39. Hilgartner S., Bosk CL. The Rise and Fall of Social Problems: A Public Arenas Model. The American Journal of Sociology. 1988;94(1) 53-78.

40. Downs A. Up and down with ecology – the issue-attention cycle. Public Interest. 1972;28 38-50.

41. Ungar S. The Rise and (Relative) Decline of Global Warming as a Social Problem. Society and Natural Resources. 1992;8(5) 443-56.

42. Ungar S. Social scares and global warming: Beyond the Rio convention. Society and Natural Resources. 1995;8(5) 443-56.

43. Lowe P., Goyder J. Environmental Groups in Politics. London: George Allen and Unwin; 1983.

44. McComas K., Shanahan J. Telling Stories about Global Change. Measuring the Impact of Narratives on Issue Cycles. Communication Research 1999;26(1) 30-57.

45. Jones BD., Baumgartner FR. The politics of attention. How government prioritizes problems. Chicago, London: University of Chicago Press; 2005.

46. Rein M., Schön D. Reframing Policy Discourse. In: Fischer F., Foster J. (eds.) The Argumentative Turn in Policy Analysis and Planing. Durham: Duke University Press; 1993. p. 145-66.

47. Rohmberg M. Mediendemokratie. die Agenda-Setting-Funktion der Massenmedien. München: Wilhelm Fink Verlag; 2008.

48. Kingdon JW. Agendas, Alternatives, and Public Policies. 2nd ed. Boston: Little Brown; 1995.

49. Jaggard L. Climate Change Politics in Europe. Germany and the International Relations of the Environment. London, New York: Tauris Academic Studies; 2007.

50. Bundesministerium für Umwelt. NuR. Nationale Nachhaltigkeitsstrategie: Mehr erneuerbare Energien und weniger Treibhausgase - Handlungsbedarf vor allem in anderen Bereichen. Berlin; 2010.

51. Gerhards J., Schäfer MS. Die Herstellung einer öffentlichen Hegemonie – Humangenomforschung in der deutschen und US-amerikanischen Presse. Wiesbaden: VS Verlag; 2006.

52. Borgatti SP., Everett MG., Freeman LC. Ucinet for Windows: Software for Social Network Analysis. Harvard: Analytical Technologies; 2002.

53. Brandes U., Wagner D. Visone - Analysis and Visualization of Social Networks. In: Jünger M., Mutzel P. (eds.) Graph Drawing Software. Berlin: Springer Verlag; 2003. p. 321-40.

54. Brandes U., Kenis P., Raab J., Schneider V., Wagner D. Explorations into the visualization of policy networks. Journal of Theoretical Politics 1999;11(1)75-106

Chapter 6

JORDAN'S WATER RESOURCES: INCREASED DEMAND WITH UNRELIABLE SUPPLY

Saad Merayyan[1], Salwa Mrayyan[2]

[1]Department of Civil Engineering, California State University, Sacramento, USA

[2]AL-Balqa Applied University, Al-Huson, Jordan

ABSTRACT

Jordan is a small county located in the Middle East. Jordan has borders with Saudi Arabia, Syria, Iraq, and Israel (Figure 1). Jordan was established in 1921 and has very limited natural resources. Jordan's current (2008) population is 6.5 million. The country has a total land area of 750,000 km^2, about one third (92,300 km^2) of which is dry land while the other two thirds (329,000 km^2) are ir- rigated land. Jordan is considered as a water poor country due to unreliable and shortages in the supply of water sector. This makes it very difficult to meet the required and steadily increasing demand. Impact of climate change adds a layer to the uncertainty on the supply side of Jordan's water portfolio. This paper addresses the water supply challenges that Jordan faces and what has been accomplished to improve supply and/or reduce demand. Many projects were undertaken or planned by the Jordanian government to increase the water supply and improve its relia- bility. Completing the proposed projects will result in Jordan meeting its water demand [1] . Otherwise, the Jordanian Government implement some or all the proposed short term solu- tions as presented in this paper.

BACKGROUND

Jordan's limited water resources supply fluctuates around a stationary [1] . The normal population growth, in ad- dition to the massive influx of refugees from neighboring countries, made the limited water supply in the coun- try even worse. The majority of water resources experts classify countries with less than 1000 m^3 per (year-ca- pita) as water poor countries. In 1997, Jordanians consumed only 883 million cubic meters (MCM), which trans- lates to about 200 m^3 per (year-capita).

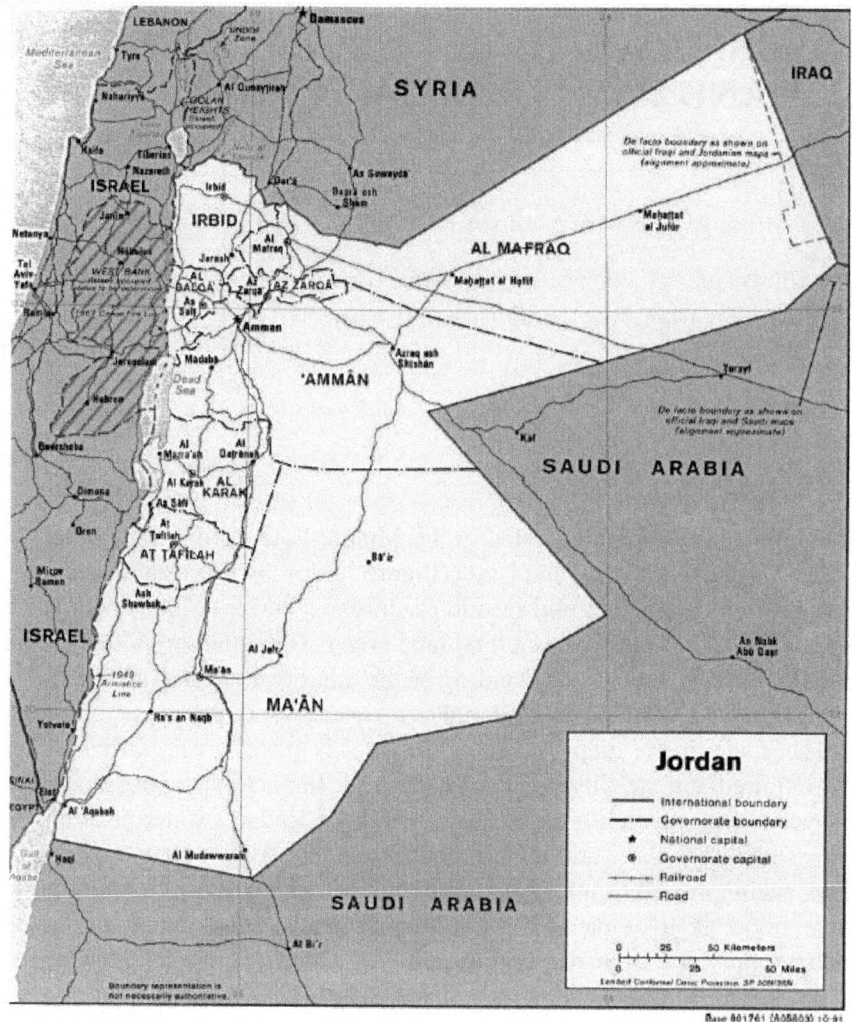

Figure 1: Map of Jordan.

The 1997 water consumption rate?when compared to the rates of neigh-boring countries?was 65% less than that consumed in Syria and Israel, and 85% less than that of Egypt [1] .

The gap between water supply and demand is expected to increase due to the steady increase in population. If the current population growth trends follow what is projected, in 2025 the annual water consumption per capita will decrease from its 1997 levels of 200 m^3 per (year-capita) to 91 m^3 per (year-capita) [1] .

In 2000, Jordan needed 1258 MCM to meet its domestic, industrial and agriculture demand. Faced with these challenges, the Jordanian government water planners devised a plan to increase the water supply, improve the delivery efficiency and reduce certain sectors demand. The increase of the water supply was partly achieved as an outcome of the peace treaty that was signed between Jordan and Israel in 1994, which ensures an additional 215 MCM of water via the building of additional dams, as well as pipelines on the Yarmouk and Jordan rivers. Other projects that the Jordanian government proposed to improve water supply are purchasing water from other countries (i.e. Turkey) and/or constructing a canal "The Two Seas Canal or the Peace Conduit" to bring water from the Red Sea to the Dead Sea. The proposed Two Seas Canal will be constructed with the participation of Israel and the financial support of the United States and the European Union. The Canal will bring 1.8 billion m^3/yea of sea water which will then be treated using desalination to provide about 850 MCM of portable water to Jordan, Israel and Palestine [1] [2] .

The water issue in the Middle East is not only a limited resource. It's also a source of conflict that might re- quire world intervention. Participation in and funding of such projects will, on the long run, provide the people of Jordan with the necessary supply of water, hence improve the overall health and quality of life and, reduce any potential for conflict.

JORDAN'S CLIMATE

Jordan's weather is classified as semi-arid to arid with hot and dry summers, and cold and wet winters. Jordan's precipitation (Figure 2) varies in magnitude, intensity, and distribution. Most of the precipitation in Jordan oc- curs between the months of October and May of each year [2] . The average annual precipitation in Jordan is 50 - 200 mm, 100 - 300 mm, and 200 - 600 mm in Jordan's Desert, Jordan's Valley, and Highlands, respectively. In a typical wet year, the total precipitation is about 1200 MCM and is half of that in a typical dry year. The evapora- tion rates in Jordan vary from 63% from the highlands to 99% from the desert. The typical long-term evaporation average is 93%.

SURFACE WATER RESOURCES

The majority water re-charge comes from precipitation, which dramatically varies from year to year [3] . The surface water, base flow, and runoff long-term averages are 713 MCM, 451 MCM, and 256 MCM, respectively. There are three major surfaces water systems in Jordan. These systems are: Jordan

River that has a catchment area of 18,194 km^2, The Yarmouk River, which has a catchment area of 67,890 km^2. About 1160 km^2 of the Yarmouk River catchment area is located in Syria. The third major river in Jordan is the Zarqa River. The Zarqa River has a catchment area of 4025 km^{2}and its water is not shared with any of Jordan's neighboring countries. The Zarqa River catchment area is densely populated, where about 65% of Jordan's population resides and about 80% of Jordan's industry located. The effluent form several wastewater treatment facilities (e.g. As-Samra,

Figure 2: Annual precipitation (M. Mohsen, 2007).

Jeresh, and Ba'qa WWTP) discharge to the Zarqa River. The Zarqa River is controlled by the King Talal Reser- voir and its water is mainly used for agriculture and livestock demands [3] . The percentage of treated sewerage that is discharged into the Zarqa River is 50% in the winter months and about 60% in the summer months. There are two surface water bodies in Jordan: The Dead Sea and the Gulf of Aqaba. Figure 3 shows the fifteen surface water basins while Table 1 provides a summary of the annual flow from each basin.

GROUNDWATER RESOURCES

Groundwater is considered to be the main source of water supply in many regions of Jordan and is the only wa- ter supply in other regions. More than half of the water supply in Jordan comes from groundwater basins. There are twelve groundwater basins in Jordan (Figure 4). The Disi and Jafer basins are considered to be "fossil" aqui- fers. The annual safe yield from these two aquifers is 12 MCM [4] . Recent studies showed that the Disi aquifer can support annual abstraction of 125 MCM for about 50 years [4] [5] . The two largest aquifers that are rechar- geable aquifers are the Amman-Zarqa basin and the Dead Sea basin. The maximum annual safe yield from these two basins is 144 MCM. Table 2 provides a summary of the annual safe yield supply from various aquifers in Jordan [6]

WASTEWATER RESOURCES

Twenty-six of Jordan's cities are served by nineteen wastewater treatment plants (WWTP) [4] . The Majority of these treatment plants are stressed. The treated wastewater is considered an integral part of Jordan's sustainable water resources system. About 70% of the population living in urban areas is serviced with sewer collection system. The annual treated wastewater effluent form theses serviced areas is about 80 MCM [1] [3] . The treated effluent is typically used to meet agriculture and livestock demands and as a groundwater recharge source. The agriculture demands are mainly directed towards the Jordan Valley, where the majorly of Jordan's agricultural is located.

The two largest wastewater treatment plants in Jordan are the As-Samra and Irbid (Wadi AL-Arab) WWTP. The design capacity of the As-Samra WWTP is 68,000 m^3/day and Irbid (Wadi Al-Arab) WWTP design capacity is 21,000 m^3/day [1] [2].

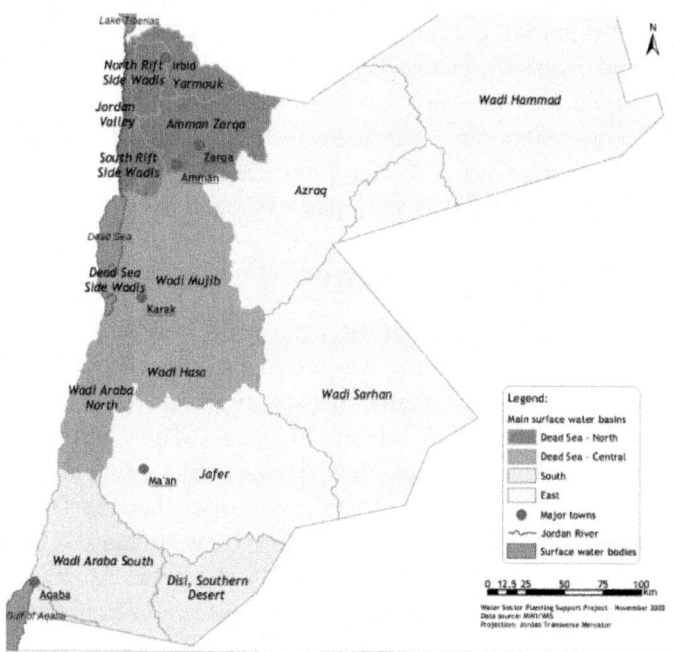

Figure 3: Surface water basins.

Table 1: Catch basin flow

Basin	Annual flow (M
Yarmouk	166
Zarqa	84
N. Side Wadi	58
S. Side Wadi	58
Jordan Valley	8
Wadi Mujib	102
Dead Sea Side Wadi	43
Wadi Hasa	43
Azraq	41
Wadi Hammad	24
Wadi Sarhan	18
Jafer	13
Disi, Southern Desert	1
N. Wadi Araba	46
S. Wadi Araba	8

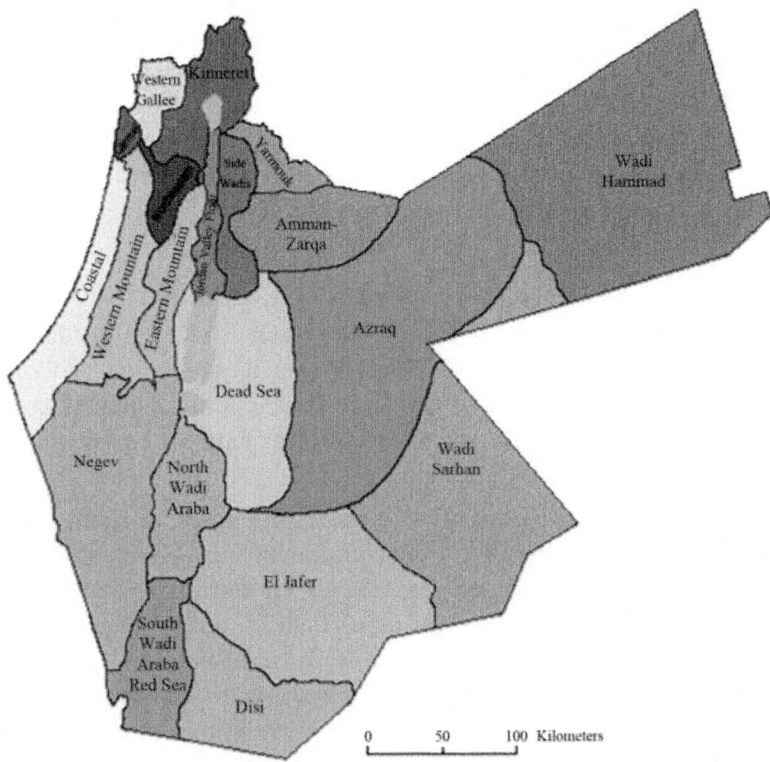

Figure 4: Groundwater basins.

Table 2. Catch basin flow

Basin	Annual flow (MCM)
Yarmouk	166
Zarqa	84
N. Side Wadi	58
S. Side Wadi	58
Jordan Valley	8
Wadi Mujib	102
Dead Sea Side Wadi	43
Wadi Hasa	43
S. Wadi Araba	6
Wadi Sarhan	5
N. Wadi Araba	4
Disi (Fossil)	3

Additional sustainable source of water in Jordan is obtained from desalination plants. There are two operating desalination plants in Jordan. They are the Abu Zeghan and Zara plants with a total combined treatment capacity of about 31 MCM (10 MCM from Abu Zeghan and 21 MCM form Zara). The desalinated water from these two plants is added to the overall water supply in Jordan. Desalination treatment plants are typically expensive to build and operate in a country with limited resources like Jordan [2] [4] .

DISCUSSION

There are several challenges facing the water sector in Jordan. These challenges include elevated levels of nor- mal population growth rates (including population migrations due to regional conflict), degrading surface and groundwater qualities, and changes in precipitation patterns due to climate change [3] [4] .

The annual average population growth rate is about 2.5%, which is relatively high when compared to other countries of Jordan's size and resources. The massive fluxes of thousands of refugees (in the span of few months at a time) from neighboring countries as a result of several regional conflicts (e.g. the first Gulf War, the conti- nuous Arab-Israeli conflict, and the second Gulf War) caused an alarming and sudden increase on the water de- mands in Jordan. In addition to the population increase, the living standards in Jordan are steadily improving and as a result the industrial and agricultural demands are increasing.

To make things worse, Jordan's population projections for the year 2020 is about 9.9 million with 65% of the populations will be served with sewer service. The projected sewerage generation is about 237 MCM, which is about three times the sewerage generation from the current population estimates [4] [7] .

In 1999, the domestic water consumption was 231.5 MCM (29% of the total consumed water in the country). About 79% of the water supply for domestic uses comes from groundwater sources with the exception of the water that is pumped from the Yarmouk River to meet Amman's water demand. The portable water infrastruc- ture in Jordan is not very efficient. Only 45% of the daily domestic consumption accounted for and about 55% of the daily domestic consumption is lost.Table 3 provides a summary of Jordan's water demands and their sources.

The effluent's salinity from WWTP in Jordan is typically higher than normal due to low domestic use levels (due water conservation efforts, and intermittent service) and the use of standard treatment technologies like stabilization ponds in the current WWTP. In addition, to the reasons mentioned above, the high evaporation rate in Jordan results in loss of liquid sewerage water from the stabilization ponds to the atmosphere, which in turn in- creases the sewerage water salinity. The use of the brackish water to recharge groundwater sources result in ad- ditional degradation of the groundwater supply.

The regulation on industrial discharge to the public sewer collection lines is loosely enforced in Jordan. Therefore, industrial effluent from various types of industries with varying industrial wastewater strengths are currently being discharged into public sewers collection lines and eventually treated by municipal WWTP using standard technologies that is used to treat domestic sewerage. As result, when released, the treated effluent from the municipal WWTP is less desirable for agricultural purposes.

Industrial demands accounts for about 5% (Table 3) of Jordan's total water supply [5] . About 94% of the in- dustrial demand comes from groundwater resources. This is projected to increase due to improvements in the industrial sector in the country, Jordan's inclusion in the World Trade Organization (WTO) and the signed free trade agreements between Jordan and the United States and the establishments of the free zones in different re- gions of Jordan. These agreements make Jordan a desirable destination for regional investors and manufacturers.

The agriculture sector in Jordan is the largest water consumer. This sector uses about 65% (seeTable 3) of the total water supply in Jordan. About 49% of the agricultural demand comes from groundwater resources. The majority of the irrigated land in Jordan is located in the highland and the desert (about 52,700 hectares), and in the Jordan Valley (about 31,600 hectares).

It's clear the importance of groundwater in Jordan's overall water resources plan to meet its current water de- mand. Yet there is a shortage of about 40% in the groundwater supply to reach the current total and required wa- ter demand. The 2005 estimates of available water supplies in Jordan from various sources is 816 MCM (275 MCM from Groundwater, 56 MCM groundwater recharge, 395 MCM surface water, 80 MCM from WWTP ef- fluent, and 10 MCM from desalinated water). The projected increase in the water demand form the three major sectors (99% of the total water demand) in Jordan are presented in Table 4.

A quick comparison of the water supply in 2005 (used as a reference for the comparison) with water demand in 1999, 2005, 2010 and 2020 (see Table 5) reveals major water deficit in all years analyzed expect in 1999. It's clearly demonstrated that the problem of water shortages in Jordan is a chronic problem for years to come. If the current domestic, agricultural and industrial practices are not addressed, the water supply in Jordan will deplete in quantity and quality with time. This will also make the problem of water shortages even worse than actually shown in Table 5. Overdraft of groundwater and surface water is creating a tremendous stress on the water supply in Jordan. Large losses from surface waters (i.e. the Dead Sea) are also a problem that the Jordan has to address in order to at least to reduce these losses.

In order for Jordan to balance its water budget, support the growth plans envisioned in the country, and pro- vide the basic element of life to Jordanians, the government needs to take certain steps to increase the water supply and reduce demands.

Increasing supply should be the most important objective of Jordan's water planners. In order to achieve this objective, the government is forging ahead with major joint projects with neighboring countries (Egypt, Syria, Israel, and Palestine) and applying water conservation practices. The steps that the government is proposing to meet and exceed the water demand are listed below.

Table 3. Water demand in MCM by sector in 1999

Use	Groundwater Sources (GWS)	% of Total GWS	Surface Water Sources (SWS)	% of Total SWS	Total Supply Sources TSS	% of Total TSS
Domestic	183	79	49	21	231	29
Industrial	35.5	94.4	2.1	5.6	37.6	5
Agriculture	256	49.1	265	50.9	521	65
Other	7.3	64.6	4	35.4	11.3	1

Table 4. Previous, current and projected water demand in MCM

Use	Year			
	1999	2005	2010	2020
Domestic	231	382	434	611
Industrial	37.6	81	99	146
Agriculture	521	858	904	890
Total	789.6	1321	1436	1647

Table 5. Water management and projections in MCM

Year	Available Water Supply	Demand	Deficit
199	801	789.6	+11.4
2005	816	1321	505
2010		1436	620
2020		1647	831

1) Building Al-Wehda dam adds 100 MCM to the water supply in Jordan. The project is on hold until a peace treaty is forged between Syria and Israel.

2) The main source of additional water supply in Jordan will come from the Dead Sea-Red Sea channel project (or the Peace Conduit). This project is in its early phases. The environmental assessment study and feasi- bility study were just awarded to French and British companies. When completed, this project will provide Jor- dan with 570 MCM in 2022. The project is in co-operation between Jordan, Egypt, Israel, and Palestine with the financial support from the world community (the United States of America, the European Union, the World).

3) Back and other financial agencies. This is a major piece in the water supply puzzle for Jordan to meet its future water supply.

4) The Disi Project is operational and was completed in 2013 after two years delay, which will increase the water supply from this basin by 100 MCM. This supply sources is unsustainable (fossil basin) which is projected to be depleted in 50 years. This is a temporary fix but it's necessary.

5) More surface water can be harvested by employing better water management practices and building small earthen dams. The exploitation of additional surface water will add 139 MCM to the current 395 MCM.

6) Employing non-traditional water management practices will increase water supply. Some of these practices include but not limited to: improvement of the effluent quality form WWTP by employing new technologies and improve water retention in the WWTP stabilization ponds. These practices will add about 89 MCM in 2020 to the current level of 80 MCM.

7) Invest in desalination technology utilizing sustainable energy sources like solar and wind energies. Jordan implemented two projects that will add 65.5 MCM of available water in 2020 to the current levels of desalinated water of 31 MCM.

8) Utilizing surface runoff to recharge groundwater will add about 26 MCM above the current recharge amount of 56 MCM.

CONCLUSIONS

Evaluation of the completed projects, proposed projects and the implementation of the strategies suggested by the Jordanian government show that it is possible that Jordan will be able to meet its water demand or close to meet it by 2020. This paper did not consider climate changes' impacts on Jordan's water resources and no adap- tation strategies were considered [7] . The authors believe that unless the peace conduit is carried out, Jordan's water supply will severely suffer under climate change scenarios. Until then, it is suggested that the Jordanian Government implement one or all of the proposed these short-term fixes:

1) Improve the portable water infrastructure which will save about 30% - 40% of the water losses (leaks and delivery efficiency).

2) Control illicit surface water use and illicit over pumping from groundwater basins.

3) Enforce existing and propose stricter laws on sewerage and industrial effluent to municipal WWTPs.

4) Reduce the agriculture water demand to a level that is appropriate with its economic impact on the county.

5) Invest in agriculture water efficiency technology and practices.

6) Invest in new water treatment technologies and sustainable water supply [2] [8] .

7) Increase outreach efforts to citizens and provide incentives to water conservation efforts especially from the agricultural sector.

8) Increase the use of recycled water to add to the reliability of the water supply.

REFERENCES

1. Mohsen, M. (2008) Water Needs in the Middle East Region, from Red Sea to Dead Sea-Water and Energy. Presentation at Sharing Knowledge across the Mediterranean (4) MAICh, Chania, 7-9 April 2008.

2. Mohsen, M. and Al-Jayyousi, O. (1999) Brackish Water Desalination: An Alternative for Water Supply in Jordan. De- salination, 124, 163-174.

3. Tutundjian, S. (2001) Water Resources in Jordan. World Bank, Amman. http://www.usaidjordan.org/upload/key-doc/water%20resources%20 in%jordan%202001.pdf

4. Al-Fataftah, A. and Abu-Taleb, M. (1992) Jordan's Action Plan. The Canadian Journal of Development Studies, 13, 153-171. http://dx.doi.org /10.1080/02255189.1992.9669488

5. Raddad, K. (2005) Water Resources and Use. Workshop on Environment Statistics, Dakar, 28 February-4 March 2008.

6. Halasah, N. and Ammary, B. (2007) Groundwater Resources in Jordan.http://www.emwis.org/WFG/GROUNDWATER%20 Resources%20Jordan%20%20Nizar%20Halasah.doc

7.   Abu-Taleb, M. (2000) Impacts of Climate Change Scenarios on Water Supply and Demand in Jordan. Water Interna- tional, 25, 457-463.http:// dx.doi.org/10.1080/02508060008686853

8.   Abdulla, F.A. and Al-Shareef, A.W. (2009) Roof Rainwater Harvesting System for Household Water Supply in Jordan. Desalination, 243, 195-207.

Chapter 7

TWO DIMENSIONAL HYDRODYNAMIC MODELLING OF NORTHERN BAY OF BENGAL COASTAL WATERS

Misbah Uddin[1], Jahir Bin Alam[1], Zahirul Haque Khan[2], G. M. Jahid Hasan[1], Tauhidur Rahman[1]

[1]Department of Civil and Environmental Engineering, Shahjalal University of Science and Technology, Sylhet, Bangladesh

[2]Coast, Port and Estuary Division, Institute of Water Modelling, Dhaka, Bangladesh

INTRODUCTION

The Bay of Bengal, including the Andaman Sea and Malacca Strait, lies roughly between latitudes 5°N to 23°N and longitudes 79.8°E to 102°E. It is bordered by eastern coast of Srilanka and India on the west, Bangladesh coast to the north, western coast of Myanmar (formerly known as Burma) and north western part of Malay Peninsula to the east [1] . The mathematical model presented in this study includes the area of northern part of the Bay of Bengal from latitude 17.65° to the coast of Bangladesh and longitude 94.57° at Gwa beach to 83.28° at Vishakhapatnam.

The estuarine along the 710 km coastline of Bangladesh is a very dynamic coastal system [2] . Here, one of the world's greatest rivers the Lower Meghna River finds its way to the Bay of Bengal. The Lower Meghna conveys the combined flows of the Brahmaputra, the Ganges and the Upper Meghna. The sediment discharge from the Lower Meghna River is the highest [3] and the water discharges the third highest, of all river systems in the world [4] . Erosion and accretion rates are high and the area is periodically subjected to severe storms and cyclones.

The hydrodynamic factors that are playing dominant role in morphological development along coast line of Bangladesh are; enormous volume of river water flow, sediment transport, strong tidal and wind actions, wave, salinity and cyclonic storm surge. These hydrodynamic factors and their interactions shape the morphology of the estuary. A complicated interplay between the forces of the river, tide and the waves creates a complex pattern of sediment displacement in the estuary. Large quantities of sediment are transferred continuously towards the shallow coastal region of Bangladesh. The displacement of sediment is a part of continuous process of the estuarine-landscape striving to achieve dynamic equilibrium between morphology and the continuously changing river discharge conditions and tidal flows.

Keeping this background in mind, in order to have the essential comprehension of the flow pattern in the Bay of Bengal which is highly affected by above mentioned natural and many other man-made activities, the authors realized to have an accurate hydrodynamic model. Scientifically based mathematical modelling is an efficient tool for establishing hydraulic and morphological conditions, reliable evaluation of coastal development plan for maximizing the benefit integrating the coastal systems incorporating upstream and downstream hydraulic conditions.

LITERATURE REVIEW

The earliest 2D Model for Bay of Bengal was developed under the project Cyclone Shelter Preparatory Study, CSPS during 1996 to 1998 [5] . The Model was later updated by Surface Water Modelling Center, SWMC during the project Meghna Estuary Study, MES in 2001 [6] .

Jacobsen et al. (2002) [7] performed numerical simulations through the two-dimensional MIKE 21 model during the Meghna Estuary Study (MWR, 1997) [8] and obtained a counterclockwise circulation with a northward flow in the Sandwip Channel and a southward flow in the Tetulia River and in the area from Hatia to Sandwip. During the study it was also observed that the residual circulation, to some extent, traps the river water inside the Meghna Estuary and thus increases the residence time, which is one of the reasons for the relatively low salinity in the estuary even during the dry season. During the study wind stress was considered not to influence the residual currents of the Meghna Estuary significantly but earlier another numerical investigation in the Northern Bay of Bengal by Ali (1995) established that south-westerly monsoon wind may increase water level in the estuary and create back water effects in the rivers. Potemra (1991) also studied the seasonal circulation in the upper Bay of Bengal.

Under the Estuary Development Programme, EDP in 2010 [9] , Institute of Water Modelling, IWM has updated the BoBM with the recently surveyed hydro-morphological data and latest satellite imagery. During EDP, the model has been upgraded from rectangular mesh to flexible mesh using MIKE 21FM modelling system. Present study mainly focuses the development and improvement of model obtained during EDP study. This article is outlined as follows: an elaborated introduction is followed by heading methodology where scientific background and detail about the data and boundary used in the model are described. Then results and discussion are pointed out and finally, the conclusions are extracted.

METHODOLOGY

A mathematical model usually describes a system by a set of variables and a set of equations that establish relationships between the variables. The BoBM is a two dimensional one layer (depth integrated) hydrodynamic model based on MIKE21FM modeling software. It is the basic module of the entire MIKE 21 system. MIKE 21 HD simulates the water level variations and flows in response to a variety of forcing functions resolved on a rectangular or triangular grid covering the area of interest when provided with the bathymetry, bed resistance coefficients, wind field, hydrographic boundary conditions etc. The module solves the vertically integrated equations of continuity and conservation of momentum in two horizontal dimensions.

Main Equations

The hydrodynamic module in the MIKE 21 Flow Model FM (MIKE 21 HD, FM) is a general numerical modeling system for the simulation of water levels and flow in estuaries, bay and coastal water areas. It simulates unsteady two dimensional flows in one layer (vertically homogeneous) fluids and has been applied in a 2-D model development of Bay of Bengal.

The following equations, the conservation of mass and momentum integrated over the vertical, describe the flow and water level variations:

The continuity equation is:

$$\frac{\partial \zeta}{\partial t} + \frac{\partial p}{\partial x} + \frac{\partial q}{\partial y} = 0$$

The momentum equation in x-direction is:

$$\frac{\partial p}{\partial t} + \frac{\partial}{\partial x}\left(\frac{p^2}{h}\right) + \frac{\partial}{\partial y}\left(\frac{pq}{h}\right) + gh\frac{\partial \zeta}{\partial x} + \frac{gp\sqrt{p^2+q^2}}{C^2.h^2} - \frac{1}{\rho_w}\left[\frac{\partial}{\partial x}(h\tau_{xx}) + \frac{\partial}{\partial y}(h\tau_{xy})\right] - \Omega_q - fVV_x + \frac{h}{\rho_w}\frac{\partial}{\partial x}(p_a) = 0$$

The momentum equation in y-direction is:

$$\frac{\partial q}{\partial t} + \frac{\partial}{\partial y}\left(\frac{q^2}{h}\right) + \frac{\partial}{\partial x}\left(\frac{pq}{h}\right) + gh\frac{\partial \zeta}{\partial y} + \frac{gq\sqrt{p^2+q^2}}{C^2.h^2} - \frac{1}{\rho_w}\left[\frac{\partial}{\partial y}(h\tau_{yy}) + \frac{\partial}{\partial x}(h\tau_{xy})\right] + \Omega_p - fVV_y + \frac{h}{\rho_w}\frac{\partial}{\partial xy}(p_a) = 0$$

where p and q (m³/s/m) are fluxes in xand y-directions respectively, t (s) is time, x and y (m) are Cartesian coordinates, h (m) is water depth, d is time varying water depth (m), g (9.81 m/s^2) acceleration due to gravity, ζ (m) is the sea surface elevation, C is a Chezy resistance parameter (m$^{1/2}$/s), f(V) wind friction factor, V, V$_x$, V$_y$ wind speed and components in xand y-directions (m/s), Ω is Coriolis parameter which is latitude dependent (s^{-1}), P$_a$ is atmospheric pressure (kg/m/s^2), ρ_w density of water (kg/m^3), τ_{xx}, τ_{xy}, τ_{yy} components of effective shear stress.

Data Used for Model Development

The modelling study has been devised in combination with data analysis and numerical modelling. The two dimensional (2-D) flexible mesh model (MIKE21 FM) has been applied for hydro-morphological investigation around the study area. Data on recent bathymetry, sediment concentration and hydrometrics of the estuary have been utilized for updating and re-calibrating the existing Bay of Bengal model and establishing baseline conditions.

Data surveyed and collected by different agencies of Bangladesh Government like BWDB, CDSPIII for executing different projects in the coastal region have been compiled in Table1

Bathymetric Data

The bathymetry data over the entire Meghna Estuary was surveyed and the data has been analysed. BWDB has checked the consistency of Bench Mark values that were used during bathymetric survey and corrected the bathymetry accordingly. MIKE-C Map which is global data base for water depth or water-land boundaries was used to generate land level data in the deep Bay of Bengal.

Table 1: List of data collected from secondary sources

Data	Source	Name of Study	Period
Bathymetry	BWDB	Meghna Estuary Study II [6]	2000
	BWDB	Feasibility Study for Haitya-Nijhum Dwip Cross Dam Project [10]	2006
	CDSP III	Survey and Modelling Study of Sandwip-Urirchar-Noakhali Cross-Dam(s) [11]	2009-10
Water Level	CDSP III	Survey and Modelling Study of Sandwip-Urirchar-Noakhali Cross-Dam(s) [11]	2009-10
Discharge	CDSP III	Survey and Modelling Study of Sandwip-Urirchar-Noakhali Cross-Dam(s) [11]	2009-10

Water Level Data

Water level data are used to estimate the variation of water depth over the year, tidal characteristics and also to calibrate the hydrodynamic model. Water level observations have been carried out with pressure gauge over 24 hours at half an hour interval. All water level data are referred to Public Works Department (PWD) datum. The measured water level data at various locations have been plotted and considerable inconsistencies have been found. Bench mark error as well as processing error may have caused these inconsistencies in water level data. The consistent water level data have been used for calibration of the model.

Discharge Data

Discharge measurement was carried out for 13 hours with one hour interval in the Lower Meghna River (at Kaliganj), Hatiya Channel, East Shahbazpur Channel, West Shahbazpur Channel, the Tetulia River and Sandwip Channels. These data have been utilised for calibration.

Mesh Generation and Bathymetry Development

Objective of mesh generation is to divide the whole model area in to a number of individual triangular flexible cells to perform the computation. To generate mesh, shoreline is essential. Shoreline can be extracted from satellite images using ARC View/ARC GIS or from MIKE-C Map. In case of BoB model shoreline is extracted from satellite images. Area included inside the shoreline represents a polygon (Figure 1). In case of BoB model one polygon should not be used. If coarser resolution is used for whole area then the estuary cannot be represented properly. Again if finer resolution is used to represent the Meghna estuary properly then the total number of computational points will be increased and simulation time will be higher. To avoid this situation total area is divided into four polygons of different resolution as shown Figure 2. The resolution varies from 5400 m × 5400 m and 1800 m × 1800 m in the Bay of Bengal, through 600 m × 600 m in the Meghna Estuary and 200 m × 200 m in the areas of special interest.

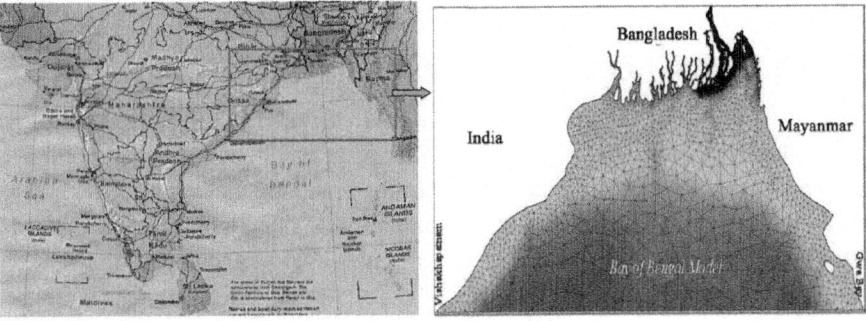

Figure 1: Model domain enclosed by shoreline of the Bay of Bengal.

For bathymetry generation the measured and generated land level data were incorporated with the mesh and by applying the interpolation method, elevation in each and every point of the modeled area were calculated. The generated bathymetry is shown in Figure 3.

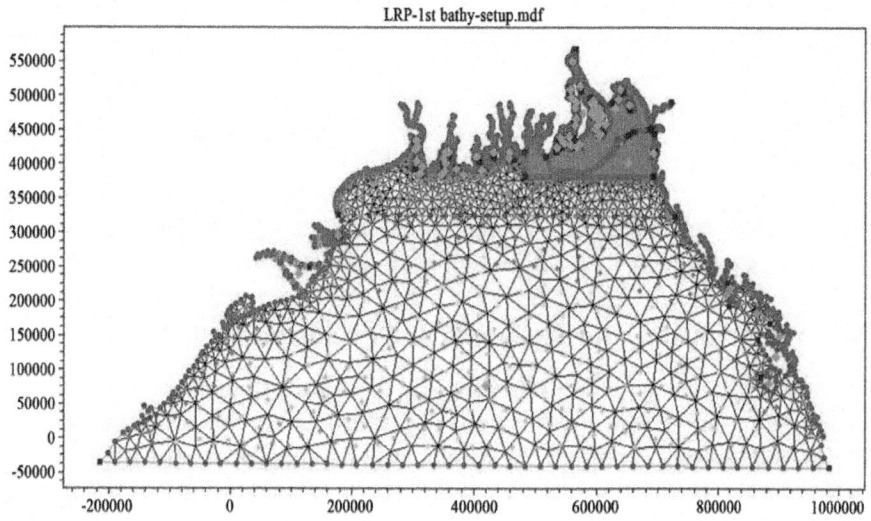

Figure 2: Different polygon and their mesh resolution in model domain.

Boundary Generation

There are two open boundaries in the model, one is in the Lower Meghna River at Chandpur, which is northern boundary and another one is in the Bay of Bengal that is southern boundary shown inFigure 3. Observed water level at Chandpur station has been used in the northern boundary. Predicted or generated water level using Global Tide Model of two stations namely Vishakhapatnam (India) and Gwa Bay (Myanmar) was used for south boundary.

RESULTS AND DISCUSSION

The two-dimensional hydrodynamic model of the Bay of Bengal have been calibrated against water level and discharge at different locations comparing the model results with field measurement to make the model performance to a satisfactory level. The validation has been made only with the available data for the year 2008. The model is based on the surveyed data of 2009-2010 and calibrated with water level and discharge measurements of 2009-2010 and found good agreement. Hydrodynamic simulation has been made for one month for both the dry and monsoon seasons in order to establish the hydraulic characteristics for low and high discharges. Locations of calibration areas have been presented in Figure 4.

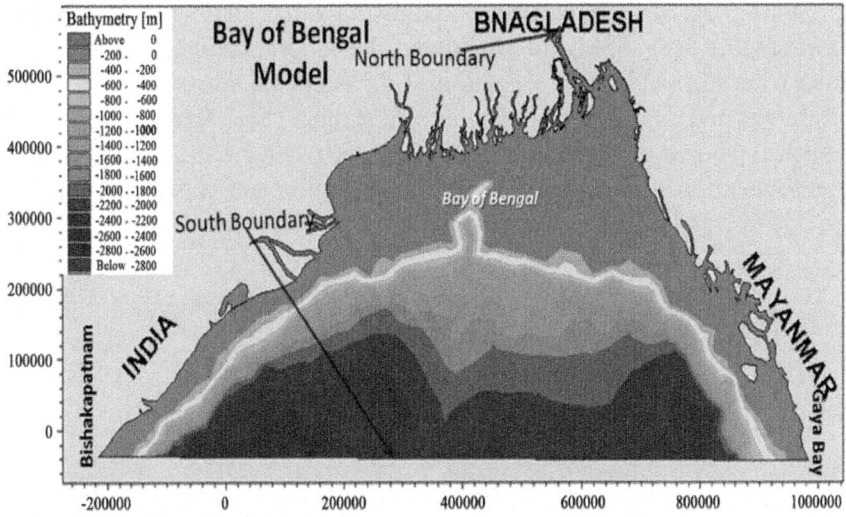

Figure 3: Bathymetry and open boundaries of BoB Model.

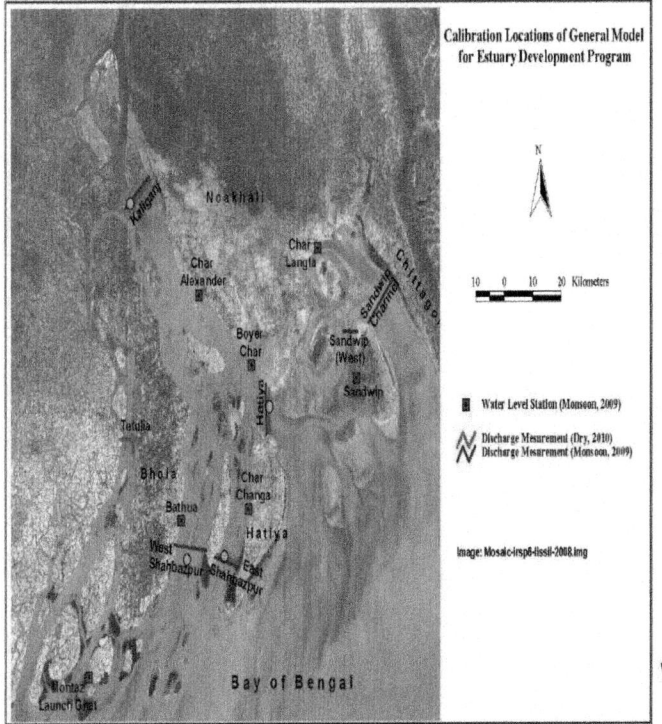

Figure 4: Calibration locations of Bay of Bengal Model.

Some plots of these calibration points have been shown in the following figure. Figure 5(a) andFigure 5(b) shows the calibration against water level at Boyer Char and Sandwip for wet season. Calibrations against observed discharge during wet and dry season at Monpura-Jahajmara (East Shabhazpur Channel) have been shown in Figure 6(a) and Figure 6(b). Figure 7(a) and Figure 7(b), represents discharge calibration at Bhola-Monpura and at Nairpur during wet season 2009 and dry season 2010 respectively.

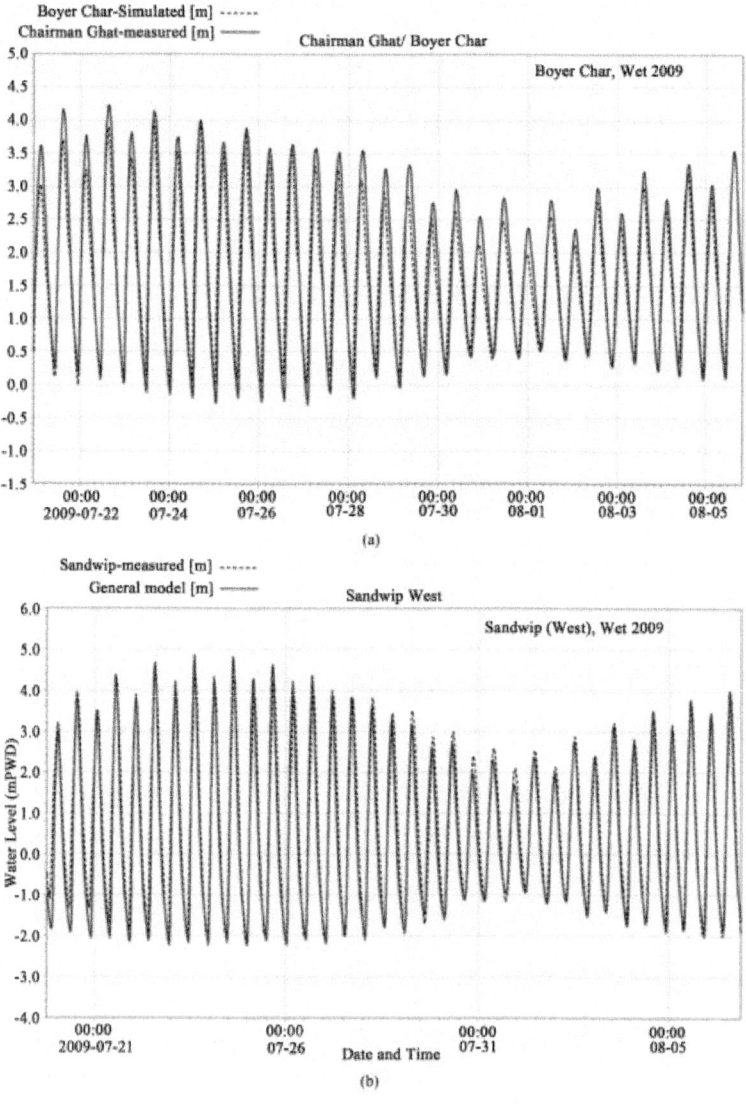

Figure 5: Calibration results of Bay of Bengal Model on water level.

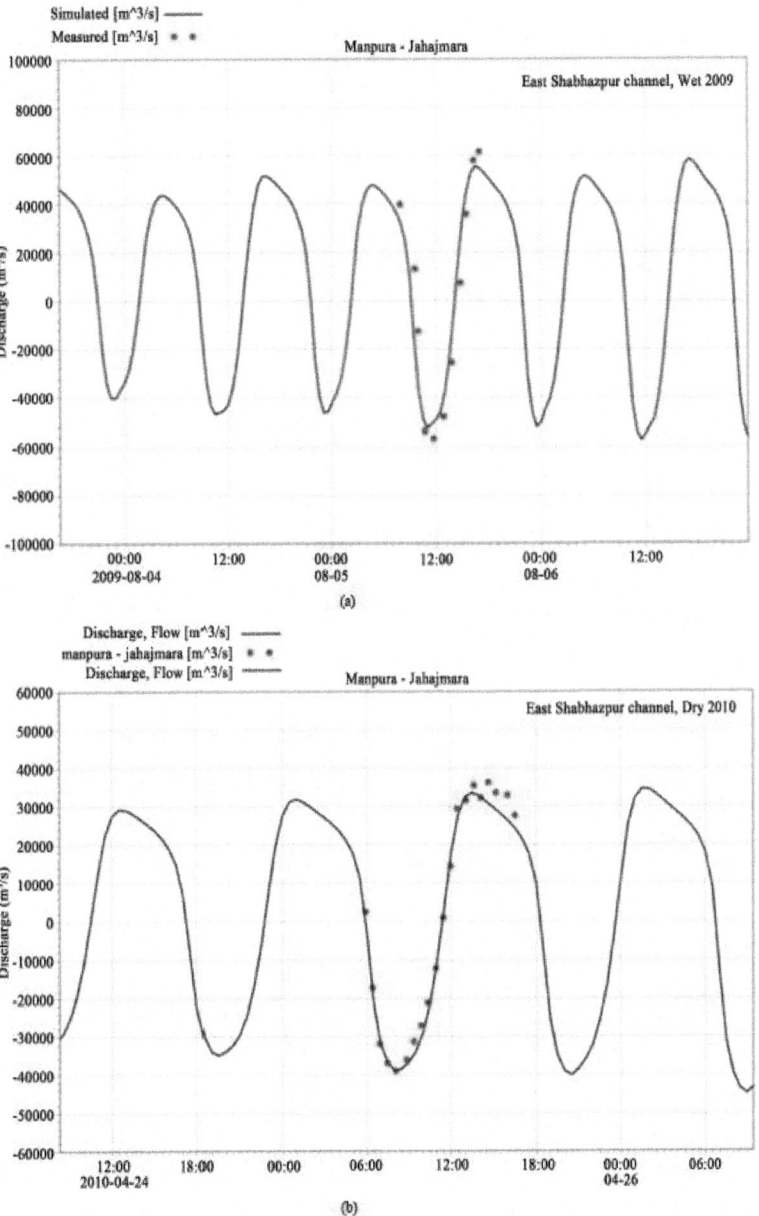

Figure 6: Calibration against discharge of Bay of Bengal Model.

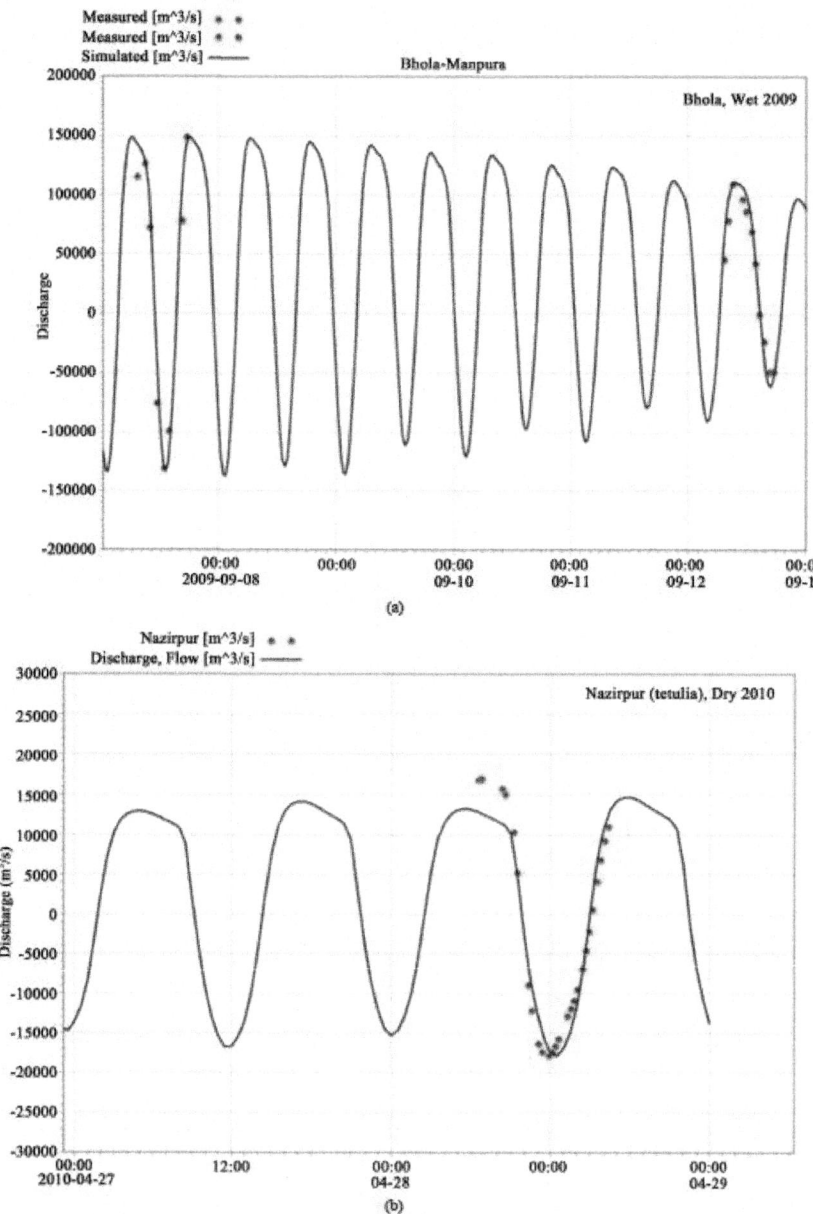

Figure 7: Calibration against discharge of Bay of Bengal Model.

All these calibration plots show good amplitude and phase agreement between the model results and observed data.

As a result of interactions of different hydrodynamic forces active in the estuary some changes like siltation in channel bed, shoreline erosion, shifting of thalweg and finally shifting of channels are occurred in the estuary. The hydraulic conditions of the Meghna estuary such as residual flow, mean current speed and tidal meeting point in the different channel of Meghna Estuary has been established and analyzed for dry and monsoon season under this study using the results of the Bay of Bengal Model.

During dry season upland fresh water flow into the Bay through the Lower Meghna River is very much lower than that of monsoon season. Tidal action becomes stronger and dominates water flow pattern in the estuary. Lowest velocities of about 0.25 - 1.0 m/s are found in the upper part of the lower Meghna River during dry period, where the tidal action is less dominant (Figure 8).

The estuary becomes morphologically very dynamic during monsoon. The upland flow is enormous during monsoon and the mean high water is considerably higher than during dry season. The distribution of flow and water level in the different channels of the estuary are governed by river discharge, the tide and the wind speed. The maximum depth average current speeds of about 1.75 m/s are found mainly in the West Shabhazpur Channel, north Hatiya channel and north of Urirchar, i.e. in the Bamni channel during monsoon. Most of the accretion and erosion occurs during monsoon. Current speed and direction during monsoon is shown in Figure 9.

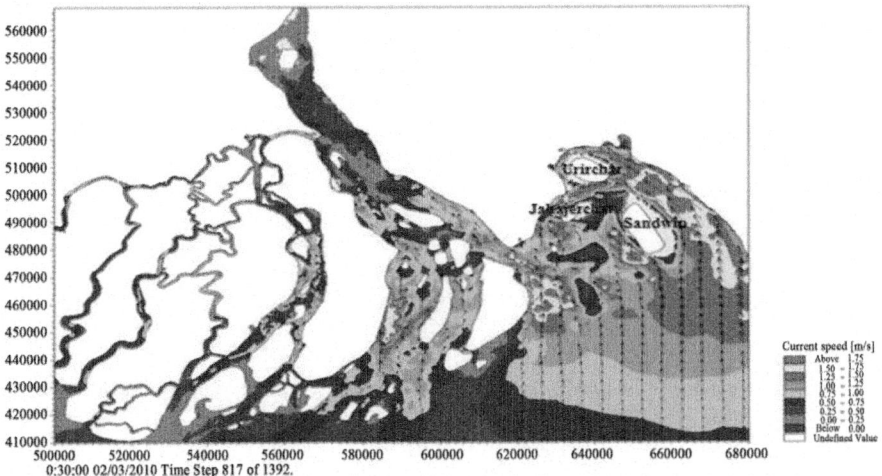

Figure 8: 2D map of flow speed/direction, covering one tidal cycle, every 2 hours (dry period).

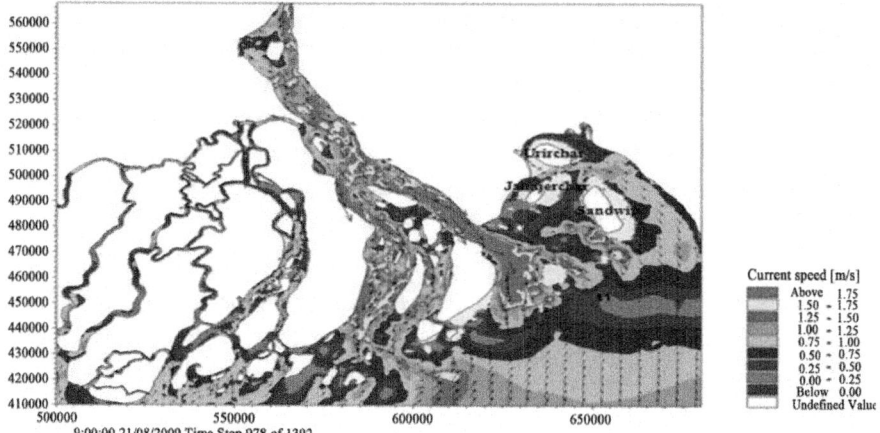

Figure 9: 2D map of flow speed/direction, covering one tidal cycle, every 2 hours (monsoon).

The residual flow has been established based on the simulation results of one month covering spring and neap tides both for dry and monsoon season. Simulation result shows a net water flow out of the Meghna estuary through West Shabhazpur channel and easterly flow outside the estuary during dry season. A prominent anticlockwise circulation is prevailing around the Sandwip Island, which is mainly forced by tide. In the Sandwip channel, the residual anti-clockwise circulation during monsoon is similar to the circulation during dry season, which implies the area is dominated by tide both in dry and monsoon seasons. The net anti-clockwise circulation traps the sediment in this area. The net flow in between Sandwip and Hatia Island is influenced by the river discharge. The residual flow and mean current speed is shown in Figure 10 and Figure 11 for dry and wet season respectively.

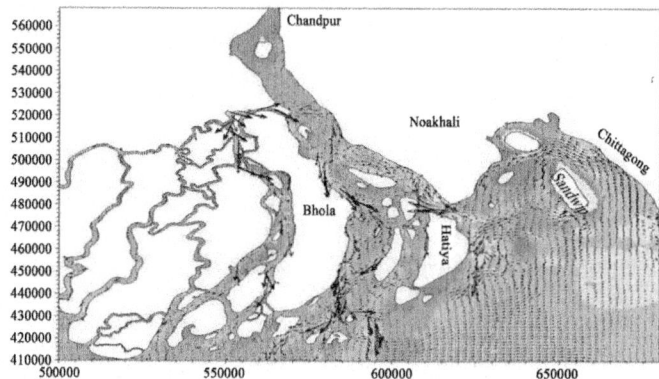

Figure 10: Residual flow pattern in the Meghna Estuary in Dry season (February, 2010).

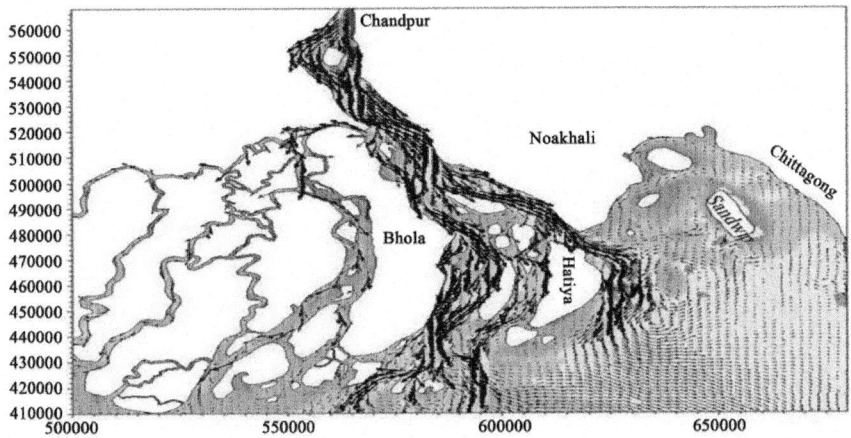

Figure 11. Residual flow pattern in the Meghna Estuary in Wet season (August, 2009).

Simulation results shows three tidal meeting points in the Sandwip, Jahazerchar and Urirchar area; one in the channel between Urirchar and Char Clark, other meeting point is in the channel between Jahazer Char and Noakhali coast near Char Bayejid and another meeting point in the channel between Jahazer Char and Sandwip, shown in Figure 12. Model results show that in this area the change of flow direction takes place over a very short distance and rapidly, which is in good agreement with the field observations.

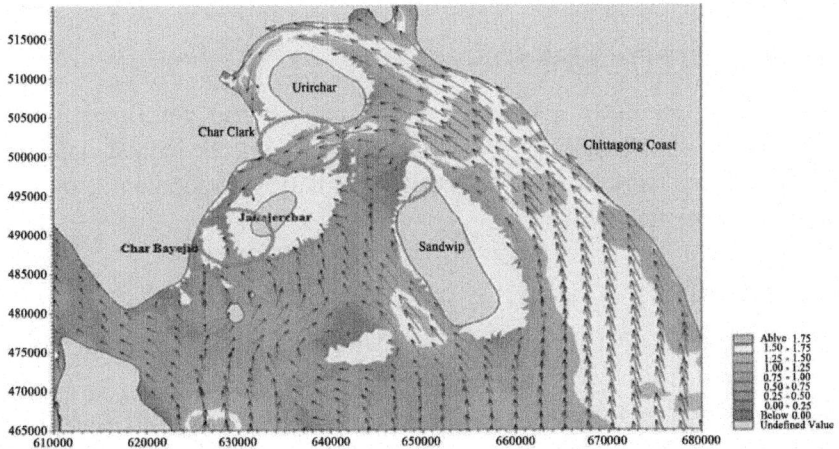

Figure 12: Tidal Meeting Points in Sandwip-Urirchar-Noakhali area (Tidal meeting points are marked with red circles).

CONCLUSIONS

During the initial development of BoB model, observed water level has been used in the upstream boundary and predicted tide has been used in the downstream boundary. To obtain optimum result, an attempt may be taken to extend the upstream boundary towards 20 km up where rating curve is available for generating discharge and hence the present boundary stations may be used as a calibration point. Use of discharge data at upstream boundary will reduce the boundary effect and improve the model performance.

Sediment movement in the Meghna estuary is very important since about 1.5 billion tons of sediment from Ganges-Brahmaputra-Meghna basin is discharged to the Bay of Bengal through this estuary. It is possible to reclaim land artificially or enhance the natural accretion process by proper management of this huge amount of sediment. Moreover for avoiding unwanted siltation efficient management of sediment is also necessary. To understand the sediment movement phenomena, to investigate the accretion and erosion process in the estuary and to calculate sediment budget, a morphology model or mud transport model along with this hydrodynamic model should be developed.

Furthermore, a sensitivity analysis can also be made to access the sensitiveness of different parameters such as roughness, critical shear stress for erosion and deposition, time step and viscosity. In addition, with more measurement stations data available, the BoBM can be calibrated and validated more effectively to increase its robustness and prediction accuracy.

ACKNOWLEDGEMENTS

The authors gratefully acknowledge the support and contributions of the Institute of Water Modelling (IWM), Dhaka. The works presented in this article is based on the work, carried out during "Estuary Development Programme" project execution in IWM.

REFERENCES

1.   Sindhu, M. (2012) Numerical Modelling of Tides and Storm Surges in the Bay of Bengal. Ph.D. Thesis, Goa University, Goa.

2.   Koen, W.D. (2011) Moving Coastlines, Emergence and Use of Land in the Ganges-Brahmaputra-Meghna Estuary. The University Press, Dhaka, 2 p.

3.   Coleman, J.M. (1969) Brahmaputra River: Channel Process and Sedimentation. Sedimentary Geology, 3, 129-239.http://dx.doi.org/10.1016/0037-0738(69)90010-4

4. Milliman, J.D. (1991) Flux and Fate of Fluvial Sediment and Water in Coastal Seas. In: Mantoura, R.F.C., Martin, J.-M. and Wollast, R., Eds., Ocean Margin Processes in Global Change, John Willey and Sons Ltd., Chichester, 69-89.

5. CSPS (1998) Cyclone Shelter Preparatory Study, Stage I: Feasibility Phase. Final Report. Mathematical Modelling of Cyclone-Surge and Related Flooding. European Commission, Directorate General External Economic Relations. Technical Unit for Asia Centre, Planning Commission, Dhaka.

6. S.W.M.C. (2001) Two-Dimensional Model of the Meghna Estuary. 2nd Update Report, Submitted to MES by Surface Water Modelling Center.

7. Jakobsen, F., Azam, M.H. and Kabir, M.M. (2002) Residual Flow in the Meghna Estuary on the Coastline of Bangladesh. Estuarine, Coastal and Shelf Science, 55, 587-597.http://dx.doi.org/10.1006/ecss.2001.0929

8. MWR (1997) Meghna Estuary Study, Salinity Distribution in the Estuary. Ministry of Water Resources, Bangladesh Water Development Board, Technical Note MES-004. Report by DHV Consultants BV for DGIS/ DANIDA and GOB.

9. EDP (2010) Estuary Development Programme, Final Report, Submitted to Ministry of Water Resources, Bangladesh Water Development Board by Institute of Water Modelling.

10. IWM (2006) Hydraulic Modelling Study in Connection with Feasibility Study for Hatiya-Nijhum Dwip Cross-Dam Project.

11. IWM (2009) Survey and Modelling Study of Sandwip-Urirchar-Noakhali Cross-Dam(s). Final Report, Submitted to Char Development and Settlement Project-III (CDSP-III) by Institute of Water Modelling.

Chapter 8

FLOOD RISK PATTERN RECOGNITION USING CHEMOMETRIC TECHNIQUE: A CASE STUDY IN MUDA RIVER BASIN

Ahmad Shakir Mohd Saudi[1,2], Hafizan Juahir[1], Azman Azid[1], Mohd Khairul Amri Kamarudin[1], Mohd Ekhwan Toriman[1], Nor Azlina Abdul Aziz[1]

[1]East Coast Environmental Research Institute, University Sultan Zainal Abidin, Kuala Terengganu, Malaysia

[2]Faculty Science and Technology, Open University Malaysia, Shah Alam, Malaysia

ABSTRACT

This study constructs downscaling statistical model in analyzing the hydrological modeling in the study area which faces the risk of flood occurrence as the impact of climate change. The combination of chemometric method and time series analysis in this study show that even during the monsoon season, rainfall and stream flow are not the major contribution towards the changing of water level in the study area. Based on Correlation Test, it shows that suspended solid and water level show high correlation with p-value < 0.05. Factor Analysis being carried out to determine the major contribution to the changes of water Level and the result show that Suspended Solid shows a strong factor pattern with value 0.829. Based on Control Chat Builder for time series analysis, the Upper Control Limit for water level and suspended solid are 7.529 m and 1947.049 tons/day and the Lower Control Limit are 6.678 m and 178.135 tons/day. This shows that human development in the area gives high impact towards climate change and risk of flood in the study area which commonly faces flood during monsoon season.

INTRODUCTION

Kedah is "the rice bowl of Malaysia" and water has always been essential to the state. The rivers have been used for irrigation, communication linkages between villages, and to transport rice grown in the vast area of paddy fields. The study area relatively is located in the district of Sik, and gentle flow from Kuala Kedah made navigation easy and Alor Setar was established as a port for the rice trade back in the 1800's.

Paddy uses a lot of water, and in spite of the reservoirs of Muda, Pedu and Ahning, water remains a scarce resource. MADA has therefore introduced water saving methods, including recycling of drainage water. Irrigation is still by far the largest consumer of water, but industrial demand is growing and the rising standard of living also means increasing domestic demand. Tourism also consumes a lot of water. Finally, Kedah supplies water to Perlis and shares Sungai Muda with Penang. The economic development thus increases the demand for a limited resource.

Water quality is the main concern. While the rivers are pristine in the upstream areas, they become increasingly polluted downstream due to discharge of urban wastewater and agricultural runoff. Solid waste often ends up in the river and this is especially a problem in urban areas. As a result, most storm water drains are very polluted and smelly, especially in the dry seasons.

Water quality status of the Sungai Muda is important to be known as to preventing potential health impacts on human beings in particular in a short period or even a long one, to protect the beneficial function of the river, and to provide relevant statistical analysis for water managers and decision makers to take effective control measures and have implementations of pollution preventing activities.

Urban wastewater and agricultural runoff are the main source of pollution in Muda River. The main functions of Sungai Muda are water supply for agriculture and also the water sources of people in Kedah and Penang. This pollution can be dangerous to human health and people rarely know about it.

Muda River basin which is located within the boundary of Kedah and Pulau Pinang with a catchment area of 4210 km^2 and 180 km length begins from Muda Dam and flows across the district of Baling, Sik and Kuala Muda. Water supply for agricultural, industrial and domestic sector for both Penang and Kedah is the key role of the river.

The catchment was often flooded during the rainy season from April to May and from September to November in every year. Many problems arise when flood keeps on worsening each year e.g., riverbank erosion, river pollution and reduction of water resources. The flood event which occurred on October 2003 was the worst compared to previous events in 1988, 1995 and 1998.

The main functions of Muda River basin are water supplies for agriculture and also the water sources of people in Kedah and Penang. It becomes sources of fresh water for Penang especially from Muda River where the total number of 17 such schemes is found in the 4 districts within the basin. These areas contain almost 3500 schemes that come from various sources, including Muda River tributaries and MADA canals.

During the rainy season, the catchment areas replenish the rivers and absorb large amount of rainwater, thereby minimizing risk of flooding. During the dry season, the catchment areas replenish the rivers and provide a continuous supply of water.

MATERIALS AND METHODS

Topograpgy of Study Area

In a two-component gel, it is easy to modify the molecular structure of either of the two components. Situated at the coordinates of 5°06'N and 100°17'E, the natural basin covers an area of 2920 km², with approximately 60 km wide and 80 km long, as well as ranges from 400 m high to the coastal plains, where they are the heirs of rice cultivation. Discovered over 250 years ago as the meeting point of Sungai Anak Bukit and Sungai Kedah, the main activity in the state capital was rice trade with the main area of the Muda Irrigation Scheme which consists of 966 km² coastal plains.

The topography of the Muda River Basin and the monitoring station by the Department of Drainage and Irrigation (DID) along the river basin is illustrated in Figure 1 and Table 1 tabulated the specific location of coordinates for monitoring stations. The secondary hydrological data were provided by the Department of Drainage and Irrigation (DID) for the year 1982-2012, which include rainfall, water level, stream flow and suspended solid.

These forests are the habitat of a diversified collection of plant and animal species. This includes the river terrapins which are threatened due to habitat destruction and excessive egg poaching. The river terrapins are protected under the Kedah Terrapin Enactment 1972. The river is also the habitat for fish species such as Channa, micropeltes (Toman), Labeo rohita (Rohu), Chitala chitala (Belida), Leptobarbus hoevenii (Jelawat) Pangasius nasutus (Patin), and Puntius gonionotus (Lampam).

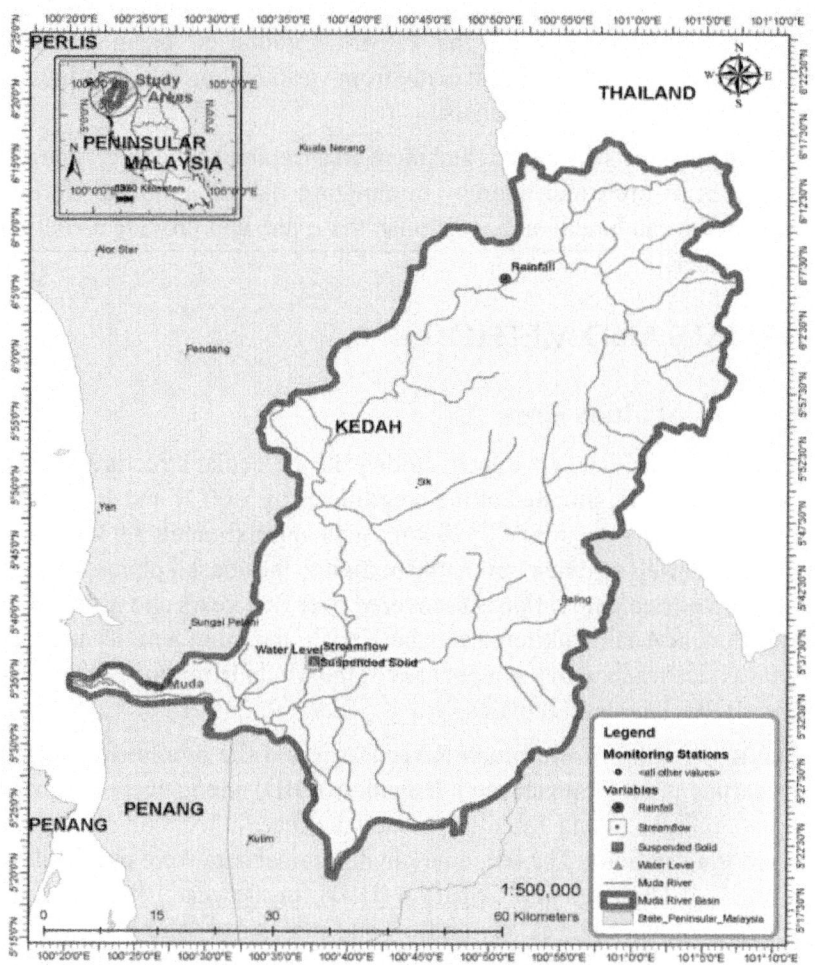

Figure 1. Location of monitoring stations in Muda River Basin.

Table 1: Location of monitoring stations in Muda River Basin

Station No.	Latitude	Longitude	Name of Station	Variables
Site 6108001	5°06'48N	100°01'55E	Kompleks Rumah Muda	Rainfall
Site 5606410	5°12'57N	100°27'16E	Sungai Muda, Jambatan Syed Omar	Suspended Solid
Site 5606510	5°17'25N	100°39'29E	Sungai Muda, Jambatan Syed Omar	Stream Flow
Site 5606410	5°18'32N	100°40'23E	Sungai Muda, Jambatan Syed omar	Water Level

However, rapid growth in certain area gives a negative impact towards rate of the surface runoff into the water body system and affecting water level at certain location in the river basin and leads to flooding. The study was being conducted to see the relationship between exceeded surface runoff to the water level of the river which leads to flood at high impact area, especially during monsoon season, the determine limitation of flood risk based on hydrological data from the year 1982-2012, and to identify suitable mitigation measure for flood prevention at the high impact area. Based on the Figure 1 and Table 1, it explains the location of monitoring stations in the study area.

METHODOLOGY IN RESEARCH

Correlation Test

For this study, the Correlation test was adapted to determine the variables with strong relationship for further analysis as the test is suitable to measure two variables that have the relationship between −1 to 1. Pearson Coefficient and Spearman Coefficient are the types of products that can be implemented in this study, but the former was widely utilized should there be an association of two variables (Moore D.S. and McCabe G.P., 1989) [1] . It was used in this study to determine the relationship between important parameters in hydrological data, as well as to determine the parameter with the strongest relationship. Upon that, the development with the biggest influence on the hydrological modelling in Muda River Basin can be determined.

Spearman's and Pearson's rank coefficients are two of the most common types of correlation types, where the former that requires ordinal data as the calculation will be based on data ranking. In addition, it also measures the degree of strength for the coefficient between variables considered in the research (Altman, 1991) [2] . There can be either positive or negative correlations for this method, where the positive correlation indicates two variables increasing together in a linear condition. On the other hand, the negative correlation indicates one variable increasing while the other decreasing in a linear condition. Meanwhile, the Pearson rank coefficient requires actual data for the calculation and all variables considered must be in the form of ratio scale. Both tests were implemented in this study, and only the best result was used for the discussion in this study.

$$r_p = \frac{\sum_{i=1}^{n}(X_i - \bar{X})(Y_i - \bar{Y})}{\sqrt{\sum_{i=1}^{n}(X_i - \bar{X})^2 \sum_{i=1}^{n}(Y_i - \bar{Y})^2}}$$

$$(1)$$

Chemometric Techniques

Chemometric technique such as application of Factor Analysis is able to see the reduction of variables into a set of factors for further analysis. Based on Floyd and Widamann (1995) [3] . Based on Floyd and Widamann (1995) [3] , the application of this method will make the researcher able to make a comparison of variables which give the highest impact towards the changes of water level with the lower cost compare to the other method.

The reduction of variables into a set of factors for further analysis can be observed using chemometric technique, such as the utilization of Factor Analysis. It is seldom that the researcher collects and analyzes data with prior knowledge regarding the relationship of the variables, but through this technique, variables with the biggest influence in the change of the hydrological modelling in the study area can be compared on a cost effective and quicker manner compared to other techniques (Gorsuch, 1990) [4] .

The utilization of this method in the study allowed the inclusion of a large number of variables into smaller set of variables, otherwise known as factors. The dimension between factor analysis variables and the measured latent construct established the dimension between these two elements and construct validity evidence of self reporting scales (Thompson, 1996) [5] . Other than that, factor analysis also examines the structure or relationship between variables, reduces the number of variables, and can be used for the detection and assessment of unidimensionality of theoretical construct (Brett W. et al., 2012) [6] . The method also considers the existence of two or more variables that are correlated (e.g., multicollinearity), which is suitable for this study. The equation implemented in this method was:

$$z_{ji} = b_j F_{1i} + b_{j2} F_{2i} \cdots + b_{jn} F_{ni} \qquad (2)$$

The common-factor approach only considers the covariation between observed variables, whereas the principal-component approach considers all variations in the observed variables.

Factor loadings represent the correlation coefficient between each factor and the observed variables.

Factor scores are the values of each observation on the factor Fk.

Time Series Analysis

Time Series Analysis is essential for the prediction of water level in the study area, where this method enables an efficient evaluation of the process from the performance by analyzing data. The method produces three important data (e.g., Upper Control Limit (UCL), Average Value (AVG) and Lower Control Limit (LCL)) for the trend and prediction of future hydrological modelling, where the Sigma is within a range value of a set of data. Control Chart can detect some trends and patterns with actual data deviations from historical baseline, be able to capture unusual resource usage, can determine the dynamic threshold, and also can become the best base lining to examine the actual data deviation from the historical baseline (Igor Trubin, 2008) [7] . The equation implemented in this analysis was:

$$\text{Moving Range} = \text{Plot}: \text{MR}_t \text{ for } t = 2, 3, \cdots, m. \tag{3}$$

Artificial Neural Network

Artificial Intelligent mimics the concept of the human brain and it has been utilized in the method for data analysis known as an Artificial Neural Network. This concept was introduced by McCulloch and Pitts in 1943, where the stimulation of structure and the performance of biological neural network in the computing system have been investigated.

An activation function is utilized to transform the weighted sum of the inputs transferred to the hidden neurons. The back propagation method is also implemented in the learning process for the purpose of error distribution, where the process can reduce the errors to the minimum level. After the error function has been minimized, the iteration is terminated when the value of the error function reached the predefined goal, thus completing the process (Juahir et al. 2009) [8] .

The function used was given by:

The process of cross validating the testing data set can be used to indicate the performance of the data, where the algorithm needs to be terminated during the process using back propagation. The architecture of the network and number of hidden units affects the learning ability of ANN. The size of the network is also important in capturing the connectivity of the data, as the degree of freedom works to capture the connection, and the size of the network must be compatible with the degree of freedom or the process will fail.

Imrie et al. (2000) [9] determined the effectiveness of ANN for rainfall—runoff modelling and flood forecasting, where the ability of ANN in predicting river flow and quality of water downstream has been highlighted. As a matter of fact, the aforementioned issues were also considered in this study.

RESULT AND DISCUSSION

Variables Which Contribute to Flood Occurrence

The correlation test being carried out for this study in order to see the relationship between variables in this study. All variables being analyzed by using Correlation Test to see whether all variables have strong correlation and based on result from Table 2 and figure 2, it shows that only water level and Suspended Solid have high correlation when the p-value for both variables is less than 0.001 (Saudi, 2014) [10] .

The result for Rainfall and Stream flow shows a weak correlation with other variables when the result of the test shows that both variables show p-value close to 1. This explains that both variables have a very weak result to show no correlation with water level and Suspended Solid.

Factors Which Contribute to Flood Occurrence

Results in table 3 and Figure 3 show that there are 2 major components which affect the most of the hydrological modelling at Sungai Muda and those components are Suspended Solid and Water Level. Both variables show the strong coefficient with value more than 0.7 in Factor 1 and the result is 0.829 for Suspended

Table 2: Correlation test

Variables	WL	Rainfall	SS	Stream Flow
WL	0	0.072	<0.001	0.061
Rainfall	0.072	0	0.230	0.026
SS	<0.0001	0.230	0	<0.001
Stream Flow	0.061	0.026	<0.001	0

Table 3: Factor analysis

	F1	F2	Initial Communality	Final Communality	Specific Variance
Stream Flow	−0.018	0.136	0.005	0.019	0.981
Rainfall	0.008	−0.471	0.014	0.222	0.778
Suspended Solid	0.829	0.085	0.460	0.695	0.305
Water Level	0.822	−0.079	0.461	0.682	0.318

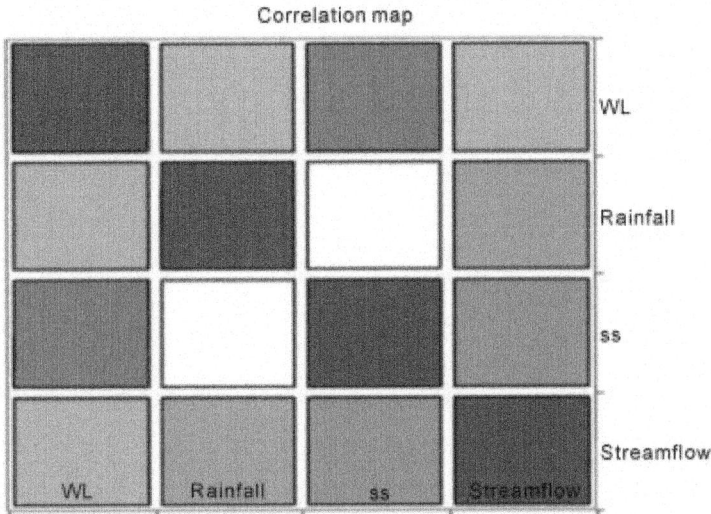

Figure 2: Correlation map.

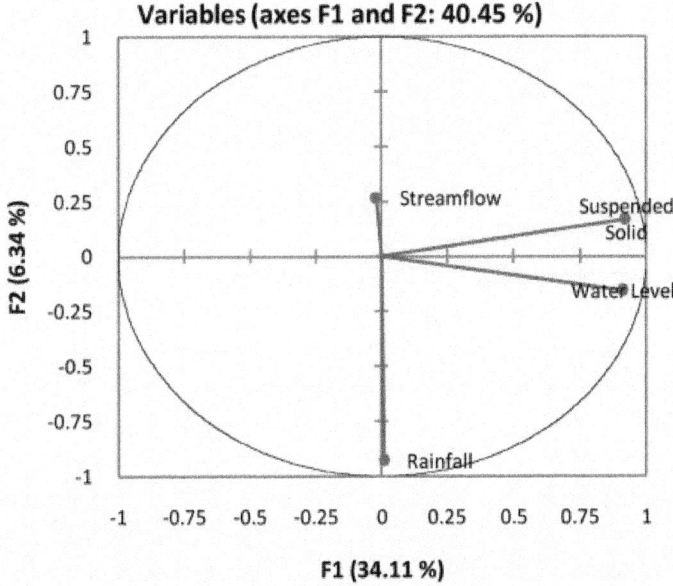

Figure 3: Correlation between variables and factors.

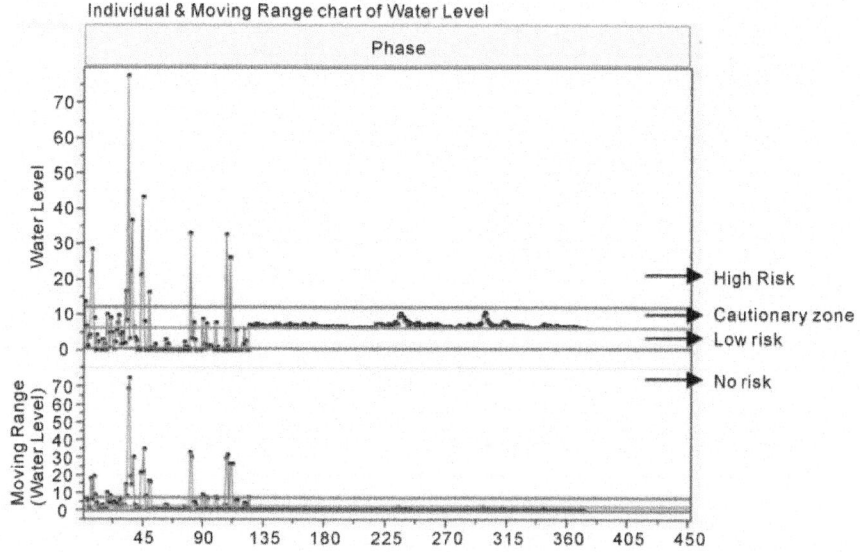

Figure 4: Result for time series based on Control Chat Builder.

Solid and 0.822 for Water Level. This concrete result shows the Rainfall is not the main factor in the changing of water level when it shows the weak coefficient with the result 0.085.

Flood Control Warning System

Based on figure 4 and table 4, the average water level for Sungai Muda from year 1982-2012 is about 6.349 m and the Lower Control Limit (LCL) is 0.417 and for the Upper Control Limit is 12.282 m. This result shows that the water level above Upper Control Limit will face the risk of flood while the value of water level, which is below Lower Control Limit considered decrement on the water level at Sungai Muda where this condition will affect the role of the river as a source of water for agriculture and the source of water for citizen of Kedah and Penang.

Based on the figure 5 and table 5, the result in Lower Control Limit for Suspended Solid from 1982 until 2012 is 178 tons ton/day, 1062 ton ton/day for average value and 1947 tons ton/day for Upper Control Limit. Result from Correlation test explains that water level and Suspended Solid show high correlation compared to Rainfall and Stream flow. This shows that when the range of Suspended Solid and water level within Upper Control Limit, the mitigating measure should be implemented in preventing flood from destroying the area even though the rate of rainfall is low within the same period. This situation can happen when the high rate of surface runoff from the water body precipitated and become the composition of the surface area of the river which cause the river turn into shallow.

The development around the study area affecting the climate in the study are not just based on the rate of rainfall anymore, but in this study area it refers to the high surface runoff will cause the high sedimentation into the river which will cause the changes of the depth of the river. This condition will cause the river easily to face flood if the heaviest rainfall occurred in a few days when the condition will cause the river become overflow and flooding.

PREDICTION OF FLOOD RISK CLASSIFICATION

Based on the result of time series analysis in figure 4 and figure 5, the risk of flood being classified into its own class based on hierarchy of risk is High Risk, Cautionary zone, Low Risk and No Risk.

The level of High Risk are classified for all data which are pointed at the above Upper Control Limit line in Control Chart graph, followed by Cautionary Zone for data which are plotted between Average line and Upper

Table 4: Result for time series based on Control Chat Builder

Points Plotted	LCL	AVG	UCL	Limit	Sigma	Sample Size
Individual	0.417	6.349	12.282	Moving	Range	1

Table 5: Result of suspended solid based on Control Chat Builder

Points Plotted	LCL	AVG	UCL	Limit	Sigma	Sample Size
Individual	288.535	1563.631	2838.726	Moving	Range	1

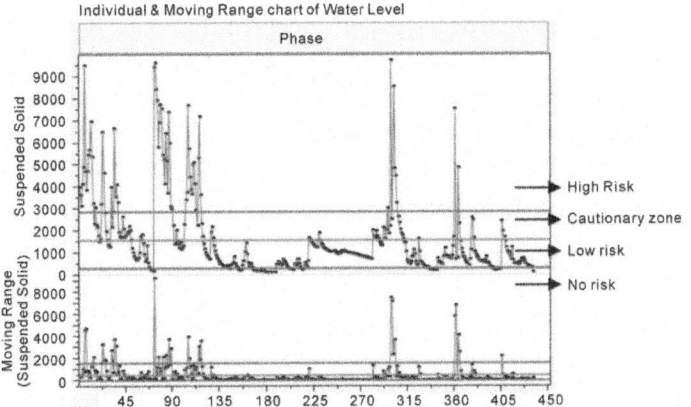

Figure 5: Suspended Solid based on Control Chart Builder (time series analysis).

Control Limit, Low Risk for data which are plotted between Lower Control Limit line and Average line and No Risk for data which are plotted below the Lower Control Limit line.

Prediction of risk hierarchy being carried out by using Artificial Neural Network and the result fromtable 6 shows that the accuracy of prediction is 0.96 which is also being considered as 96% and this explains that the prediction is accurate and also can be used for future prediction in risk assessment for flood occurrence.

CONCLUSIONS

Local Authority should give a strong commitment in controlling excessive amount of surface runoff into the river. They must fit with a few conditions which are information management and performance monitoring, integrated policy and strategies, constitution legislation and standard, Erosion and Sediment Control Plan (ESCP) to control erosion and sediment, an effective enforcement by the Department of Drainage and Irrigation (DID) referring to the regulation of Environmental Quality Act 1974 (Act 127) & Subsidiary Legislation, Waters Act 1920 (Act 418) & Water Supply (Federal Territory of Kuala Lumpur) Act 1998 (Act 581), and Water Act 1989—Chapter 15. This action and legislation will be able to control the uncontrolled development along the river bank, being carried out by an irresponsible developer not following the guideline which has been set up by the government. This condition will be able to reduce the risk of flooding in the study area.

Other mitigating measure that has been implemented such as construction of the Barrage, River Bund, Pump House, Diversion, Pond, Dam and River Improvement work at study area should be well maintained and improvised from time to time. The effectiveness of these mitigating measures also depends on the awareness and strong legal enforcement in controlling rate of surface runoff, which comes from uncontrolled human development in the study area and if it is not being configured well, all the structure mitigating measure means nothing in preventing of flood occurrence. Time Series Analysis is able to identify the limitation for all factors which affect the most of the changing of water level based on the results from Correlation.

Table 6: Prediction for hierarchy of flood risk

OUT_1	Accuracy	Total
Train	1	173
Test	0.96	75

Test and Factor Analysis, and this will not only reduce the cost of operation but also reduce the total lost from flood destruction and save lives. The application of Artificial Neural Network (ANN) is able to trigger earlier warning for citizens to take precaution for flood prevention based on level of risk from the prediction.

ACKNOWLEDGEMENTS

I am grateful to the Ministry of Higher education for scholarship through my Ph.D. Scholarship for this research where I completely identified source and formulation in preventing flood occurrence in the study area. I would like also to thank my supervisor, Hafizan Juahir for advising me until this research completely done.

REFEENCES

1. Moore, D.S. and McCabe, G.P. (1989) Introduction to the Practice of Statistics. W. H. Freeman, New York.

2. Altman, D.G. (1991) Practical Statistics for Medical Research. Chapman & Hall, London, 285-288.

3. Floyd, F.J. and Widaman, K.F. (1995) Factor Analysis in the Development and Refinement of Clinical Assessment Instruments. Psychological Assessment, 7, 286-299.http://dx.doi.org/10.1037/1040-3590.7.3.286

4. Gorsuch, R.L. (1990) Common Factor-Analysis versus Component Analysis: Some Well and Little Known Facts. Multivariate Behavioral Research, 25, 33-39.http://dx.doi.org/10.1207/s15327906mbr2501_3

5. Thompson, B. and Daniel, L.G. (1996) Factor Analytic Evidence for the Construct Validity of Scores: A Historical Overview and Some Guidelines. Educational and Psychological Measurement, 56, 197-208. http://dx.doi.org/10.1177/0013164496056002001

6. William, B., Brown, T. and Onsman, A. (2012) Exploratory Factor Analysis: A Five-Step Guide for Novices. Australasian Journal of Paramedicine, 8.

7. Trubin, I.A. (2008) Exception Based Modelling and Forecasting. Proceedings of the Computer Measurement Group, Nevada, 7-12 December 2008, 353-364.

8. Juahir, H., Sharifuddin, M.Z., Ahmad, Z.A, Mohd, K.Y. and Mazlin, M. (2009) Spatial Assessment of Langat RIVER Water Quality Using Chemometrics. Journal of Environmental Monitoring, 12, 287-295. http://

dx.doi.org/10.1039/b907306j

9. Imrie, C.E., Durucan, S. and Korea A. (2000) River Flow Prediction by Using Artificial Neural Networks: Generalisation beyond Calibration Range. Journal of Hydrology, 233,138-153. http://dx.doi.org/10.1016/S0022-1694(00)00228-6

10. Saudi, A.S.M., Juahir, H., Azid, A., Yusof, K.M.K.K., Zainuddinc, S.F.M. and Osman, M.R. (2014) Spatial Assessment of Water Quality Due to Land-Use Changes along Kuantan River Basin. From Sources to Solution 2014, 297-300.

Chapter 9

SIMULATION OF LONG TERM CHARACTERISTICS OF ANNUAL RAINFALL IN SELECTED AREAS IN SAUDI ARABIA

Nidhal Saada

Civil Engineering Department, AL Ahliyya Amman University, Amman, Jordan

ABSTRACT

Simulation experiments with different stochastic models were conducted to investigate the long term characteristics of rainfall in Saudi Arabia using selected Autoregressive Moving Average (ARMA) models. The results of the study indicated that the ARMA models were able to capture the long term statistics for one of the rainfall records investigated (Surat Obeida). However, the other rainfall record investigated in this study (Malaki) was characterized with a slow and long decaying structure and a high Hurst coefficient indicating the possibility of non-stationarity of the data. Trend analysis (Pettitt test) of the data revealed that a break point or a shift in the record happened around 1983 at Malaki. As a result, ARMA models should not be used in modeling the rainfall data at that station.

INTRODUCTION

Stochastic modeling of hydrologic time series has been widely used for planning and management of water resources systems. Stochastic models are used in operational hydrology to generate synthetic time series which exhibit similar statistical characteristics as the observed data. One of the crucial problems in stochastic modeling of hydrologic time series is to find a model which is capable of preserving the historical statistical characteristics that affect the variability of the data. Furthermore, the model should be capable of reproducing certain statistics that are related to the intended use of the model [1] . Generally, the properties of a process include the mean, variance, skewness, and the correlation structure of the data.

Additional properties related to the long term characteristics such as storage and drought related statistics may also be included, depending on the particular problem at hand [1] . In this era of possible adverse effects of climate change, preservation of such long term characteristics is important. The "Hurst" behavior of a time series is one of these characteristics, which is related to the long term persistence of that series [2] . Besides persistence, other reasons could explain the Hurst behavior such as the non-stationarity in the mean, which could be one of manifestation of the effects of climate change [2] [3]

Rehman [4] analyzed rainfall data at 10 locations in Saudi Arabia and showed that the Hurst exponent value for all stations was >0.5, indicating the existence of a persistence behavior of the rainfall data in Saudi Arabia.

Elfeki, Al-Amri, and Bahrawi [5] used a spectral density function (SDF) approach to analyze annual rainfall signals in the southwestern part of the Kingdom of Saudi Arabia. Results showed that multiple cyclic components with significant variances existed and that a common cycle of 26 years existed in all annual rainfall data studied [5] .

Almazroui, Nazrul Islam, Athar, Jones and Ashfaqur Rahman [6] compared temperature and rainfall data from 3 gridded datasets (CRU, CMAP and TRMM) with the observed temperature and rainfall data for Saudi Arabia. Results showed that the observed annual rainfall showed a significant decreasing trend (47.8 mm per decade) in the last 15 years with a relatively large inter-annual variability, while the maximum, mean and minimum temperatures had increased significantly at a rate of 0.71°C, 0.60°C, and 0.48°C per decade, respectively [6] .

The objective of this study is to investigate the use of ARMA models in modeling and simulation of annual rainfall data in Saudi Arabia and their ability to capture the long term statistics observed in the historical records. In this paper, three univariate (single site) models will be used in this study, namely, AR (1), ARMA (1,1), and ARMA (2,1) models.

MATERIALS AND METHODS

Data Used

The historical annual rainfall amounts in, two stations (Surat Obeida and Malaki) in Saudi Arabia were used in this study. The data used for Surat Obeida was for 30 years for the period of 1981 through 2010 while at Malaki, 27 years of data (1967-1993) was used. It is noted here that the record at Malaki has data between 2001 and 2010 but missing records for the period of 1994-2000.

As a result, it was decided to use only the continuous record (1967-1993) in this study. Figure 1shows a time series plot of the annual rainfalls at the two stations

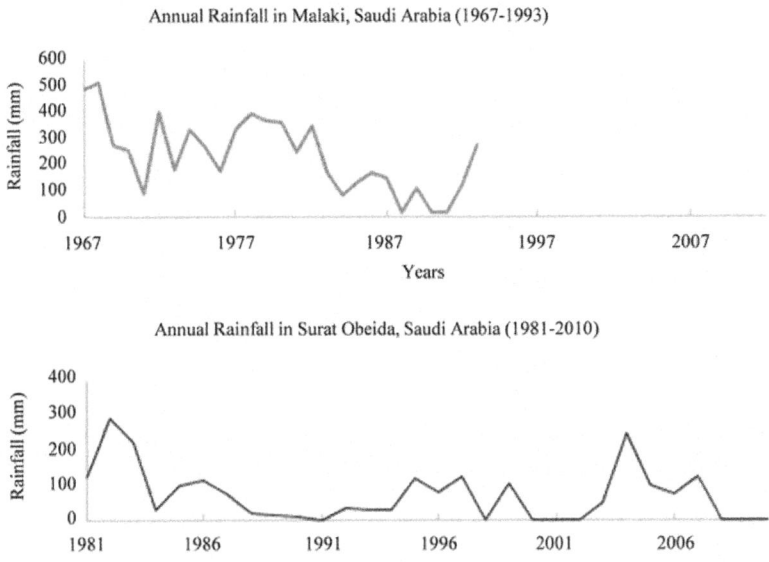

Figure 1: Annual rainfall in Malaki and Surat Obeida, Saudi Arabia.

Models Used

The univariate Autoregressive Moving average (ARMA) model may be written as [1] :

$$y_t = \sum_{i=1}^{p}\phi_i y_{t-i} + e_t - \sum_{j=1}^{q}\theta_j e_{t-j} \qquad (1)$$

where y_t represents the standardized process for year t, it has a mean = 0 and variance σ_y^2 and is normally distributed; e_t is the uncorrelated noise term with mean = 0 and variance σ_e^2 and is also normally distributed. $\phi_1, \cdots, \phi_p$ are the autoregressive parameters; $\theta_1, \cdots, \theta_q$ are the moving average parameters. For example, for p = q = 1, the ARMA (1,1) model becomes:

$$y_t = \phi_1 y_{t-1} + e_t - \theta_1 e_{t-1} \qquad (2)$$

Long Term Related Statistics

Consider the above time series (with length N) $y_i, i = 1, \cdots, N$ and a subsample $y_1, \cdots, y_n$ with n < N. If one forms the sequence of partial sum S_i as:

$$S_i = S_{i-1} + (y_i - \bar{y}_n) \quad i = 1, \cdots, n \tag{3}$$

where $S_0 = 0$ and $\bar{y}_n$ is the sample mean of $y_1, \cdots, y_n$, then, the adjusted range R_n^* and the rescaled adjusted range R_n^{**} can be calculated as:

$$R_n^* = \max(S_0, S_1, \cdots, S_n) - \min(S_0, S_1, \cdots, S_n) \tag{4}$$

And

$$R_n^{**} = \frac{R_n^*}{V_n} \tag{5}$$

In which V_n is the standard deviation of $y_1, \cdots, y_n$. Likewise, the Hurst coefficient (K) is then estimated by [7] :

$$K = \frac{\ln(R_n^{**})}{\ln(n/2)} \quad \text{for } n > 2 \tag{6}$$

For the series $y_i, i = 1, \cdots, N$ and for a demand level = $\bar{y}$ (the mean), a deficit with duration (L) occurs when $y_i < \bar{y}$ consecutively during one or more years until $y_i > \bar{y}$ again. Assume that m deficits occur in a given sample, then the maximum deficit duration, or longest drought (D), is given by [7] :

$$D = \max(L_1, \cdots, L_m) - \min(L_1, \cdots, L_m) \tag{7}$$

Similarly, a surplus with duration (P) occurs when $y_i > \bar{y}$ consecutively during one or more years until $y_i < \bar{y}$ again. Assume that j surpluses occur in a given sample, then the longest surplus duration (U) is given by [7] :

$$U = \max(P_1, \cdots, P_j) - \min(P_1, \cdots, P_j) \tag{8}$$

Simulation Experiments

Monte Carlo simulation experiments were conducted with AR (1), ARMA (1,1), and ARMA (2,1). The purpose of such experiments were to test the capability of such models to preserve the long term statistics of the historical rainfall data used in this study. A software package, Stochastic Analysis Modeling and Simulation (SAMS), was used in this study to conduct the simulation experiments [8] .

To apply and use the models mentioned above, the observed data must be normally distributed. The analysis of the skewness coefficient revealed that the observed data at Surat Obeida is not normal. SAMS provide option to transform the data in order to normalize it. Logaritmic, Box-Cox, and power transformations options are available in SAMS. The data at Surat obeida was normalized by using a logarithmic transformation. The skewness coefficient of the transformed data revealed that the transformed data was Norma. The data at Malaki was shown to be Normal and no transformation was done for Malaki.

Each ARMA model was then fitted to the normalized data. Simulation experiments were then conducted by generating synthetic time series data using the fitted models. In each experiment 100 samples, each with length equal to the historical length of the series (i.e. 30 years for Surat Obeida and 27 for Malaki) were generated from the fitted models. The average statistics calculated from these generated series were then compared with the historical statistics. These include the annual Correlogram, longest drought, longest surplus, and Hurst coefficient.

RESULTS

The autocorrelogram of the historical annual data at Surat Obeida as well as the genertaed autocorrelogram of AR (1), ARMA (1,1), and ARMA (2,1) models are shown in Figures 2-4 respectively. Results indicate that the autocorrelogram was well preserved by all three models.Table 1 shows the results of the simulation experiments in preserving the long term statistics of the historical data. The three models were capable of reproducing the historical Hurst coefficient. The models, in general, also performed well in preserving the longest drought and the longest surplus.

Results at Malaki were not as good. All three models underestimated the historical autocorrelogram as shown in Figures 5-7. Table 1 shows the results of the simulation experiments in preserving the long term statistics. These statistics were not well preserved for at Malaki. The hurst coefficient was consistently underestimated by all the models used in this study.

The Pettit test was used to detect possible change points in the annual rainfall data at the two stations used in this study. Results for Malaki revealed that a change point (shift) in the data happens around 1983, with a 95% and 99% significance levels, as shown in Figure 8. On the other hand, for Surat Obeida, the pettitt test revealed that no break point exist in the record.

Table 1: Historical and generated long term statistics at Surat Obeida and Malaki, Saudi Arabia

	Longest Drough (D)		Longest Surplus (U)		Hurst Coefficient (K)	
	Surat Obeida	Malaki	Surat Obeida	Malaki	Surat Obeida	Malaki
Historical	7.00	10.00	4.00	6.00	0.70	0.88
AR (1)	5.46	6.13	5.03	5.69	0.71	0.77
ARMA (1,1)	5.28	6.06	5.03	6.55	0.72	0.78
ARMA (2,1)	5.24	6.04	4.99	6.42	0.71	0.78

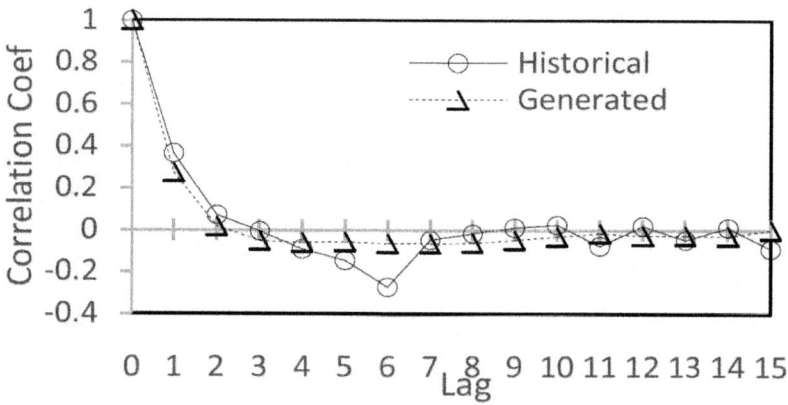

Figure 2: Historical and generated annual correlogram for AR (1) model at Surat Obeida, Saudi Arabia.

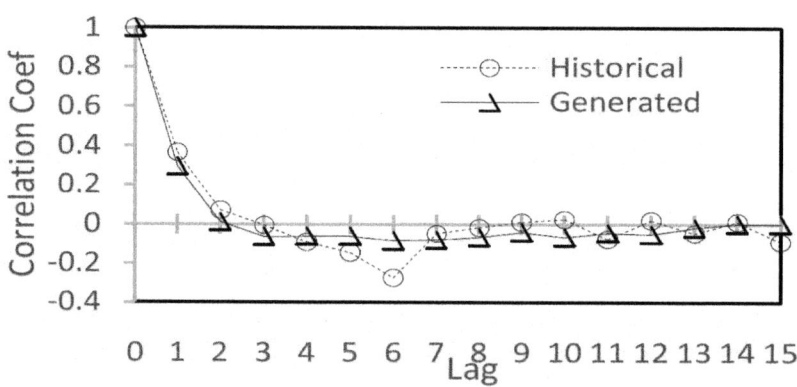

Figure 3: Historical and generated annual correlogram for ARMA (1,1) model at Surat Obeida, Saudi Arabia.

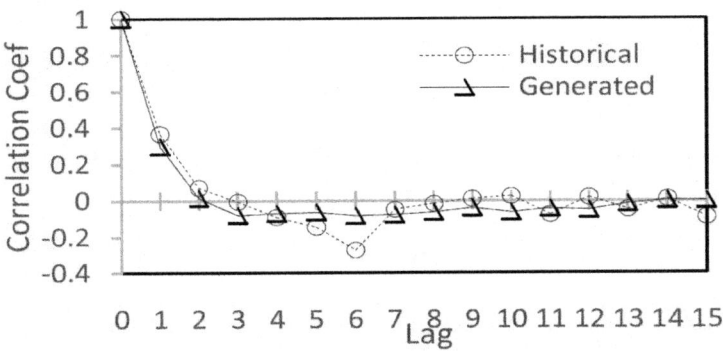

Figure 4: Historical and generated annual correlogram for ARMA (2,1) model at Surat Obeida, Saudi Arabia.

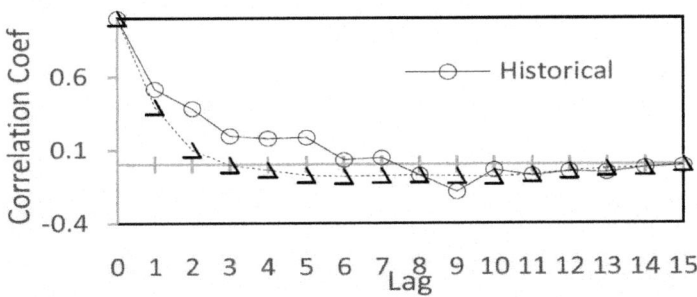

Figure 5. Historical and generated annual correlogram for AR (1) model at Malaki, Saudi Arabia.

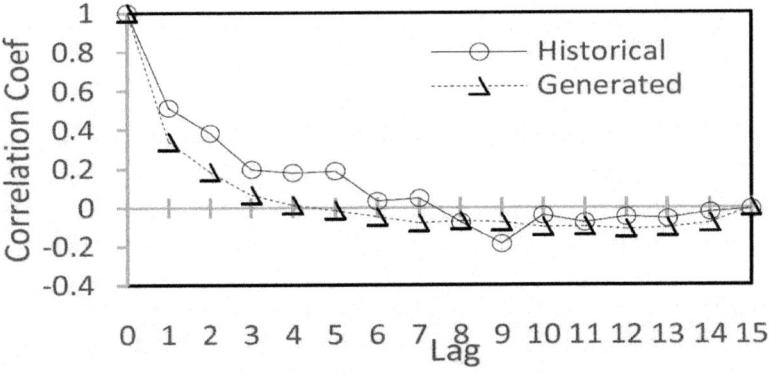

Figure 6. Historical and generated annual correlogram for ARMA (1,1) model at Malaki, Saudi Arabia.

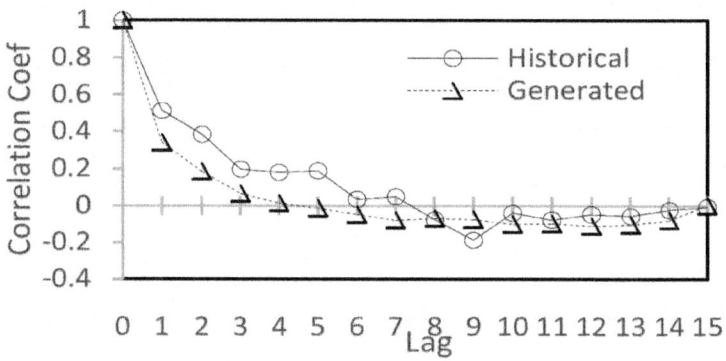

Figure 7. Historical and generated annual correlogram for ARMA (2,1) model at Malaki, Saudi Arabia

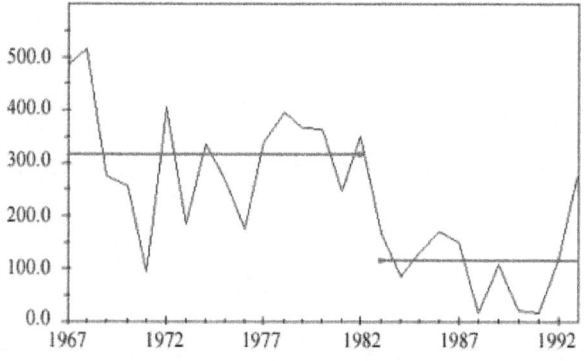

Figure 8. Results of Pettitt test for Malaki rainfall data (1967-1993) where a break point around 1983 is identified.

DISCUSSION

Auto regressive moving average (ARMA) models have been widely used in stochastic hydrology to model annual time series where the mean, variance, and the correlation structure do not depend on time. These models are capable of preserving the basic statistical characteristics of annual historical time series such as the mean and variance and also can preserve the long term related statistics [3] . However, the simulation experiments in this study revealed that the tested ARMA models did not perform well at Malaki, but performed reasonably well at Surat Obeida.

A time series Hurst Coefficient in the range 0.5 - 1 would indicate a series with a long decaying positive autocorrelation, meaning both that a high value will probably be followed by another high value resulting in having periods or clusters of high values [9] A value in the range 0 - 0.5 indicates a switching behavior between high and low values, meaning that a high value will probably be followed by a low value and vice versa and a value of 0.5 indicates a completely uncorrelated behavior [9] . The historical autocorrelgram at Malaki is characterized by slow long decaying correlation structure whereas the autocorrelogram at Surat Obeida is fast decaying type. The Hurst coefficient at Malaki was high (K = 0.88) indicating a strong peristance features. As was shown earlier, results of the pettit test indicate an existance of a break point or a shift in the record that could have happened around 1983 at Malaki. This could suggest the possibility of non-stationarity of the historical record of Malaki.

CONCLUSIONS

The ARMA models were used for modeling and simulation of rainfall in arid and semi-arid regions. The ARMA models were able to preserve the long term statistics at Surat Obeida, but failed to do so at Malaki. Malaki historical data reveled a slow and long decaying structure and a high Hurst coeficient indicating the possibility of non-stationarity of the data. Pettitt test revealed that a break point or a shift in the record of Malaki happened around 1983, which could indicate that the data were non-stationary. Thus, it is recommended that ARMA models should not be used in modeling the annual rainfall at Malaki.

Finally, future investigation of other stochastic models that are capable of preserving long term characteristics is needed.

ACKNOWLEDGEMENTS

The researcher would like to extend his gratitude and appreciation to Dr. Mohammed Al Zahrani, Associate Professor of Civil Engineering at King Fahd University of Petroleum and Minerals, Dhahran, Saudi Arabia for providing the historical rainfall data.

REFERENCES

1. Salas, J.D., Saada, N.M. and Chung, C.H. (1995) Stochastic Modeling and Simulation of the Nile River System Monthly Flows. Tech.Rep.5, Colo.State Univ., Fort Collins.

2. Golder, J., Joelson, M., Neel, M. and DI Pietro, L. (2014) A Time Fractional Model to Represent Rainfall Process. Water Science and Engineering, 7, 32-40.

3. Fortin, V., Perreault, L. and Salas, J.D. (2004) Retrospective Analysis and Forecasting of Streamflows Using a Shifting Level Model. Journal of Hydrology, 296, 135-163.http://dx.doi.org/10.1016/j.jhydrol.2004.03.016

4. Rehman, S. (2009) Study of Saudi Arabian Climatic Conditions Using Hurst Exponent and Climatic Predictability Index. Chaos, Solitons and Fractals Journal, 39, 499-509.http://dx.doi.org/10.1016/j.chaos.2007.01.079

5. Elfeki, A., Al-Amri, N. and Bahrawi, J. (2013) Analysis of Annual Rainfall Climate Variability in Saudi Arabia by Using Spectral Density Function. International Journal of Water Resources and Arid Environments, 4, 205-212.

6. Almazroui, M., Nazrul Islam, M., Athar, H., Jones, P.D. and Ashfaqur Rahman, M. (2012) Recent Climate Change in the Arabian Peninsula: Annual Rainfall and Temperature Analysis of Saudi Arabia for 1978-2009. International Journal of Climatology, 32, 953-966. http://dx.doi.org/10.1002/joc.3446

7. Salas, J.D., Saada, N.M., Chung, C.H., Lane, W.L. and Frevert, D.K. (2000) Stochastic Analysis, Modeling and Simulation (SAMS) Version 2000—User's Manual. Colorado State University, Fort Collins.

8. Sveinsson, O.G.B., Salas, J.D., Lane, W.L. and Frevert, D.K. (2007) Stochastic Analysis, Modeling, and Simulation (SAMS Version 2007) User's Manual. Department of Civil and Environmental Engineering, Colorado State University, Fort Collins.

9. Mesa, O.J., Gupta, V.K. and O'Connell, P.E. (2012) Dynamical System Exploration of the Hurst Phenomenon in Simple Climate Models. American Geophysical Union, 196, 209-230.

Chapter 10

INFLUENCE OF FLY ASH ON BRICK PROPERTIES AND THE IMPACT OF FLY ASH BRICK WALLS ON THE INDOOR THERMAL COMFORT FOR PASSIVE SOLAR ENERGY EFFICIENT HOUSE

Golden Makaka

University of Fort Hare, Alice, South Africa

ABSTRACT

In quest for quality and sustainable development, it is necessary to find alternative materials, methods of brick making and house design. Bricks made in open kilns using locally available materials usually do not meet the requirements of the South African Bureau of Standards; hence it needs to add some ingredients such as fly ash to produce better quality bricks. This paper reports the effects of fly ash on properties of clay bricks that can improve the thermal performance of buildings. Bricks of different clay and fly ash mixing proportions were molded. A passive solar house was designed and constructed using fly ash bricks. Results indicate that thermal conductivity and water absorption decrease with increase in fly ash. Compressive strength was found to increases with increase in amount of fly ash. A mixing proportion of 50% of fly ash to 50% clay by volume produced a brick with the highest compressive strength, lowest thermal conductivity and minimum water absorption. The bricks were observed to have uniform size as they experience minimal burning shrinkage. These properties were found to have a significant impact on the thermal performance of the house. The mean indoor temperature swing was found to be 11°C.

INTRODUCTION

In spite of being cheap, the locally made bricks possess some grave disadvantages, i.e. high water absorption, great shrinkage, high thermal conductivity, low compressive strength and non-uniformity in size and shape. The aim is to decrease these disadvantageous characteristics. Simple intervention such as north orientation (in the southern hemisphere) of the house and placement of bigger windows on the north wall plays a crucial role in creating a more comfortable indoor environment [1] . Raw materials used and the firing temperature influence the brick properties. The South Africa Building Standard (SABS) recommends a minimum compressive strength of 5 Mpa and a thermal conductivity less than 0.69 W/km [2] . It is desirable to add some more ingredients to produce bricks of good quality. Fly ash is one such material that can improve brick quality [3] . Bricks with low thermal conductivity, low water absorption, high compressive strength and high sound damping are most preferable. This paper investigates the effects of fly ash on brick properties and the thermal comfort of a passive solar fly ash brick house. The burning of harder, older anthracite and bituminous coal typically produces Class F fly ash. Fly ash produced from the burning of younger lignite or sub-bituminous coal, also has some self-cementing properties. In the presence of water, Class C fly ash will harden and gain strength over time. Class C fly ash generally contains more than 20% lime (CaO) [2] .

BRICK PROPERTIES

Brick properties are affected by composition of the raw materials and the manufacturing processes[4] . For the production of bricks raw materials must possess some specific properties and characteristics. Clays must have plasticity, which permits them to be shaped or moulded when mixed with water; and they must have sufficient wet and air-dried tensile strength to maintain their shape after forming [5] . The properties that determine brick quality include the following: durability, colour, texture, size variation, compressive strength, water absorption, thermal conductivity and sound damping. High thermal conductive bricks result in houses with high temperature swings with rapid response to the outdoor temperature variations while bricks with less sound damping results in house with poor privacy. The addition of fly ash to clay can significantly improve brick properties. The chemical composition and the energy content of the fly ash determine these properties. The chemical and physical properties of fly ash depend upon many parameters such as coal quality, type of coal pulverization and combustion process followed nature of ash collection and disposal technique adopted, etc. [6] .

METHODOLOGY

A mesh sieve was used to separate the unburnt coal from the fine ash; the mesh sieve was inclined at an angle of about 60 degrees and supported by two iron rods. The fly ash was first dried by heating to a temperature of about 80˚C. The fine ash was used to mould bricks and the incomplete burned coal (coarse fly ash) was used to burn the bricks. Bricks of size 230 mm × 67 mm × 118 mm and of different mixing proportions of clay soil to fly ash were moulded. The amount of fly ash was increased in steps of 10% by volume up to 70%. The bricks were dried in shade for a period of six days and then fired in an oven up to a temperature of ~1300°C. Compressive strength, thermal conductivity and water absorption of the bricks were measured.

The brick compressive strength was measured using the compressive strength machine. The brick was first prepared by removing all the unevenness from all the brick faces and the dimensions were also measured. The brick was placed horizontally between the flat plates of the machine. An axially load was applied until brick failure occurs. Then the compressive strength was calculated by dividing the maximum load by the average area of the brick face. The brick thermal conductivity was measured the Hot Wire Method [7] . The standard water absorption method was used [8] .

A passive solar energy-efficient house was designed and constructed in Somerset East. Ecotect building design software was used to simulate the thermal performance of the house. Material properties were also specified. Cost efficiency decisions were taken in the design and construction of the house. On completing constructing the house monitoring sensors were installed, i.e., thermocouples, HMP50 temperature-humidity probe, model 03001-wind sentry anemometer and vane, and pyranometer were installed. Thermocouples were installed inside and outside the building to measure air and wall temperatures. 26 thermocouples were distributed in the house to map-out the temperature pattern. Six thermocouples were installed to measure surface wall temperatures. Thermocouples to measure indoor and outdoor wall surface temperatures were placed and glued at the center of the walls. A wind anemometer and a vane, temperature-humidity probe, and a LI-COR pyranometer were placed on the top of the roof. A second temperature humidity probe was place in the center of the house at height of about 2 m above the floor. All these sensors were then connected to a RC1000 datalogger, powered by a 12 V battery, which was charged by a 20-watt solar panel.

Chemical Properties of Fly Ash

The properties of fly ash bricks depend mainly on two factors: 1) the energy content of the fly ash used and 2) the chemical composition of the fly ash. Table 1 shows the chemical composition of fly ash from two different sites. The X-Ray powder diffraction (XRD) method was used to determine the fly ash chemical composition. Site A was cloth manufacturing company, and site B was a soft drink producing company. The two companies were selected because of the different technology of the boiler system used. The boiler efficiency will then determine the characteristics of the coal ash. The other constituents include FeO, Na_2O, K_2O and unburnt carbon that form the bulk part of the fly ash. The South African fly ash has high-energy content, which makes it excellent in brick making. The fly ash from site A was used to make the bricks. Chemical composition of the fly ash and the temperature attained during burning determine the brick colour. High-energy content fly ash has a great influence on thermal conductivity. Figure 1 shows the variation of thermal conductivity of bricks for different proportions of fly ash. From Figure 1 it can be observed that thermal conductivity of fly ash bricks generally decreases with increase in the amount of fly ash. The minimum thermal conductivity (0.0564 W/mk) corresponds to a mixing proportion of 50% of fly ash by volume. This means the mixing proportion of 50% fly ash will result in 93% reduction of the thermal conductivity. On burning the bricks, the incompletely burned coal in the brick acts as a fuel to enhance uniform burning of the bricks with a lower level of external energy required while at the same time producing uniform bricks.

The end product is very light in weight (density 400 - 1190 $kg·m^{-3}$) as the burnt ash leaves small-unconnected cavities that give the bricks effective heat insulating properties (low conductivity). On the other hand, low energy content fly ash mainly influences the water retention properties without much modification to thermal conductivity. Low energy content fly ash can therefore be added to soil to increase its water retention capacity thereby increasing agricultural production.

Figure 1 also shows the variation of water absorption by fly ash bricks. Water absorption of fly ash bricks decreases with increase in fly ash. A mixing proportion of 50% fly ash produces a brick with minimum water absorption. Since the bricks contain incompletely burned coal, on burning, the brick is burnt from the outside inward thus melting metallic elements (reaching temperatures between 980°C and 1300°C) sintering the brick to form a ceramic material with minimal water absorption (less than 15% by weight). According to the South African Building Standard (SABS), the brick water absorption must be less than 20% by weight.

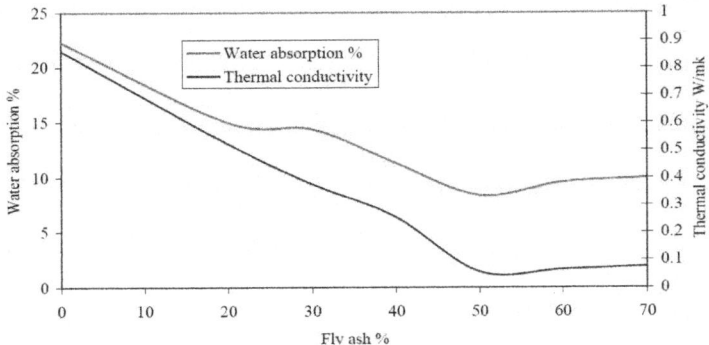

Figure 1: Variation of thermal conductivity and water absorption at 25°C.

Table 1: Chemical composition of fly ash, %

Site	Sulphate (as SO_4)	Phosphate (as PO_4)	Silicate (SiO_2)	Calcium (Ca)	Magnesium (Mg)	Potassium (K)	Aluminium (Al)
A	0.3	<0.1	20.3	0.36	0.05	1.63	14.1
B	0.2	0.1	20.9	2.15	0.12	2.68	19.6

With reference to Figure 1, the locally made bricks do not meet the minimum requirements, and the addition of fly ash has shown to improve the water absorption of the locally made bricks. The addition of 20% of fly ash by volume reduced the water absorption by 32%, while the addition of 50% fly ash lowered the water absorption by 62%. Since the created cavities are unconnected, it implies that permeability and porosity are reduced. The reduction in permeability and porosity implies the reduction in freezing/thawing damage of the brick, since there will be small amount of water in the brick.

Figure 2 illustrates the variation of compressive strength of fly ash bricks for different proportions of fly ash. The compressive strength increases with the amount of fly ash, attaining a maximum compressive strength at 50% fly ash. The SABS specify the minimum brick compressive strength to be 5,000,000 Pa. From Figure 2, it can be seen that the locally made bricks (with no fly ash) do not meet the minimum compressive strength requirement. The addition of fly ash significantly improves the compressive strength of bricks. 7000 fly ash bricks with a mixing proportion of 50% fly ash were made and used to construct a passive solar house.

Experimental Solar House Plan

The floor plan measures 6880 mm by 6580 mm, giving an approximate area of 45.27 m². Figure 3 shows the floor plan of the experimental passive solar and Figure 4 shows the completed passive solar energy efficient house

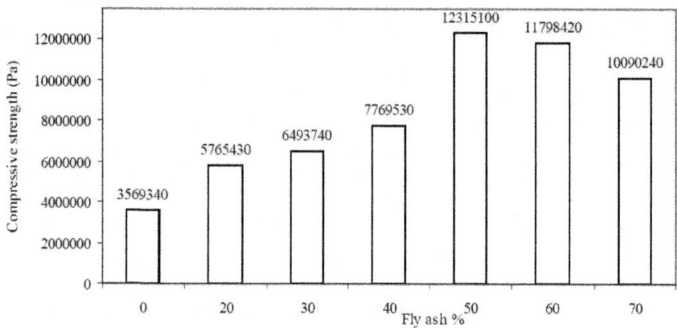

Figure 2: Compressive strength of fly ash bricks.

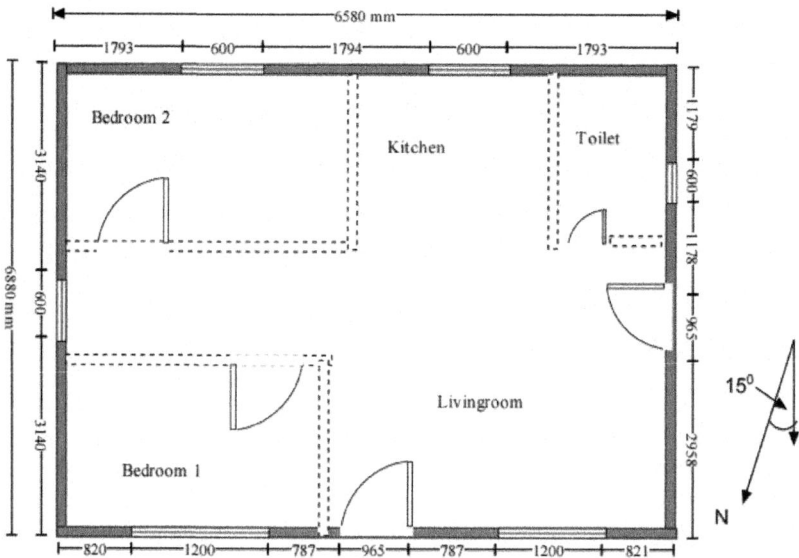

Figure 3: The floor plan of the passive solar house.

The house was made to face N15°W. This orientation was to optimize the solar radiation that penetrates indoor in winter. The house has two standard doors and a total of six windows, i.e., two large windows on the north (front wall), two medium windows on the south wall, and one small window on the west wall and another small window on the eastern wall. On the north-facing roof there are four clerestory windows. One of the doors is on the north wall and the other on the west wall. An open plan layout was adopted in order to optimize natural ventilation. Mechanical ventilation systems were avoided in order to keep the running cost of the house low. The location of the experiences prevailing westerly winds (W60° ± 15°N), and the placement of a door on the west and a small window on the east wall allows to control the ventilation rate, making it possible to regulate the indoor temperature. The small window on the west serves to capture westerly prevailing winds and this minimizes the indoor moisture condensation.

Figure 4. Completed passive solar energy efficient house.

The roof was split into two, the lower and upper roof. The lower roof faces north while the higher roof faces south. This was done in order to enable the insertion of clerestory windows making it possible to direct solar radiation to the desired rear zone (floor and southern wall) and to maximize day lighting, thus minimizing the use of electricity during the day. The northern roof is ideal for mounting photovoltaic modules as active solar energy converters.

Operation of the Passive Solar House

In summer, the sun almost rises from the east and sets in the west. In this case, the roof overhangs were made long enough (simulation was done ECOTECT) to eliminate the possibility of sunrays penetrating indoor. With reference to the clerestory windows, the upper roof was extended out by 200 mm while the lower roof was extended in by 100 mm. This eliminates the possible direct penetration of the solar radiation in summer while allowing maximum penetration in winter. In winter (May to August) the sun rises almost northeast, but following a low northern path in the sky and then set in the northwest.

From May to August the daily maximum angle of the sun ranges from 34° to 48° and this maximum angle occurs at around 12 h15 with June 21st having the smallest angle. Thus, the north facing windows allows solar radiation to penetrate indoor, while the clerestory windows allow the south wall and the far south floor (thermal mass) to receive solar radiation. The thermal masses of high heat capacity (i.e. concrete floor of 100 mm thickness and the wall made from fly ash bricks) absorb solar radiation during the day. Figure 5 shows how the lower winter sun penetrates indoors and heats the thermal mass. This penetration of solar radiation is prevented in summer by overhangs.

Post Construction Thermal Performance

The passive solar house was monitored for six days, i.e., March 1st, 2006 to March 6th, 2006.Figure 6 shows the indoor and outdoor temperatures patterns for the period tested. The house was being used as the kitchen, a sitting room and at the same time used as a bedroom. The house was subdivided using cloth. A stove was situated at the west wall close to the door. The indoor temperature follows the outdoor temperature, but with thermal time lag of about 3 hours. The indoor experience a temperature swing of about 11°C while for the outdoor it was 14°C.

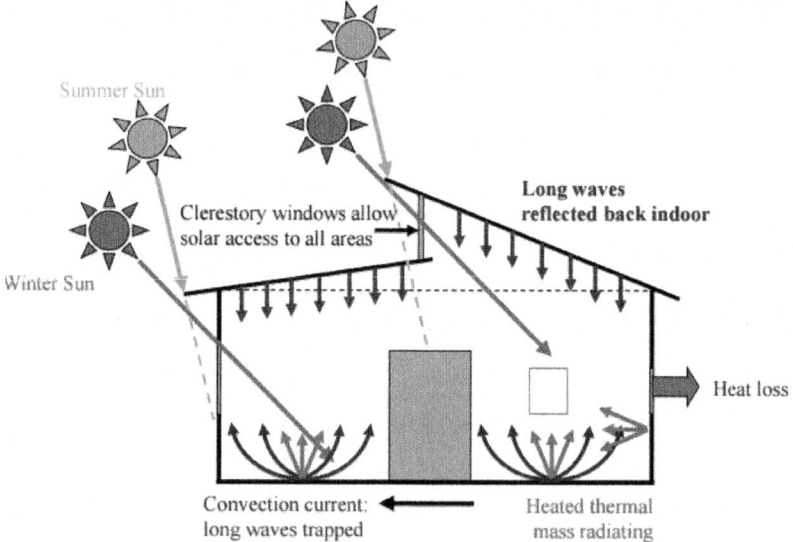

Figure 5: Operation of the passive solar house.

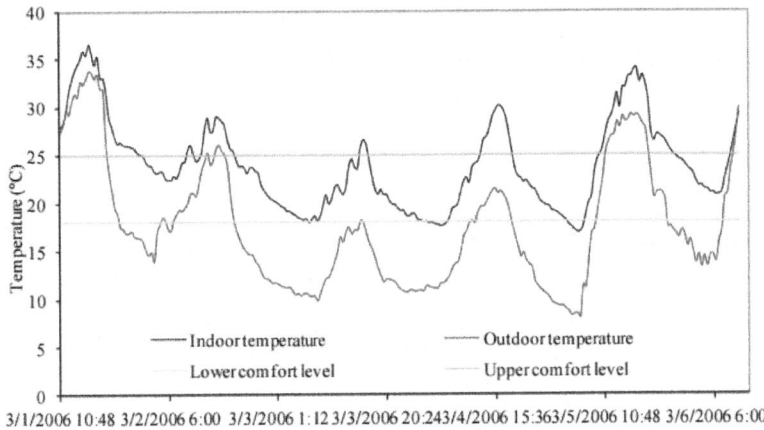

Figure 6: Indoor and outdoor temperature.

The outdoor attained a maximum temperature of about 33.66°C while the indoor experienced a higher temperature of about 36.63°C, however, the indoor experienced a low temperature swing.Figure 6 also shows the lower and upper limits of the thermal comfort zone. The indoor was observed to be within the comfort level for 60% of the total time tested while the outdoor was only 30%. The performance of the house could have been higher if the occupants were operating the house properly, i.e., opening and closing of windows and doors when necessary.

For most of the time windows were closed, thus preventing the prevailing winds to cool the house. On March 3rd, 2006 at around 08:30 local time, the outdoor temperature dropped to 9.75°C but the indoor temperature only dropped to about 18°C. Surviving this strenuous test shows that the passive solar heating design can be freeze resistant. The ordinary windows were not opened for approximately 60% of the time they were supposed to be opened, this then contributed negatively to the thermal performance of the house.

Figure 7 shows the variation of the outdoor and indoor relative humidity.

It can be seen that Somerset East experience high humidity levels reaching a maximum value of ~97.75% while the indoor attained ~67% with the external and internal coinciding and being equal to ~22.86%. The high

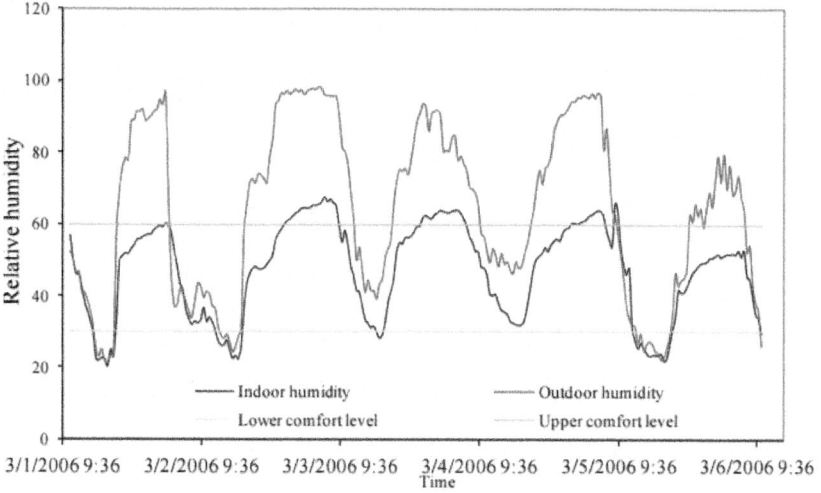

Figure 7: Indoor and outdoor humidity.

indoor humidity is mainly due to cooking while the windows and doors are both closed. The peak outdoor and indoor humidity were out of phase by about 3 hours. The mean ratio of the outdoor to indoor relative humidityi.e., $\left(\dfrac{RH_{out}}{RH_{in}}\right)$ was observed to be approximately 1.4 while the mean indoor-outdoor humidity difference was

~26.09%. The indoor humidity was observed to be within the comfort levels (30% - 60%), for about 75% of the total time tested. For proper operation of the house windows and doors needs to be opened or closed when necessary to allow air circulation and unnecessary heat loss/gain. However windows were closed for most of the time and clerestory windows were never opened, thus minimizing the rate of air exchange.

Vertical Temperature Variation

Figure 8 shows the mean indoor vertical temperature variation at three different heights. With reference to Figure 8, it can be observed that there is a systematic variation in temperature within the house that is related to the height above the floor.

From Figure 8 it is evident that temperature increases with height during the day. However, at night there was a tendency of uniform temperature distribution. At night the roof (iron corrugated sheets) will be at a lower temperature than the indoor air temperature, and the tendency that warm air raises results in the warm rising air to be cooled by the roof making it to descend again. This results in a more uniform temperature distribution. However, during the day, the roof will be at a higher temperature than the indoor air, so the warm air that rises is trapped closer to the roof, and as the roof temperature continue to increase, the temperature of the air closer to the roof also continue to rise. This therefore creates an almost stagnant vertical temperature gradient. By opening the clerestory windows, the warm air can therefore be exhausted out, thus allowing cooler outdoor air to penetrate indoor. The vertical temperature variation was seen to depend with the side of the building, i.e. north, south, east or west. From a height of 1.9 m to 2.4 m the mean temperature gradient was found to be 2.5°C/m while from a height of 0.9 m to 1.9 m the mean temperature gradient was found to be 2.1°C/m. The vertical temperature variation creates a density gradient resulting in vertical air movement (stake effect). Figure 9 and Figure 10 show the vertical temperature variation at a particular time for summer and winter respectively.

From Figure 9 it can seen that the north and west sides are at a relatively higher temperatures than the south and east and this results in horizontal airflows towards the south and the east. FromFigure 10 it can be seen that there is minimal horizontal airflows as the air temperatures are almost the same.

During the day, temperature increases with height. From literature, it has also been found that the type of heating employed within the house impacts on the temperature distribution (Inard, et al., 1996; Howarth, 1985). That work was however conducted in laboratories and considered static situations when a particular heater was operating.

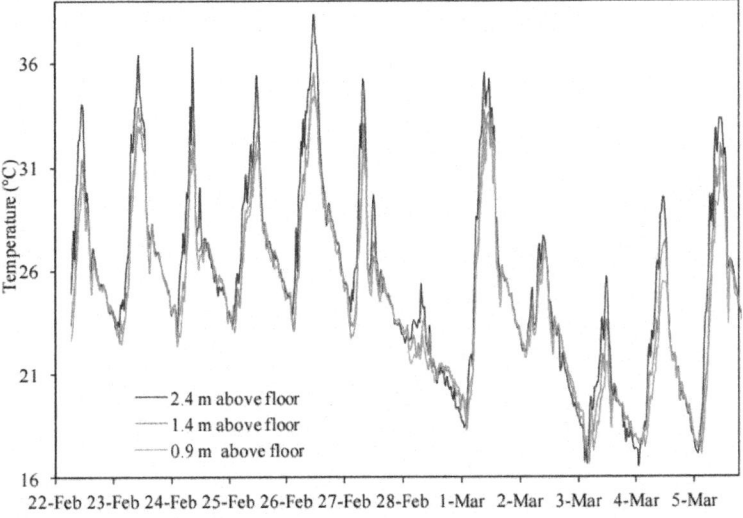

Figure 8: Vertical temperature variations.

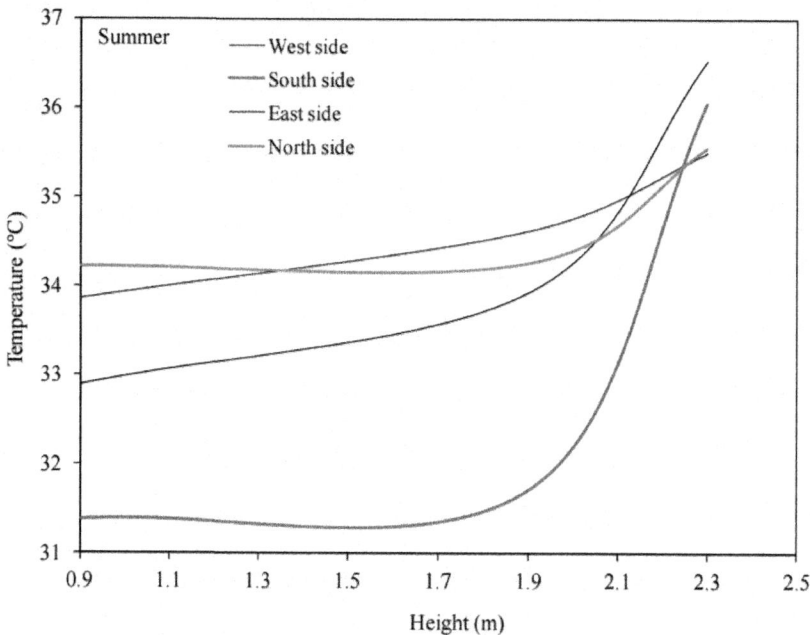

Figure 9: Summer: Indoor vertical temperature variation (23/2/2006 15:30).

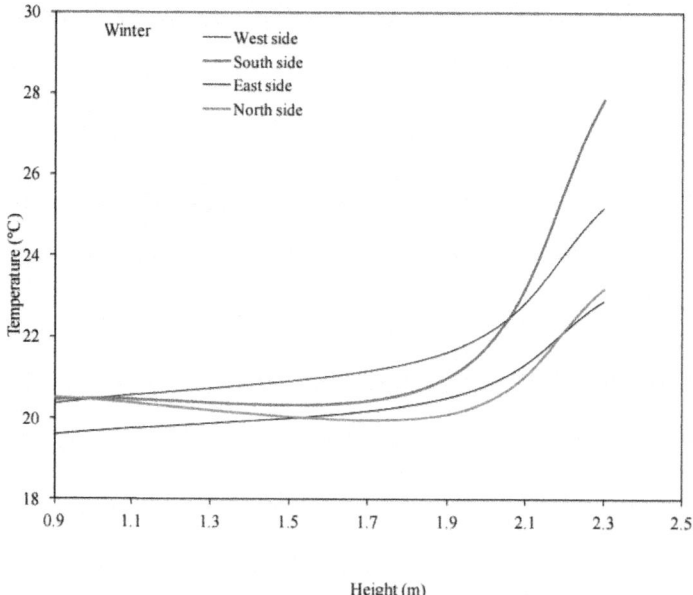

Figure 10: Winter: Indoor vertical temperature variation (5/7/2006 15:30).

Heating due to solar radiation is time dependent. The mean temperature difference for 1 m height difference was found to be approximately 0.7°C. This means the roof has a significant influence on the indoor temperature distribution. It can also be noted from Figure 8 that as the indoor environment is heated, there is an increase in the vertical temperature difference. When the indoor cools (due to conduction through the walls and infiltration heat losses) the vertical temperature difference shrinks. It was also observed that the lowest vertical temperature difference occurred during 9 pm to 2 am.

CONCLUSION

Fly ash can significantly improve brick properties. Water absorption and thermal conductivity decrease with the amount of fly ash, and the mixing proportion of 50% fly ash produces a brick with the least thermal conductivity (0.0564 W/mk) and water absorption (8.4%) but with the highest compressive strength ~12 MPa. Outside wall plastering could be avoided as these bricks are smooth and their shape and size are uniform. These properties can lead to construction of low-cost modern houses that have better thermal comfort and reduce the cost of the house. A passive solar house was designed and constructed in Somerset East. The indoor temperature swing was found to be 11°C while the outdoor was 14°C. The indoor humidity swing was 44.14% while outdoor was 74.89%.

ACKNOWLEDGEMENTS

The authors would like to acknowledge Tuskegee University and the USAID for financial support.

REFERENCES

1. Givoni, B. (1998) Effectiveness of Mass and Night Ventilation in Lowering the Indoor Daytime Temperatures. Energy and Buildings, 28, 25-32.http://dx.doi.org/10.1016/S0378-7788(97)00056-X

2. South Africa Building Manual, 2005.

3. Dondi, M., et al. (2002) Orimulsion Fly Ash in Clay Bricks—Part 2: Technical Behaviour of Clay/Ash Mixtures. European Ceramics Society, 22, 1737-1747.http://dx.doi.org/10.1016/S0955-2219(01)00494-0 http://en.wikipedia.org/wiki/Fly_ash

4. Smith, M.W., et al. (2001) Thermal Performance Analysis of a High-Mass Residential Building. National Renewable energy Laboratory.

5. Adam, A.A. (2009) Strength and Durability Properties of Alkali Activated

Slag and Fly Ash-Based Geopolymer Concrete. Ph.D. Thesis, RMIT University, Melbourne.

6. Yoshitoshi, S., et al. (2009) Measurement of Thermal Conductivity of Magnesia Brick with Straight Brick Specimens by Hot Wire Method. Materials Transactions, 50, 2623-2630.http://dx.doi.org/10.2320/matertrans.M2009098

7. Pel, L. (1995) Moisture Transport in Porous Building Materials. Ph.D. Thesis, Eindhoven University of Technology, Eindhoven.

Chapter 11

APPLICATION OF WATER QUALITY MODEL QUAL2K TO MODEL THE DISPERSION OF POLLUTANTS IN RIVER NDARUGU, KENYA

Letensie Tseggai Hadgu[1], Maurice Omondi Nyadawa[2], John Kimani Mwangi[1], Purity Muthoni Kibetu[1], Beraki Bahre Mehari[1]

[1]Department of Civil, Construction, and Environmental Engineering, Jomo Kenyatta University of Agriculture and Technology, Nairobi, Kenya

[2]Jaramogi Oginga Odinga University of Science and Technology, Bondo, Kenya

ABSTRACT

Ndarugu River, Kenya, during its course through the different agricultural and industrial areas of Gatundu, Gachororo and Juja farms, receives untreated industrial, domestic and agricultural waste of point source discharges from coffee and tea factories. During wet season the water is also polluted by non-point (diffuse) sources created by runoff carrying soil, fertilizer and pesticide residues from the catchment area. This study involved the calibration of water quality model QUAL2K to predict the water quality of this segment of the river. The model was calibrated and validated for flow discharge (Q), temperature (T°), flow velocity (V), biochemical oxygen demand (BOD5), dissolved oxygen (DO) and nitrate (NO3-N), using data collected and analyzed during field and laboratory measurements done in July and November-December 2013. The model was then used in simulation and its performance was evaluated using statistical criteria based on correlation coefficient (R^2) and standard errors (SE) between the observed and simulated data. The model reflected the field data quite well with minor exceptions. In spite of these minor differences between the measured and simulated data set at some points, the calibration and validation results are acceptable especially for developing countries where the financial resources for frequent monitoring works and higher accuracy data analysis are very limited. The water is being polluted by the human activities in the catchment. There is need for proper control of wastewater by various techniques, and preliminary treatment of waste discharges prior to effluent disposal. Management of the watershed is necessary so as to protect the river from the adverse impacts of agricultural activities and save it from further deterioration.

INTRODUCTION

Intensive developments of industry, agricultural production and ever intensive urbanization have led to the in- crease in number of pollutants and the amount of wastewater which pollute water flows. On the contrary, the need for water of satisfying quality continuously grows. A big amount of agricultural, municipal and industrial wastewater discharges to water bodies around the world. The discharging of degradable wastewater in water bo- dies result in decrease in water quality generally and particularly DO (Dissolved Oxygen) concentrations [1] . Disposal of municipal, agricultural and industrial wastewater into the rivers with little or no treatment prior to discharge is a common practice in many developing countries. This has caused a serious concern over the dete- rioration of river water quality. Many big and small rivers in Kenya such as Nairobi River, Athi River and their tributaries are under threat due to influx of pollutants without prior treatment. Therefore, it is important and timely that a rigorous approach to the water quality modeling of such water-courses be undertaken. Ndarugu River is a tributary of Athi River which is the second longest river in Kenya. During its flow through the differ- ent agricultural and industrial areas of Gatundu, Gachororo and Juja farms, it receives untreated industrial and agricultural waste discharges such as effluent from coffee and tea factories in the catchment area and the neigh- boring small settlements situated on the bank of the river. This river is a main source of fresh water for domestic use to the villages along the river bank and Nairobi City. Considering the implications of water pollution on hu- man and aquatic health, the effective management of this segment of the river is important. QUAL2E model, developed by United States Environmental Protection Agency (USEPA), is the most widely used mathematical model for conventional pollutant impact evaluation [2] . However, several limitations of the QUAL2E have been reported. One of the major inadequacies is the lack of provision for conversion of algal death to carbonaceous biochemical oxygen demand (CBOD). Another limitation of this model is that the river section has to be seg- mented in to equal lengths reaches and equal the number of elements in each reach. Park and Lee [3] developed QUAL2K after modification of QUAL2E. The modifications include the expansion of computational structures and addition of more parameters. An enhanced and modernized version of QUAL2E, QUAL2K version 2.11, was developed as a continuation of modification and simplification of the model [4] . In most of the studies car- ried out in the application of QUAL2K model, it was observed that the model represented the field data quite well and this reasonable modeling guarantees the use of QUAL2K for future river water quality options [5] . For example Q2K was applied for water quality modeling in the Baghmati River and this application showed that, the model represented the field data pretty accurate. In this study, various water quality management options are taken into account to control DO, such as pollution loads modification and local oxygenation (by affixing weirs). Apparently local oxygenation is effective in raising DO levels [6] . The water environmental capacity of the Hongqi River (China) was simulated by Q2K. In this study Q2K was calibrated and confirmed using data from field monitoring carried out during the winter of 2009 and spring of 2010. The simulated results correlated with the measured data precisely [7] . The aim of the study was to model the water quality of the polluted segment of Ndarugu River by the comprehensive application of water quality model QUAL2K and evaluate the perfor- mance of the model using statistics based on correlation coefficient (R^2) and standard error (SE).

MATERIALS AND METHODS

Study Area and Sampling Sites

Ndarugu River is one of the tributaries of Athi River. Ndarugu sub-catchment extends from Kieni and Kinale forest eastwards and parts of ridges of Aberdares to Juja farm all the way to Munyu where it is joined by River Komu before it joins Athi River. Ndarugu River is a perennial river with its source in the Kikuyu escarpments. It flows in south east direction and meanders through farmed slopes of Gatundu and Thika District before joining Athi River at Munyu near Kilimambogo. The tributaries of Ndarugu River are Ruabora, Githobokoni and Kara- kuta rivers. The study area has a catchment area of 135 km² bounded between latitudes 1°00'36" South and 1°08'59" South and longitudes 36°53'33" East and 37°10'25" East. Its UTM zone is 37S, at an average altitude of 1560 meters above sea level. The drainage area is coded by the Government of Kenya as 3CB sub catchments in Athi Basin. The study area covers a 15.5 km stretch of the river starting 2.4 km downstream of Munya estate and 0.5 km downstream of Karakuta estate (upstream boundary of the stretch) to the upstream side of Juja farm (downstream boundary of the stretch). Figure 1shows the location of catchment area in Kiambu County and Figure 2 shows the sampling sites for water quality testing along the river stretch selected for the study.

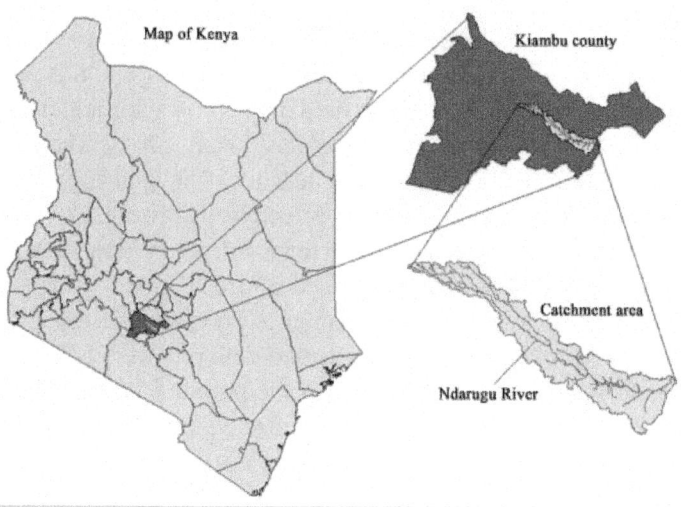

Figure 1: Location map of the study area

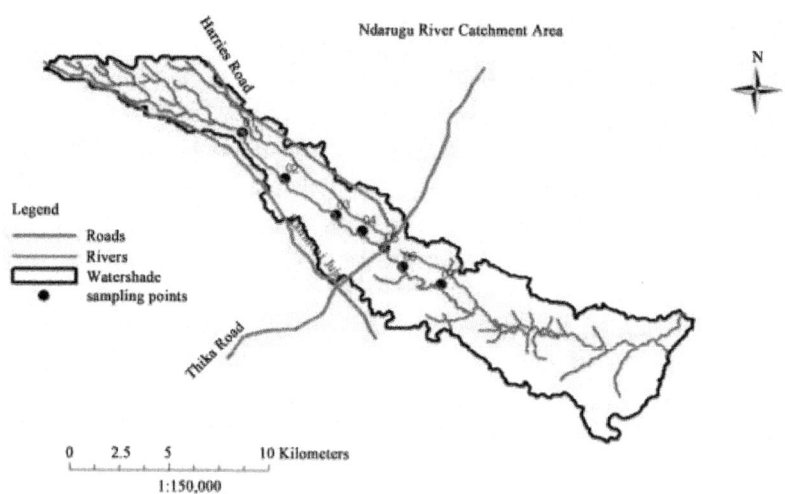

Figure 2: Catchment area and sampling sites.

Water Sampling and Analysis

Water quality data of the river and wastewater of some coffee estates were collected in July 2013 for calibration during dry season and in November and December 2013 for validation during the short rainy season. Water samples were collected at 15 cm depth (to avoid floating material) from seven points along the river. Sampling bottles made of plastic each 500 ml volume and Global Positioning System (GPS) were used to collect the water samples and determine the locations of sampling points. A sampling pole was used to get samples from the run- ning water in the main body of the river since it gives the best overall sample unlike the water at the edges or that held in pools, which is likely to have a variation from the running water. The bottles were rinsed few times from the river water before taking the samples and touching the rim was avoided to prevent from any contami- nation of the samples. The bottle was placed inverted in the body of water 10 to 20 cm below the surface and then turned horizontal facing in to the flow of water. The samples analysis for eight parameters namely pH, temperature, conductivity, turbidity, DO, BOD, nitrates and phosphates were done in situ and at Jomo Kenyatta Environmental Laboratory. Sampling was done for ten days in each season and average values of the results were presented in Table 1. These values were used in the calibration and validation of the model. In addition weather data from Meteorology Department and stream flow and velocity data from Water Resources Manage- ment Authority (WRMA) were used as input values. The river section (area of study) was divided in to six seg- ments based on the presence of point and non-point source pollutions and the uniformity in the river characteris- tics such as dimensions and slopes.

Description of the Model

QUAL2K (Version 2.11) is a modeling framework for simulating river and stream water quality developed by Chapra et al. [4] . It is implemented within the Microsoft Windows environment. Numerical computations were programmed in Fortran 90. Microsoft Office Excel is used as the graphical user interface. All interface opera- tions are programmed in the Microsoft Office macro language: Visual Basic for Applications (VBA). It divides the system into reaches and elements. In contrast to QUAL2E, the element size for QUAL2K can vary from reach to reach and reach lengths can vary. In addition, multiple loadings and withdrawals can be input to any element. QUAL2K model is suitable for simulating the hydrological and water quality conditions of a small river. It is a simple one-dimensional model that simulates basic stream transport and mixing processes. It allows spe- cifying many of the kinetic parameters on a reach-specific basis. A complete discussion of the model theory is described in QUAL2K Documentation and User's Manual [4] . Therefore, QUAL2K was selected as the tool to develop the Ndarugu River water quality model.

Table 1: Physicochemical parameters of the water samples in dry and wet seasons

Sampling Points	Season	pH	Temp. (°C)	EC (μS/cm)	Turbidity (NTU)	DO (mg/l)	BOD₅ (mg/l)	NO₃ (mg/l)
01	Dry	7.18	18.42	61.5	25.37	5.79	19.15	18.6
	Wet	7.22	17.9	52.7	49.02	7.64	16.9	25.3
02	Dry	7.1	18.24	58.6	25.11	5.57	19.19	20.4
	Wet	7.28	18.02	56.5	51.45	7.12	17.32	27.7
03	Dry	7.09	18.45	63.7	24.85	5.17	19.45	19.3
	Wet	7.13	17.8	61.1	58.36	6.4	17.45	26.1
04	Dry	7.06	18.4	66.9	25.53	4.82	19.78	17.6
	Wet	7.15	18.1	58.3	52.27	6.45	18.08	28
05	Dry	7.2	18.44	66.1	22.09	4.14	19.84	19.2
	Wet	7.04	18.54	56.9	63.45	6.14	17.67	28.2
06	Dry	6.93	18.37	67.7	28.21	4.21	20.24	18.1
	Wet	7.06	18.6	62.7	57.9	6.12	19.32	28.5
07	Dry	6.98	18.24	72.1	27.23	3.82	20.99	22.2
	Wet	6.93	17.1	64.3	58.6	5.93	18.99	28.9
WHO		6.5 - 8.5	-	500 - 5000	5	>5	-	50

Implementation of the Model

The QUAL2K model has greater flexibility, which can follow the specific circumstances of users to set the pa-rameter values and transform the simulation equation, satisfying user requirements for water quality simulation. The water quality data analyzed for dry season were applied for calibration. The calculation time step was set in such a way that ensures the model was maintained in the steady-state. The model was run (for validation) with another set of data (wet season) which was done without altering the calibrated parameters. This way the ability of the calibrated model to forecast the component concentration under different circumstances could be exa- mined.

RESULTS AND DISCUSSION

The calibration and validation results of Ndarugu River are presented in Figure 3 and Figure 4. The physico- chemical parameters of the water sample used for calibration and validation of the model are shown in Table 1. These results also show the variation of the pollution levels as we move downstream along the river. Phosphates were below the detectable limits in all sampling points for the entire study period. Temperature and pH values did not show much variation between the dry and wet seasons. Turbidity was higher in wet season than in dry season. This was due to the runoff of soil and other particles from the catchment area. Electrical conductivity, an indirect measure of the total dissolved solids (TDS), was slightly lower in wet season than in dry season.

This could be due to the dilution of the salt particles (TDS) by the input of rainfall and runoff. Nutrients, particularly nitrate values showed a considerable increment in wet season due to the runoff of fertilizers from the agricultural areas in the catchment. The flow and velocity patterns were varying significantly because of the low flow in the dry season and the input of runoff in the rainy season and discharge of effluents from drains across the reach. The QUAL2K generated results are in the form of graphs combined with the observed data set as shown in Fig- ures 3(a)-(f) and Figures 4(a)-(f). The correlation between simulated and observed values for all the parameters showed a high correlation coefficient (R^2) and lower standard errors as shown in Table 2.

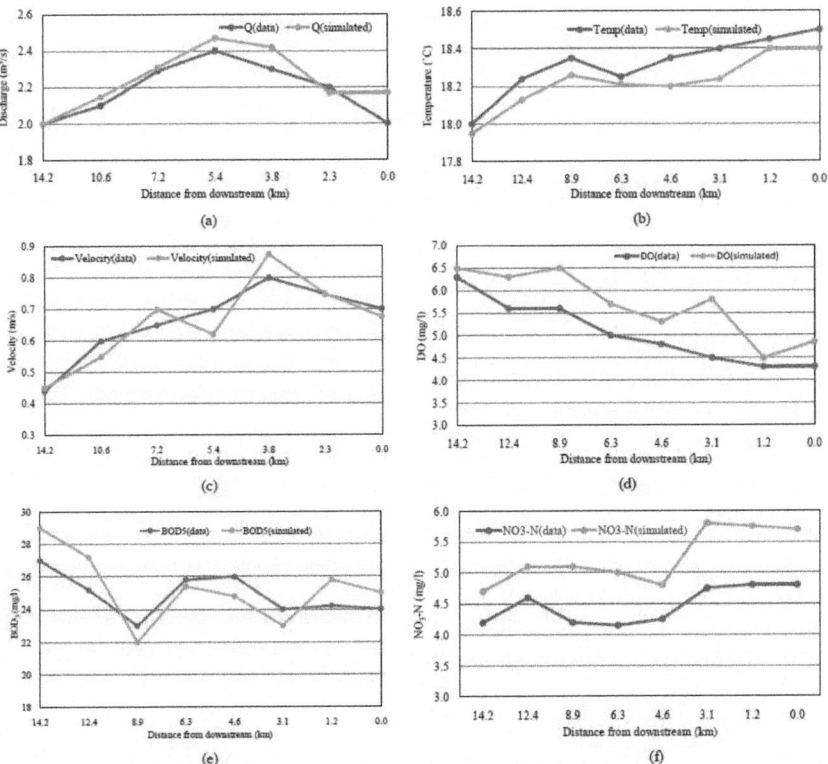

Figure 3: (a)-(f) Calibration results. (a) Calibration graph of discharge; (b) Calibration graph of temperature; (c) Cali- bration graph of velocity; (d) Calibration graph of DO; (e) Calibration graph of BOD_5; (f) Calibration graph of NO_3-N.

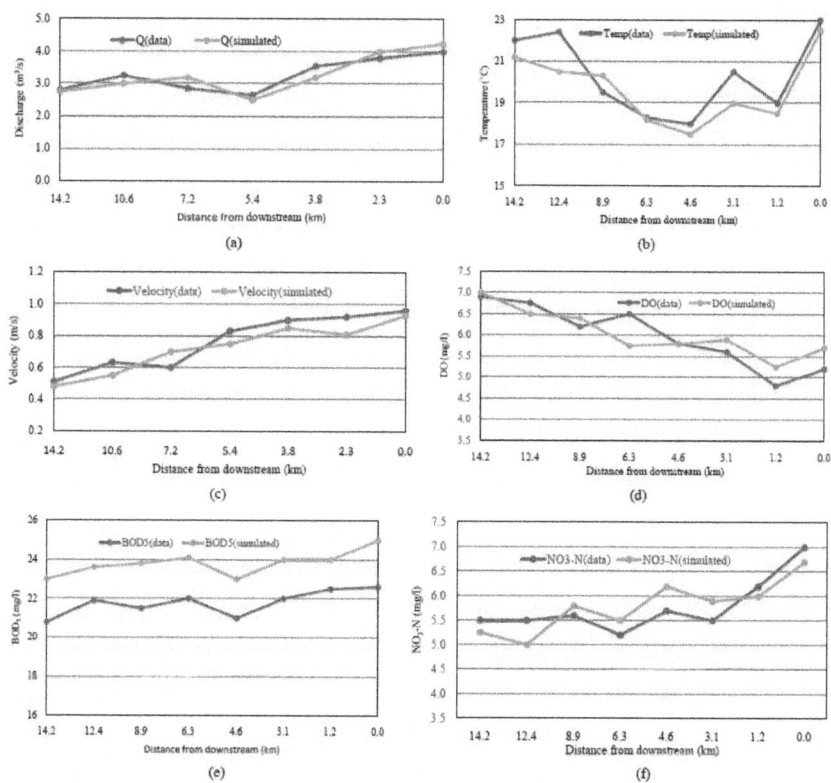

Figure 4: (a)-(f) Validation results. (a) Validation graph of discharge; (b) Validation graph of temperature; (c) Validation graph of velocity; (d) Validation graph of DO; (e) Validation graph of BOD_5; (f) Validation graph of NO_3-N.

These high correlation coefficients between observed and simulated values show that this model is perfectly reliable in modeling streams when detailed and complex data are not available due to an acceptable match of the simulated data with meas- ured data. Thus the QUAL2K model can be used to predict the effect of point and non-point diffusion and ab- straction at any section of the river on its water quality on the downstream side. This saves time and capital which would otherwise be required for frequent monitoring works. The method applied in this study can provide a basis for water environmental management in decision making for the future.

The results indicated that most of the physico-chemical water quality parameters for Ndarugu River were

Table 2: Correlation between observed and simulated values

Parameter	Calibration		Validation	
	R^2	SE	R^2	SE
Discharge	0.904	0.077	0.923	0.277
Temperature	0.956	0.047	0.907	0.759
Velocity	0.925	0.059	0.916	0.071
DO	0.882	0.382	0.855	0.318
BOD₅	0.812	1.394	0.864	0.354
NO₃-N	0.902	0.491	0.805	0.538

within the WHO and Kenya Bureau of Standards (KEBS) limits for drinking water and the water is therefore suitable for domestic purposes. The pH of the river was within the range of 6.5 to 8.5 both in the dry and rainy seasons. The DO in dry season was significantly depleted and reduced due to the mixing of these wastewaters even though re-aeration (replenishing with dilution and surface re-aeration) is expected at the surface of the river. Turbidity level varied from wet to dry season with higher levels in the wet season and was higher than the WHO and KEBS limits for drinking water throughout the study period. Electrical conductivity level reduced slightly during wet season due to the dilution from the runoff. Nutrient levels were generally low during the study pe- riod although they were higher during the wet season than the dry season. However, despite these low levels care should be taken in the application of inorganic fertilizers in order to protect the river from eutrophication.

CONCLUSION AND RECOMMENDATIONS

Observed values of physical-chemical parameters of water samples showed that the river is being polluted by the human activities in the catchment. Values of discharge (Q), flow velocity (V), NO_3-N, electrical conductivity, BOD_5 and of dissolved oxygen (DO), acquired by calibrating the water quality model, QUAL2K, are similar to observed values of physical-chemical analysis of water samples. Therefore, the water quality model QUAL2K is calibrated successfully. The performance of this model was evaluated using statistics based on correlation coef- ficient (R^2). The results of the evaluation revealed that there is not much variation in these values between cali- bration and validation. The model reflected the field data quite well with some minor exceptions. In spite of these minor differences between the measured and simulated data sets at some points, the calibration and valida- tion results are acceptable especially for developing countries where the financial resources for frequent moni- toring works and higher accuracy data analysis are very limited. There is also a need for proper control of wastewater by various techniques, and preliminary treatment of waste discharges prior to effluent disposal. Ma- nagement of the watershed is necessary so as to protect the river from the adverse impacts of agricultural activi- ties and save it from further deterioration.

REFERENCES

1. Nakhaei, N. and Shahidi, A.E. (2010) Waste Water Discharge Impact Modeling with QUAL2K, Case Study: The Zayandeh-Rood River. International Environmental Modelling and Software Society (IEMSS), Ottawa.

2. Brown, L.C. and Barnwell, T.O. (1987) The Enhanced Stream Water Quality Models QUAL2E and QUAL2E-UNCAS (EPA/600/3-87-007). US Environmental Protection Agency, Athens.

3. Park, S.S. and Lee, Y.S. (2002) A Water Quality Modeling Study of the Nakdong River Korea. Ecological Modeling, 152, 65-75.

4. http://dx.doi.org/10.1016/S0304-3800(01)00489-6

5. Chapra, S.C., Pelletier, G.J. and Tao, H. (2008) QUAL2K: A Modeling Framework for Simulating River and Stream Water Quality, Version 2.11. USA: Documentation and User's Manual. Civil and Environmental Engineering Department, Tufts University, Medford.

6. Kalburgi, P.B., Shivayogimath, C.B. and Purandara, B.K. (2010) Application of QUAL2K for Water Quality Modeling of River Ghataprabha (India). Journal of Environmental Science and Engineering, 4, 6-11.

7. Kanne, P.R., Lee, S., Lee, Y.S., Kanel, S.R. and Pelletier, G.J. (2007) Application of Automated QUAL2Kw for Water Quality Modeling and Management in the Bagmati River. Elsevier, Amsterdam, 503-517.

8. http://www.sciencedirect.com/

9. Zhang, R.B., Qian, X., Yuan, X.C., Ye, R., Xia, B.S. and Wang, Y.L. (2012) Simulation of Water Environmental Capacity and Pollution Load Reduction Using QUAL2K for Water Environmental Management. International Journal of Environmental Research and Public Health, 9, 4504-4521. http://dx.doi.org/10.3390/ijerph9124504

Chapter 12

MODELING OF DISCHARGE DISTRIBUTION IN BEND OF GANGA RIVER AT VARANASI

Manvendra Singh Chauhan, Prabhat Kumar Singh Dikshit, Shyam Bihari Dwivedi

Department of Civil Engineering, Indian Institute of Technology, IIT (BHU), Varanasi, India

ABSTRACT

Dynamics of river behavior play a great role in meandering, sediment transporting, scouring, etc. of river at bend, which solely depends on hydraulics properties such as horizontal and vertical stress, spatial and temporal variation of discharge. Therefore understanding of discharge distribution of river Ganga is essential to apprehend the behavior of river cross section at bend particularly. The measurement of discharge is not very simple as there is no instrument that can measure the discharge directly, but velocity measurement at a section can be made. Velocity distribution at different cross sections at a time is also not easy with single measurement with the help of any instrument and method, so it required repetitions of the measurement. Velocity near the end of bank, top and bottom layer of natural streams is difficult to be measured, yet velocity distribution at these regions plays important role in characterizing the behavior of river. This paper deals with the new advanced discharge measurement technique and measured discharge data has been used for modelling at river bend. To carry out the distribution of discharge and velocity with depth in river Ganga, the length of river in study area was distributed into 14 different cross sections, M-1 to M-14, measured downstream to upstream and the measurement was done by using of ADCP (Acoustic Doppler Current Profiler). At each cross section, profiles were measured independently by an ADCP and data acquired from ADCP were further used for the regression modeling. A multiple linear regression model was developed, which showed a high correlation among the discharge, depth and velocity parameters with the root mean square error (R^2) value of 0.8624.

INTRODUCTION

Flow in open channel and a natural river is often described by simplifying cross section. But in reality cross section of river and its bend is so complicated that it needs vast practical experience to understand the hydrodynamics. On the other hand the most dangerous natural disaster, worldwide known as flood, can cause a huge economic losses as well as losses of life and livelihood. Therefore understanding the flow behavior and estimation of discharge for open channel flow (most commonly natural rivers) is very vital hence it had keen interest for the researchers for decades. Many researchers develop various methods for the discharge estimation; however, some enhance the accuracy of the previously available methods. Researcher develops regression based models [1] - [3] and some develop soft computing methods [4] -[6] The main river flow (discharge of river) might be changed at a very large scale as human interruption takes place in term of occupying the place along the river bank or within the river basin [7] . Regression based approach is most commonly very useful for the ungauged sites for discharge estimation [8] - [10] .

The increase in the flow of any river is caused by the large volume of rainfall in its basin, which would probably change the physical parameters of the river involving the changes in the depth due to bank erosion' taking place and width of flow [11] - [14] . The river hydrodynamics could directly affect the flow pattern of river and may change in river morphology. [15] - [18] observed that the meandering is one of the most common pattern followed by fluvial rivers. A lot of research work is completed by researchers for the study of bankfull discharge and bankfull velocity of river, but there is a lack of research about the natural flow and natural velocity of river.

It is to understand that, the discharge of river is a function of river meandering wavelength and amplitude, as the higher the value of river meandering wavelength and amplitude, higher will be the discharge and vice versa. The above understanding gives a way to go forward with this research in the direction that the river parameters must naturally have a relationship with each other. The main purpose of this paper is to develop a correlation among parameters of Varanasi bend of holy River Ganga which are directly related to the physical parameters of river i.e. discharge, depth and velocity. The complete measurement on the Varanasi bend was done in the month of November 2013.

With the progresses of measuring discharge and understanding behavior of Natural River, Acoustic Doppler Current Profiler (ADCP) technologies, a moving boat discharge measurement technique is gradually replacing the classic procedure using mechanical meters when the water is sufficiently deep for ADCP applications. While measuring discharge through ADCP, the transducers of an ADCP are mounted facing down and barely submerged under the water surface. They ping continuously while the boat is traversing from bank to bank. The boat motion is monitored by bottom tracking acoustic pings or by a global positioning system (GPS). The water flux crossing the vertical plane of the boat path is computed, which is the same as the river discharge. The ADCP can be used for measuring a velocity profile in the vertical when the ADCP is held at a fixed position for taking a large number of the single ping velocity measurements. The averaged single ping velocity profiles reduce the measurement errors so that a meaningful mean velocity profile can be obtained.

This paper is designed to address the following objectives by using data generated from the ADCP

Quantify the discharge and velocity distribution for the different cross section along the bend of river Ganga

Identify relationships between different hydraulic parameters and thus perform regression analysis.

Organization of paper includes:

- an overview of the discharge measurement and regression modeling
- description of study area (Section 2)
- descriptions of the methods used (Section 3)
- the data analysis and model development (Section 4)
- discussion of the results (Section 5); and
- conclusions (Section 6).

STUDY AREA

Varanasi (25°20'N and 83°7'E) is located in the middle Ganges valley of North India, in the Eastern part of the Uttar Pradesh, along the left crescent-shaped bank of the Ganges, averaging between 50 feet (15 m) and 70 feet (21 m) above the river. It is oldest city situated on the convex bank of holy River Gangaas shown in Figure 1. It is called the longest river of India, having its total length 2525 KM from Gangotri to Ganga Sagar. Being located in the Indo-Gangetic Plains of North India, the land is very fertile because low level floods in the Ganges continually replenish the soil. Varanasi is often said to be located between two confluences: one of the Ganges and Varuna, and other of the Ganges and Assi, although the latter has always been a rivulet rather than a river. The distance between the two confluences is around 2.5 miles (4.0 km). Rarely has any river gathered in itself so much meaning and reverence as the Ganga has over three millennia in the Indian subcontinent. The land-water interface on the Ganga's banks is fashioned out of the need to access the rising and falling water levels in the monsoon and dry seasons. The cultural landscape of this interface a ghat (steps and landings) lined by temples and other public buildings, pavilions, kunds (tanks), streets and plazas is layered and kinetic, and responsive to the river's flow. At Varanasi, where the Ganga reverses its flow northwards, the ghats describe a crescent sweep in a 7.6 km stretch.

Figure 1: Study area: natural bend in river Ganga at varanasi.

The climate of the city, as of Northern India on the whole, is of tropical nature with extremes of temperature, varying from a minimum of 5°C in winter to a maximum of 45°C in summer. The annual rainfall varies from 680 mm to 1500 mm, with a large proportion occurring during the monsoon season, in the months of July to September.

METHODOLOGY

To achieve the objective of measuring velocity distribution and understanding the behavior of velocity distribution with depth of river in the river cross-section, an ADCP, was used. The whole study river length was divided into the 14 distinct cross-sections for discharge measurement, named as M-14 to M-1 respectively from upstream of flow to downstream of flow. Further with the help of ADCP, complete profiling for depth and discharge of each cross-section had been done. Recorded ADCP data have been extracted by using the supporting software of ADCP i.e. Win River-II, for analysis purpose. Excel sheets for each cross-section (from M-1 to M-14) of distance from bank, velocity and depth was prepared for calibration of regression based model.

For preparation of data, shortest width cross-section was selected and divided it into 4 uniform parts (width wise), the width of shortest cross-section was 281 meters after dividing it, the division width was 74.25 meters, average the velocity and depth parameters of each part as V1, V2, V3, V4 & D1, D2, D3, D4 for the cross-section M-7. The area of each part was also calculated by using AutoCAD software termed as A1, A2, A3, and A4 respectively.

Similarly by applying this process on all the data of each cross-section from M1 to M14 was estimated and listed in Table 1. M-1 has been divided into 7 parts having the average velocity from V1 to V7, average depth from D1 to D7 and the area from A1 to A7 and Cross-sectional view with reduced level is also shown in Figure 2, M-2 has been divided into 6 parts having the average velocity from V1 to V6, average depth from D1 to D6 and the area from A1 to A6, M-3 has been divided into 5 parts having the average velocity from V1 to V5,

Table 1: Description of cross sectional data

Sr. No.	Profile No.	Total Width in (m)	No. of Division	Name of Average Depths	Average Depth in (m)/74.25 m width	Name of Average Velocities	Average Velocity in (m)/74.25 m width	Area of each divided section in m²	Discharge at each cross section
				D1	11.16	V1	0.113	1116.6769	126.18449
				D2	19.98	V2	0.221	1475.8199	326.1562
				D3	18.11	V3	0.3	1351.3572	405.40716
1	M-1	460	6.1953	D4	16.53	V4	0.34	1228.3031	417.62305
				D5	11.07	V5	0.34	830.8381	282.48495
				D6	6.95	V6	0.14	526.488	73.70832
				D7	4.82	V7	0.08	15.2103	1.216824
				D1	9	V1	0.192	879.809	168.92333
				D2	15.72	V2	0.43	1163.7889	500.42923
2	M-2	434	5.8451	D3	14.62	V3	0.371	1094.5191	406.06659
				D4	12.98	V4	0.252	967.9261	243.91738
				D5	9.7	V5	0.163	731.0454	119.1604
				D6	5.58	V6	0.098	210.2677	20.606235
				D1	15.44	V1	0.32	1226.6425	392.5256
				D2	18.22	V2	0.29	1349.0814	391.23361
3	M-3	357	4.8081	D3	14.97	V3	0.3	1112.5433	333.76299
				D4	12.86	V4	0.22	950.3806	209.08373
				D5	6.8	V5	0.17	391.39	66.5363
				D1	15.1	V1	0.173	1271.3793	219.94862
				D2	16.31	V2	0.489	1204.4832	588.99228
4	M-4	378	5.0909	D3	12.52	V3	0.507	931.5344	472.28794
				D4	11.06	V4	0.295	821.6697	242.39256
				D5	7.62	V5	0.097	560.2798	54.347141
				D1	8.57	V1	0.122	864.4567	105.46372
				D2	17.87	V2	0.479	1323.7853	634.09316
5	M-5	386	5.1987	D3	15.75	V3	0.537	1170.9607	628.8059

No	M			D		V			
				D4	15.41	V4	0.248	1144.988	283.95702
				D5	8.31	V5	0.101	661.7477	66.836518
				D1	13.88	V1	0.504	1054.9663	531.70302
6	M-6	297	4	D2	13.1	V2	0.641	1062.1626	680.84623
				D3	12.71	V3	0.313	949.9967	297.34897
				D4	6.51	V4	0.137	488.8122	66.967271
				D1	11.5	V1	0.744	944.3512	702.59729
7	M-7	281	3.7845	D2	9.41	V2	0.75	707.7555	530.81663
				D3	8.54	V3	0.356	628.8874	223.88391
				D4	4.61	V4	0.111	252.4331	28.020074
				D1	8.4	V1	0.656	728.3134	477.77359
8	M-8	312	4.202	D2	10.27	V2	0.896	765.2419	685.65674
				D3	5.95	V3	0.344	443.2548	152.47965
				D4	2.9	V4	0.068	222.3736	15.121405
				D1	9.05	V1	0.899	678.7563	610.20191
				D2	6.47	V2	0.873	479.2026	418.34387
9	M-9	307	4.1347	D3	3.46	V3	0.684	255.54	174.78936
				D4	2.25	V4	0.506	169.8287	85.933322
				D5	1.45	V5	0.329	5.3285	1.7530765
				D1	10.1	V1	0.678	761.6668	516.41009
				D2	6.02	V2	0.798	447.5982	357.18336
10	M-10	392	5.2795	D3	4	V3	0.776	346.3682	268.78172
				D4	3.105	V4	0.581	228.551	132.78813
				D5	1.9	V5	0.412	142.8722	58.863346
				D6	1.12	V6	0.279	12.7513	3.5576127
				D1	4.56	V1	0.671	337.0713	226.17484
				D2	4.66	V2	0.771	346.8669	267.43438
11	M-11	422	5.6835	D3	5.57	V3	0.835	414.4396	346.05707
				D4	4.4	V4	0.797	328.4041	261.73807
				D5	3.51	V5	0.571	259.5272	148.19003
				D6	1.88	V6	0.461	78.0174	35.966021
				D1	4.19	V1	0.817	311.7137	254.67009
				D2	4.45	V2	0.79	330.417	261.02943
				D3	4.21	V3	0.8	313.0432	250.43456
12	M-12	559	7.5286	D4	4.55	V4	0.771	338.2732	260.80864
				D5	3.5	V5	0.704	257.2048	181.07218
				D6	2.66	V6	0.666	196.848	131.10077
				D7	1.81	V7	0.494	135.492	66.933048
				D8	1.2	V8	0.253	37.3966	9.4613398

				D1	3.74	V1	0.524	279.8455	146.63904
				D2	4.48	V2	0.56	332.7477	186.33871
				D3	5.92	V3	0.526	436.2374	229.46087
				D4	6.12	V4	0.516	453.295	233.90022
				D5	4.48	V5	0.521	333.9429	173.98425
13	M-13	814	10.963	D6	3.48	V6	0.497	255.3419	126.90492
				D7	2.27	V7	0.484	170.2859	82.418376
				D8	1.84	V8	0.534	135.8387	72.537866
				D9	2.2	V9	0.436	163.9797	71.495149
				D10	2.42	V10	0.356	179.3759	63.85782
				D11	2.38	V11	0.314	152.1876	47.786906
				D1	7.82	V1	0.287	594.5108	170.6246
				D2	7.566	V2	0.319	560.7728	178.88652
				D3	5.27	V3	0.337	392.2132	132.17585
				D4	3.71	V4	0.357	274.2688	97.913962
14	M-14	694	9.3468	D5	3.95	V5	0.354	291.6898	103.25819
				D6	4.14	V6	0.46	307.0175	141.22805
				D7	5.21	V7	0.459	386.965	177.61694
				D8	5.72	V8	0.455	424.994	193.37227
				D9	5.48	V9	0.43	410.8616	176.67049
				D10	4.5	V10	0.39	102.7781	40.083459

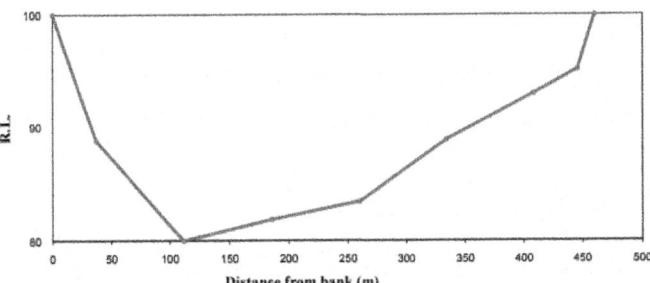

Figure 2: Typical behavior of river cross section w.r.t. Reduced Level (R.L.) at M-1.

average depth from D1 to D5 and the area from A1 to A5, M-4 has been divided into 5 parts having the average velocity from V1 to V5, average depth from D1 to D5 and the area from A1 to A5, M-5 has been divided into 5 parts having the average velocity from V1 to V5, average depth from D1 to D5 and the area from A1 to A5, M-6 has been divided into 4 parts having the average velocity from V1 to V4, average depth from D1 to D4 and the area from A1 to A4, M-8 has been divided into 4 parts having the average velocity from V1 to V4, average depth from D1 to D4 and the area from A1 to A4, M-9 has been divided into 5 parts having the average velocity from V1 to V5, average depth from D1 to D5 and the area from A1 to A5, M-10 has been divided into 6 parts having the average velocity from V1 to V6,average depth from D1 to D6 and the area from A1 to A6, M-11 has been divided into 6 parts having the average velocity from V1 to V6,average depth from D1 to D6 and the area from A1 to A6, M-12 has been divided into 8 parts having the average velocity from V1 to V8, average depth from D1 to D8 and the area from A1 to A8, M-13 has been divided into 11 parts having the average velocity from V1 to V11, average depth from D1 to D11 and the area from A1 to A11, M-14 has been divided into 10 parts having the average velocity from V1 to V10, average depth from D1 to D10 and the area from A1 to A10.

DATA ANALYSIS AND MODELING

a) Data Analysis

Before the development of the models of regression, it is the most important to check whether the variables in data have any correlation or not. Therefore, each cross-sectional data of discharge, depth and velocity are checked for the multiple regression, the R^2, Adjusted R^2, Standard error of estimates, standard error, t value and p value for each cross-section are listed in Table 2 which shows there is a strong correlation between discharge, depth and velocity data, this analysis gives an clear idea to develop a multiple linear regression model.

R-squared (R^2) is a statistical measure of how close the data are to the fitted regression line. It is also known as the coefficient of determination, or the coefficient of multiple determinations for multiple regressions. The value (R^2) should always between 0% and 100%:

0% indicates that the model explains none of the variability of the response data around its mean.

100% indicates that the model explains all the variability of the response data around its mean.

For any regression model first indicator of generalizability is the adjusted (R^2) value, which is adjusted for the number of variables included in the regression equation. This is used to estimate the expected shrinkage in (R^2) that would not generalize to the variable because our solution is over-fitted to the data set by including too many independent variables. If the adjusted (R^2) value is much lower than the (R^2) value, it is an indication that our regression equation may be over-fitted to the sample, and of limited generalizability.

The R^2 method is a useful linear regression tool for exploratory model building as it assists in finding subsets of independent variables that best predict a dependent variable in a given sample (SAS Institute, Inc., 1994). This algorithm examines all of the possible combinations of the independent variables and ranks them according to decreasing order of R^2 (fraction of the variance explained by the regression) magnitude for the given sample. Using this output of ranked R^2, the best combination of independent variables was selected for further testing for inclusion in the final regression equations. The type of regression equation that is most suitable to describe the relation depends naturally on the variables considered and with respect to hydrology on the physics of the processes driving the variables. Furthermore, it also depends on the range of the data one is interested in.

b) Development of Regression Models

(i)   Multiple Regression model for 8 Cross-Sections: For development of the regression model, the complete data set of all cross-section were analyzed separately. Three cross-sections from both ends of the bend and two cross-sections from center location have been selected for model development (as shown in Figure 3). Selected cross-section gives a complete picture of the Varanasi bend of River Ganga. For calibration of the regression model complete 55 data (about 65% of total) and remaining 31 data (about 35% of total) are used for the validation of the model. As shown in theTable 3 the value of R^2 is 0.8674 of the calibrated model which shown a strong correlation between discharge, depth and velocity data of the complete data set.

Thus developed discharge equation from regression analysis is $Q = Y_o + a \times V + b \times D$, where Q is Discharge V is Velocity and D is depth, Y_o, a and b are constants which has to determined by regression analysis.

(ii)   Partial regression model: For analyzing the fact that whether the discharge is more dependent on which parameter, depth or velocity, a partial regression model has been studied by keeping depth and velocity constant. For the modeling purpose (keeping depth constant) the data had shorted in a manner that the depth ranging in

Table 2: Cross-sectional data analysis

Sr. No.	C-S	R	R²	AdjR²	Y₄	a	b	SEE	t	P	Std Error
									-4.72	0.01	29.66
1	M-1	0.9892	0.9784	0.9677	-140.1343	823.845	14.8447	28.9603	5.87	0.00	140.24
									5.51	0.01	2.69
									-3.37	0.04	31.52
2	M-2	0.9973	0.9946	0.991	-106.0717	1451.219	-2.2241	16.7843	7.15	0.01	202.95
									-0.33	0.76	6.72
									-12.73	0.01	18.22
3	M-3	0.999	0.9981	0.9961	-232.0105	1086.547	16.5426	8.4102	7.77	0.02	139.79
									8.06	0.02	2.05
									-3.45	0.07	54.18
4	M-4	0.995	0.9901	0.9802	-187.1077	892.8937	17.3317	28.7211	9.54	0.01	93.58
									3.46	0.07	5.01
									-2.38	0.14	54.09
5	M-5	0.9974	0.9948	0.9896	-128.5444	1178.466	9.0698	27.8852	8.24	0.01	142.98
									1.40	0.30	6.50
									-4.41	0.14	27.67
6	M-6	0.9997	0.9993	0.9979	-121.947	1170.01	4.4053	12.27	20.82	0.03	56.20
									1.21	0.44	3.65
									-1.03	0.49	253.87
7	M-7	0.9785	0.9576	0.8727	-261.565	587.393	40.15	106.93	1.23	0.44	479.24
									0.77	0.58	51.99
									2.84	0.22	97.24
8	M-8	0.9994	0.9988	0.9964	275.978	2038.62	-137.83	18.05	6.18	0.10	329.67
									-3.69	0.17	37.36
									-10.03	0.01	12.59
9	M-9	0.9998	0.9997	0.9994	-126.307	78.1056	73.449	6.3376	2.31	0.15	33.87
									28.23	0.00	2.60
									-5.64	0.01	23.44
10	M-10	0.9976	0.9953	0.9921	-132.187	258.619	46.329	17.09	5.26	0.01	49.13
									14.97	0.00	3.09
									-12.53	0.00	16.27
11	M-11	0.999	0.9979	0.9965	-203.784	306.824	50.422	6.3165	5.72	0.01	53.61
									8.21	0.00	6.14
									-12.60	<0.0001	6.96
12	M-12	0.9992	0.9984	0.9978	-87.7579	53.9182	69.184	4.6753	2.00	0.10	26.95
									16.86	<0.0001	4.10
									-14.09	<0.0001	7.36
13	M-13	0.9988	0.9976	0.9969	-103.664	202.286	37.935	3.7203	11.52	<0.0001	17.55
									41.81	<0.0001	0.91
									-2.86	0.02	55.97
14	M-14	0.9201	0.8465	0.8026	-159.944	244.71	28.683	17.493	2.36	0.05	103.59
									6.19	0.00	4.63

C-S: cross-section; R: Correlation constant; R^2: square of the correlation constant; Adj R^2: adjusted value of R^2, Y_o, a and b are the intercept constants; SEE: Standard error of estimate, t: test value for each constants; p: test value for each constants and Std Error: standard error for each constants.

Figure 3: Bifurcation of cross-sections at Varanasi Bend of River Ganga.

Table 3: Statistical parameters of the calibrated model

Profile	R	Rsqr	AdjRsqr	Y₀	a	b	SEE	t	P	Std Error
								-8.5517	<0.0001	23.9917
8-C-S	0.934	0.8723	0.8674	-205.169	519.233	27.3198	59.0513	13.1601	<0.0001	39.4551
								16.096	<0.0001	1.6973

1 m - 5 m, 5 m - 10 m, 13 m - 15 m and finally 15 m - 20 m, the model gives the R^2 value as follows 0.816, 0.802, 0.947 and 0.966 respectively. For the model (keeping velocity constant) average the velocity in previously shorted data, it ranged up to 0.3168 m/s, 0.3645 m/s and 0.5 m/s, the model gives the R^2 value as follows 0.897, 0.998 and 0.988 respectively as listed in Table 4 below. These values concluded that the discharge is more depending upon the depth of the flow as the R^2 value for the model when velocity is constant is more except once i.e. 0.897.

c) Validation of Model

From whole data set remaining 31 data used for the validation of the model. The detailed calculation for each data is shown in Table 5. From this it is clearly noticeable that at very low discharge values, depth below 5 m and low velocities, the model doesn't works properly and it gives unreasoned results.

RESULT AND DISCUSSION

River hydraulics is quite complex in natural channels and rivers. For practical and engineering purposes, the flows in river channel are often characterized by depth averaged or cross-sectional averaged properties. While these simplifications might be justifiable and necessary for practical reasons, it is important to be cognizant about the complex nature of the three-dimensional free-surface flows in rivers and open channels. A better understanding of the hydraulic properties in natural rivers would give rise to a more accurate approximation in practical applications. In this study, the velocity distribution in a river cross-section has been investigated in detail.

Also as we know that atmospheric and human intervention affects the hydrology of any area which influences the flow behavior of river. To understand these effects on main governing parameters of hydraulics on river flow characteristics, 14 different cross section shad marked along the river which lies between 7500 m. The river flow velocity, width and depth has been computed and plotted to compare each other and identified their relationship among the above said parameters. The results showed that river depth is almost having increasing trend except the cross section (M-2), second last from downstream as clearly shown in Figure 4(a).

Table 4: Statistical parameters of the partial regression model

Sr. No.	R^2	Std. deviation	MSE	RMSE	Y_o	A	Range	Constant Parameter
1	0.816	36.261	1205.293	34.717	−44.442	342.204	1 - 5 m	
2	0.802	59.932	3169.251	56.296	−35.172	633.193	5 - 10 m	
3	0.947	53.104	2115.015	45.989	62.655	769.011	10 - 15 m	Depth
4	0.966	11.597	89.657	9.469	150.237	805.866	15 - 20 m	
5	0.897	45.307	1824.616	42.716	−60.380	22.605	0.3168 m/s	
6	0.998	5.656	25.593	5.059	1.750	25.202	0.3645 m/s	Velocity
7	0.988	17.128	234.689	15.320	19.174	28.876	0.5 m/s	

Table 5: Discharge data for validation of model

Sr. No.	Velocity	Depth	Observed Discharge	A × Velocity	B × Depth	Modeled discharge	% error
1	0.173	15.1	219.9486189	89.827309	412.529	297.186989	35.11655
2	0.489	16.31	588.9922848	253.904937	445.5859	494.321575	−16.0733
3	0.507	12.52	472.2879408	263.251131	342.0439	400.125727	−15.2793
4	0.295	11.06	242.3925615	153.173735	302.157	250.161423	3.205074
5	0.097	7.62	54.3471406	50.365601	208.1769	53.373177	−1.79212
6	0.122	8.57	105.4637174	63.346426	234.1307	92.307812	−12.4743
7	0.479	17.87	634.0931587	248.712607	488.2048	531.748133	−16.1404
8	0.537	15.75	628.8058959	278.828121	430.2869	503.945671	−19.8567
9	0.248	15.41	283.957024	128.769784	420.9981	344.598602	21.3559
10	0.101	8.31	66.8365177	52.442533	227.0275	74.300771	11.16793
11	0.504	13.88	531.7030152	261.693432	379.1988	435.722956	−18.0514
12	0.641	13.1	680.8462266	332.828353	357.8894	485.548433	−28.6846
13	0.313	12.71	297.3489671	162.519929	347.2347	304.585287	2.433612
14	0.137	6.51	66.9672714	71.134921	177.8519	43.817519	−34.5688
15	0.899	9.05	610.2019137	466.790467	247.2442	508.865357	−16.6071
16	0.873	6.47	418.3438698	453.290409	176.7591	424.880215	1.562434
17	0.684	3.46	174.78936	355.155372	94.52651	244.51258	39.88985
18	0.506	2.25	85.9333222	262.731898	61.46955	119.032148	38.51687
19	0.329	1.45	1.7530765	170.827657	39.61371	5.272067	200.7323
20	0.678	10.1	516.4100904	352.039974	275.93	422.800654	−18.127
21	0.798	6.02	357.1833636	414.347934	164.4652	373.64383	4.608408
22	0.776	4	268.7817232	402.924808	109.2792	307.034708	14.23199
23	0.581	3.105	132.788131	301.674373	84.82798	181.333052	36.55818
24	0.412	1.9	58.8633464	213.923996	51.90762	60.662316	3.05618
25	0.279	1.12	3.5576127	144.866007	30.59818	−29.705117	−934.973
26	0.671	4.56	226.1748423	348.405343	124.5783	267.814331	18.41031
27	0.771	4.66	267.4343799	400.328643	127.3103	322.469611	20.57897
28	0.835	5.57	346.057066	433.559555	152.1713	380.561541	9.970747
29	0.797	4.4	261.7380677	413.828701	120.2071	328.866521	25.64719
30	0.571	3.51	148.1900312	296.482043	95.8925	187.205241	26.32782
31	0.461	1.88	35.9660214	239.366413	51.36122	85.558337	137.8866

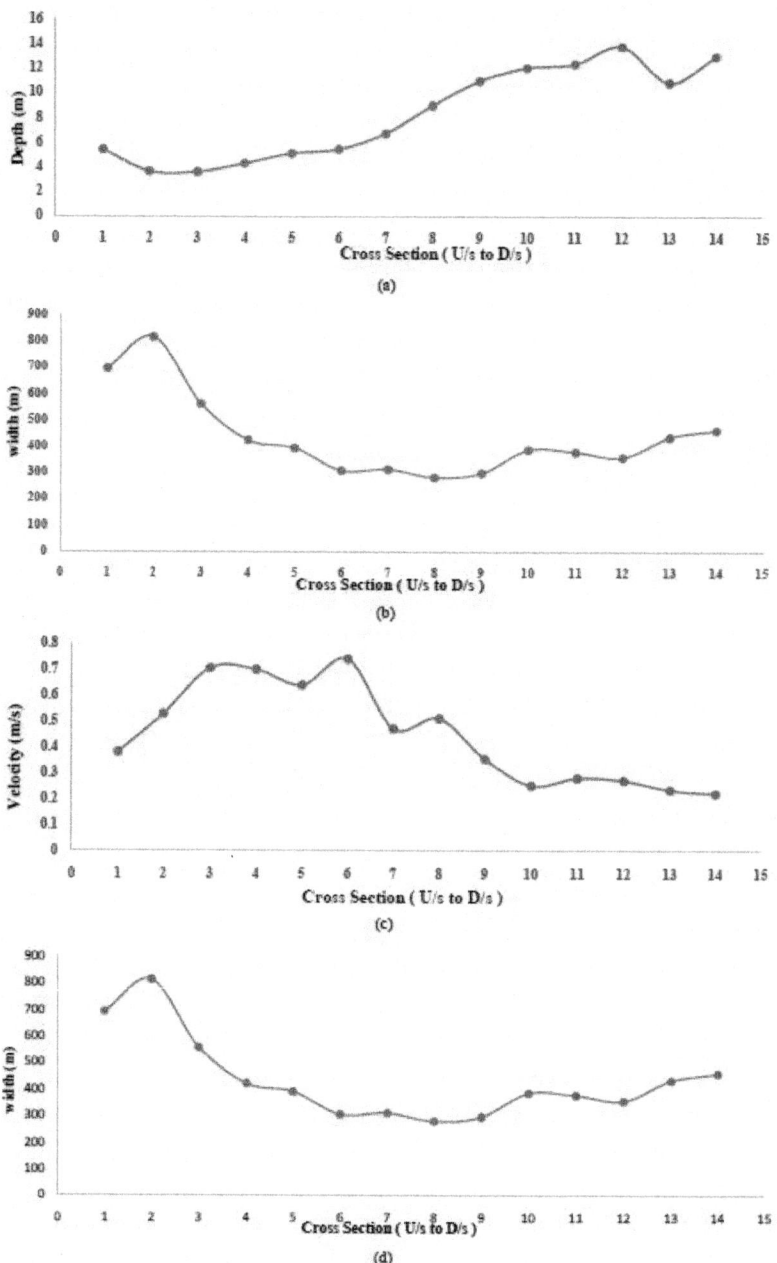

Figure 4: (a) Depth Variation along with cross section from M-14 to M-1; (b) Width variation with cross section from M-14 to M-1; (c) Velocity distribution with cross section from M-14 to M-1; (d) Discharge variation with cross section from M-14 to M-1.

Leaving the starting upstream station (M-14) the width of river is almost having decreasing trend till (M-6) cross section as shown in Figure 4(b). This shows a varying average depth and average width of flow at its different cross section due to its meandering and sinusoidal characteristics.

Generally long profile gradient of river is decreases in downstream due to increasing hydraulic radius (cross section efficiency) but here at Varanasi bend the velocity is increasing up to M-12 cross section after which the inconsistent bend for velocity obtained although from the M-7 cross section the average velocity of flow is continuously decreasing as shown in Figure 4(c). The above theory of increasing velocity in downstream seems to be not valid for River Ganga at Varanasi bend. The decreasing of velocity had resulted by the increasing the depth of flow in downstream consistency. Discharge variation with the cross sections is also shown in Figure 4(d).

Any stream with having changing volume may assume a meandering course, alternatively eroding sediments from the outside of a bend and depositing them on the inside. This meandering characteristic and sinuosity along Ganga river course had showed non uniformity in the river width and uneven depth of water. The main cause of overall these parameters is heavy rainfall from which the runoff in term of river flow depends. The rainfall intensity mainly governs the amount of erosion and the geological parameters decide the deposition of eroded sediment in the river which leads to variation in the geometry of any river. Human activity is also the indisputable cause for high flow which involved of building of impervious structures, deforestation, and caused to the higher surface runoff and decrease the time of concentration by which even on small rainfall leads to change in depth of river flow. On the other hand, forestation such as pine tree and other tree which can increase infiltration so that time of peak can delay and its harmful effects can be reduced. To minimize the impact of surface runoff directly to the river flood mitigations concept should be undertaken and another river training work to be adopted along the Ganga River in order to minimize erosion as well as sedimentation enter to the river.

CONCLUSIONS

The monitoring of discharge and velocity distribution was conducted using consistent protocols designed to ensure the scientific validity of the data. These stream flow datasets will aid in the management of water resources in a sustainable manner for the benefit of water users and the environment.

A multiple linear regression model was developed by using the measured discharge, depth and velocity through ADCP of Varanasi bend of river Ganga. The regression equation shows a high correlation between the discharge, depth and velocity parameters with the R^2 value of 0.8624. Among the validation set of 31 data's, 9 data's of discharge were in the range of 100m³/s and out of which 6 data's gave more errors, as well as the average velocity lies in the range of 0.101 m/s to 0.279 m/s in validation set which gave more errors in validation of model. The proposed model is validated for the average velocity greater than 0.279 m/s up to 0.899 m/s. The developed model also shows variation when the depth of flow is less than 5 m, so this model is suitable for the depth above 5 m up to the maximum of 19.98 m at the Varanasi bend of River Ganga. The equations developed for this study are not applicable for ungaged sites in which the basin characteristics are not in the range of those used to develop the regression equations.

ACKNOWLEDGEMENTS

Authors are thankful to Department of Civil Engineering, Indian Institute of technology (BHU) for providing the infrastructure and computational facilities to complete this work.

REFERENCES

1.   Engeland, K. and Hisdal, H. (2009) A Comparison of Low Flow Estimates in Ungauged Catchments Using Regional Regression and the HBV-Model. Water Resources Management, 23, 2567-2586. http://dx.doi.org/10.1007/s11269-008-9397-7

2.   Eslamian, S., Ghasemizadeh, M., Biabanaki, M. and Talebizadeh, M. (2010) A Principal Component Regression Method for Estimating Low Flow Index. Water Resources Management, 24, 2553-2566. http://dx.doi.org/10.1007/s11269-009-9567-2

3.   Tayfur, G. and Singh, V.P. (2011) Predicting Mean and Bank Full Discharge from Channel Cross Sectional Area by Expert and Regression Methods. Water Resources Management, 25, 1253-1267. http://dx.doi.org/10.1007/s11269-010-9741-6

4.   Zhu, Y.-Y. and Zhou, H.-C. (2009) Rough Fuzzy Inference Model and Its Application in Multi-Factor Medium and Long-Term Hydrological Forecast. Water Resources Management, 23, 493-507. http://dx.doi.org/10.1007/s11269-008-9285-1

5.   Akbari, F.M., Afshar, A. and Sadrabadi, M.R. (2009) Fuzzy Rule Based Models Modification by New Data: Application to Flood Flow Forecasting. Water Resources Management, 23, 2491-2504. http://dx.doi.org/10.1007/s11269-008-9392-z

6.   Chen, C.H., Chou, F.N.-F. and Chen, B.P-.T. (2010) Spatial Information-

Based Back-Propagation Neural Network Modeling for Outflow Estimation of Ungauged Catchment. Water Resources Management, 24, 4175-4197. http://dx.doi.org/10.1007/s11269-010-9652-6

7. Walter, C. and Tullos, D.D. (2010) Downstream Channel Changes after a Small Dam Removal: Using Aerial Photos and Measurement Error for Context; Calapooia River, Oregon. River Research and Applications, 26, 1220-1245.http://dx.doi.org/10.1002/rra.1323

8. Wharton, G., Arnell, N.W., Gregory, K.J. and Gurnell, A.M. (1989) River Discharge Estimated from River Channel Dimensions. Journal of Hydrology, 106, 365-376.http://dx.doi.org/10.1016/0022-1694(89)90080-2

9. Wharton, G. and Tomlinson, J.J. (1999) Flood Discharge Estimation from River Channel Dimensions: Results of Applications in Java, Burundi, Ghana and Tanzania. Hydrological Sciences Journal, 44, 97-111. http://dx.doi.org/10.1080/02626669909492205

10. Bhatt, V.K. and Tiwari, A.K. (2008) Estimation of Peak Stream Flows through Channel Geometry. Hydrological Sciences Journal, 53, 401-408. http://dx.doi.org/10.1623/hysj.53.2.401

11. Andersson, L., Wilk, J., Todd, M.C., Hughes, D.A. and Earle, A. (2006) Impact of Climate Change and Development Scenarios on Flow Patterns in the Okavango River. Journal of Hydrology, 331, 43-57. http://dx.doi.org/10.1016/j.jhydrol.2006.04.039

12. Kamarudin, M.K.A., Toriman, M.E., Mastura, S., Idrisand, M. and Jamil, N.R. (2009) Temporal Variability on Lowland River Sediment Properties and Yield. American Journal of Environmental Sciences, 5, 657-663. http://dx.doi.org/10.3844/ajessp.2009.657.663

13. Jung, I.W., Chang, H. and Moradkhani, H. (2011) Quantifying Uncertainty in Urban Flooding Analysis Considering Hydro-Climatic Projection and Urban Development Effects. Hydrology and Earth System Sciences, 15, 617-633. http://dx.doi.org/10.5194/hess-15-617-2011

14. Hoyle, J., Brooks, A. and Spencer, J. (2012) Modelling Reach-Scale Variability in Sediment Mobility: An Approach for Within-Reach Prioritization of River Rehabilitation Works. River Research and Applications, 28, 609-629. http://dx.doi.org/10.1002/rra.1472

15. Schwendel, A.C., Fuller I.C. and Death, R.G. (2012) Assessing DEM Interpolation Methods for Effective Representation of Upland Stream Morphology for Rapid Appraisal of Bed Stability. River Research and Applications, 28, 567- 584.http://dx.doi.org/10.1002/rra.1475

16. Chitale, S.V. (1970) River Channel Patterns. Journal of Hydraulic

Division American Society Civil Engineering, 96, 201-221.

17. Allen, J.R.L. (1982) Sedimentary Structures, Their Character and Physical Basis. Elsevier Science, New York, 633.

18. Howard, A.D. (1992) Modeling Channel Migration and Floodplain Sedimentation in Meandering Streams. In: Carling, P. and Petts, G.E., Eds., Lowland Floodplain Rivers: Geomorphological Perspectives, John Wiley, Hoboken, 41-41.

Chapter 13

RUNOFF HARVESTING AND STORAGE FOR RICE CROP AT HAMELMALO, SEMIARID REGION OF ERITREA

Ramesh Prasad Tripathi, Woldeselassie Ogbazghi, Semere Amlsom, Simon Measho

Department of Land Resources and Environment, Hamelmalo Agricultural College, Keren

ABSTRACT

Rice is staple food in Eritrea but it is not cultivated in the country due to semiarid conditions. However, possibilities exist for growing rice using runoff produced from nonagricultural hilly lands, which occupy >50% - 80% area of all agricultural watersheds in Eritrea. Study was undertaken in 6 ha watershed at Hamelmalo to design and develop waterway for safe harvesting of runoff from 5.5 ha catchment into a pond for facilitating runoff farming of rice in 0.5 ha field at the outlet end receiving recurrent floods. Slope of the catchment ranged from 1% to 6%. The waterway was designed to intercept and carry runoff from two major drains in nonagricultural land together and delivering into a pond made adjacent to rice field. The waterway was about 323 m long with 3 m top width, 1.5 m bottom width and 0.3 - 0.8 m depth from surface. The pond was 60 m long, 9 m wide and 1 m deep with 1.5 m high earthen dam towards rice field using soil excavated from the pond. Embankment on the remaining sides was mango orchard land slopping towards pond. The dam base width was 4 m, top width was 2.5 m and height was 1.5 m from the ground surface. Two spill ways were provided in the dam at ends at the ground level to facilitate release of runoff from the pond for irrigation. Gross capacity of pond was >1000 m^3. Combined effect of river water, percolation from pond and wetter rice field raised groundwater table from 3.25 m depth in June to 1.4 m by mid-crop season. This resulted in soil wetness exceeding field capacity below 0.7 m depth and greener crop. Rice yields exceeding 2000 kg·ha^{-1} **were harvested under runoff farming conditions. Soil** puddling was more conducive to rice crop than compaction under available soil and water resources.

INTRODUCTION

Rice is not a semiarid region crop because of its high water requirements. However, in Eritrea significant runoff produced from nonagricultural lands, which occupy >50% - 80% area of all agricultural watersheds, can be harvested and used as additional rainwater to facilitate rice farming [1] - [3] . Farmers cultivate crops on un-terraced sloppy lands that also contribute 60% - 70% of rainfall as runoff [2] [4] , which flows forming numerous channels and gullies damaging downstream field crops. In watersheds of size more than 5 ha, runoff frequently floods the downstream fields at the outlet rendering them unfit for raising a good crop during rainy season [3] . The same runoff could be boon to rice production if diverted through designed waterways into small ponds constructed on upstream side of such fields [3] . Crop yields in Eritrea are low (<0.2 - 0.6 t·ha^{-1}) because >70% of the total rainfall is lost as runoff from traditional sloppy fields [1] [2] [5] . Storage of even part of the runoff in watersheds, which is also rich in nutrients and soil colloidal fractions, can serve as not only extra rainwater but also nutrients for crops [3] [6] - [8] . Rice cultivation through water harvesting in terraced fields was successfully demonstrated in Uganda [9] . Although rice likes wetter regime in the root zone, irrigations applied at 8 days interval were also optimum for rice growth in Sudan [10] .

Rice needs water to meet its evapotranspiration (ET) requirements and to satisfy percolation from the soil moisture regime of field capacity to submergence necessary for optimum yields [11] [12] . High percolation from rice fields and ponds made for harvesting runoff to meet water requirements of rice will also recharge ground water table that may serve as source of sub-irrigation directly into the crop root zone [3] [13] . Tesfamichael [14] observed development of water table that fluctuated from 0.4 - 1.2 m depth from surface during October 2006 to March 2007 in about 6 ha valley farmland in Akriya, Asmara region of Eritrea, surrounded by 36.5 ha hilly terrain. Contribution of groundwater table to ET requirements of wheat was 90% from the water table fluctuating from 0.4 - 0.55 m depth from surface, which declined to 38% as water table dropped to 1.2 m depth by milk stage [14] .

Runoff farming has been traditionally practiced as spate irrigation in Eritrea since more than 100 years ago [7] . Spate irrigation is a pre-planting system of irrigation by diverting seasonal rivers producing flash floods from highlands and mountainous areas to recharge soil profile in the lowlands. More than 50% of total irrigation in Eritrea is through spate irrigation practiced from eastern to western lowlands along the Red sea coast [15] [16] . Depending upon rainfall, catchment size and runoff from highlands, diversion dams are constructed in the rivers to channelize runoff into cultivated fields to recharge soil profile in several flash floods before planting a crop. Water use efficiency of spate irrigation systems in Eritrea has been low due to wild flooding of uncropped fields leading to over or inadequate watering [7] [16] . All spate irrigated fields in Eritrea could prove boon to rice production with high water use efficiency through improvements in existing diversion structures to regulate flood water supply and field level management. Rice could be grown during preplanting irrigation period from the third week of June to October with little intervention in the existing cropping system and flood water management. Objective of this study was to design and develop runoff harvesting system in agricultural watersheds associated with nonagricultural lands to facilitate runoff farming of rice in semiarid environments of Hamelmalo, Anseba region of Eritrea.

MATERIALS AND METHODS

Study was undertaken in 2013 and 2014 in a 6 ha partly cultivated watershed at Hamelmalo Agricultural College (15°52'20.6"N and 38°27'57.6"E at 1280 msl), located in the semiarid region of Eritrea. Annual rainfall in the past seven years ranged from 370 - 663.1 mm with a mean of 488 mm and average annual pan evaporation of 1931 mm. Highest mean monthly temperature occurred in May (35.7°C) and lowest in January (11.1°C). Total rainfall was 388 mm in 2013 and 429 mm in 2014 (Figure 1). Monsoon season in the two cropping years started from third week of June and ended by third week of September. Highest storm rainfall was 57.7 mm in July, 2013 and 48.5 mm in September, 2014.

Watershed Characteristics

The watershed land was surveyed using total station and GPS to determine slope, contour lines, ridge lines, boundary, area and location of natural drains. About 2.5 ha watershed land was cultivated and was on the downstream side and 3.5 ha uncultivated. The uncultivated land comprised of wild bushes and trees, dominated by

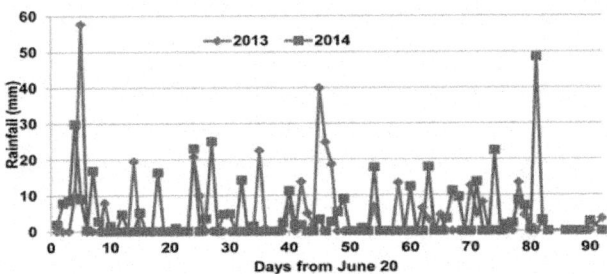

Figure 1: Rainfall during the crop season of 2013 and 2014.

Acacia and Zizyphus species and 4 buildings, each having cemented plan area of 800 m², together with an old mango orchard and a small building (40 m²) adjacent to the selected rice field. Runoff from buildings was 100% of the rainfall. Soil texture was stony sandy loam in uncultivated portion, sandy loam in cultivated portion and loam in the rice field. Slope of the catchment contributing runoff for use by rice crop ranged from 1% - 6% and that of rice field was <1%. Average bulk density of soil was 1.69 Mg·m⁻³ in the uncultivated field and 1.4 Mg m⁻³ in the rice field. Presence of stones, beating action of raindrops and trampling by grazing animals over the years increased density of the uncultivated land. Two major drains were carrying runoff from uncultivated land on the upstream and leading to gully formation in downstream cropped fields before entering into Anseba river through the selected rice field. Anseba river is normally dry but flows whenever rainfall occurs in its catchment.

Design Considerations

Waterway was designed to intercept and carry runoff from the two drains of the uncultivated land together about 3 m upstream of the cultivated land into a pond constructed adjacent to the rice field (Figure 2). The two drains at the junction of cultivated and uncultivated lands were about 10 m wide and 40 m apart. Slope of the land, about 3 m upstream from the cultivated land, was <1% - 1.5%. The waterway was thus proposed from 3 m on upstream side of the cultivated land running about 50 m across the slope dividing the cultivated and uncultivated land before turning downslope towards the pond. Minor bending was allowed to accommodate local land surface conditions. To ensure non-erosive velocity of flow, channel bed slope was maintained zero by providing successive drops in bed level at the points of sudden change in slope. Equation (1) guided vertical interval between the successive channels [17] .

$$VI = 0.12\,s + 0.3 \tag{1}$$

where VI is vertical interval (m) between channel beds, and s is per cent slope.

Flow in the waterway between successive drops was due to hydraulic gradient established by runoff. Length of each level section of the waterway and drop height was variable. Stones and gravels were placed at the drop section of the waterway to reduce scouring. Due to lack of data on rainfall intensity and duration, waterway and pond dimensions were decided from rainfall-runoff relationship developed from measurements made in another 4.29 ha model watershed at Hamelmalo farm under similar soil and vegetation conditions [2] and peak storm rainfall in the last 10 years. Peak storm rainfall in 2010 at Hamelmalo was 77.2 mm, therefore, waterway and pond were designed for the runoff that can be produced from a peak storm rainfall of 80 mm. Most storm rainfall amounts in the area are <30 mm. Following was the rainfall-runoff relationship on 3% slope used in the design.

$$\text{Runoff} = 0.52 \times \text{rainfall} - 2.0,$$ \hfill (2)

Both rainfall and runoff are in mm. The waterway length was 323 m, top width 3 m, bottom width 1.5 m and depth 0.3 - 0.8 m from surface. The pond length was 60 m, width 9 m and depth 1 m with 1.5 m high earthen dam on the downstream side adjacent to the rice field. Soil excavated from the pond was acceptable for construction of the dam. Embankment on the other three sides was not disturbed to allow natural local inflow of runoff from the adjoining mango orchard. Base width of the dam was 4 m, top width 2.5 m and height 1.5 m from the ground surface. Top width of 2.5 m was to facilitate thorough passage of farm machineries and ensure stability of the dam. Provision of two spill ways was made at ends of the dam at the ground level to allow irrigation.

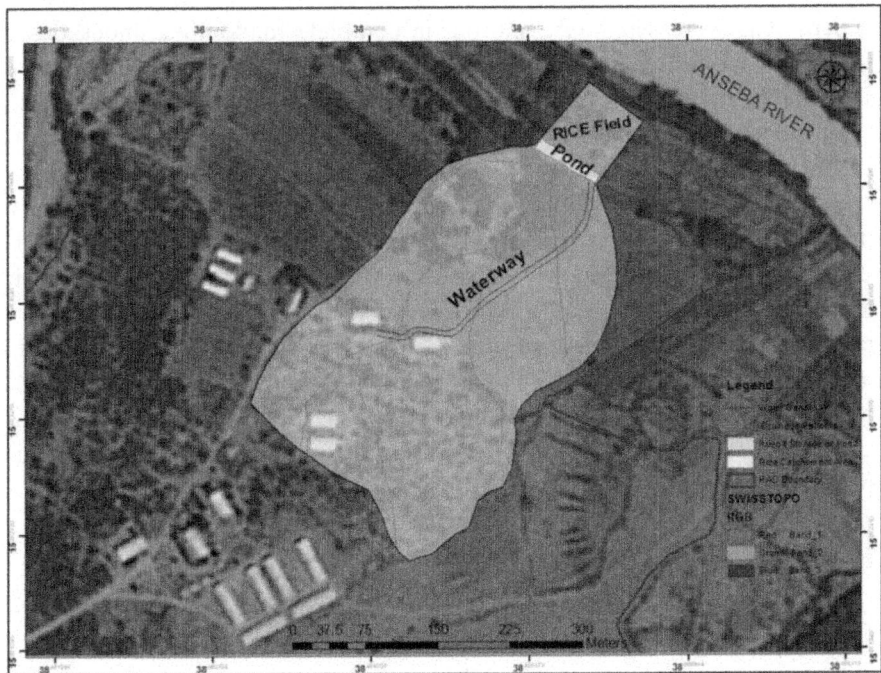

Figure 2: Rice field and catchment area contributing runoff.

and release of extra runoff from the pond. Gross capacity of the pond was >1000 m^3. Location of buildings, natural drains, Anseba River, rice field and designed waterway can be seen in Figure 2.

Crop Response Evaluation

Effect of water harvesting on runoff farming of rice was evaluated through field experiments in 2013 and 2014. Since percolation is a major water management problem in rice farming, emphasis on treatment selections was to optimize level of soil puddling or compaction done to control percolation without affecting rice growth. Rice variety NERICA 4 was transplanted at 5 levels of puddling and 5 levels of compaction. Rice seedlings were raised 21 days before transplanting in the field. Puddling treatments were no puddling (T_0), puddling by one (T_1), two (T_2), three (T_3) and four (T_4) passes of puddler replicated 4 times [18] . To simulate puddling by local bullock drawn plough, wet tillage in the plots was done manually by spade followed by churning-cum-leveling by small wooden plank. Soil compaction treatments were no compaction (T_0), compaction by one (T_1), two (T_2), three (T_3) and four (T_4) passes of 600 kg tractor driven roller in 4 replications [19] . Both the experiments were separately laid out in complete randomized block design. Irrigations were given to maintain 20 - 50 mm submergence for 10 days for seedling establishment after transplanting in puddled and compacted fields. Fertilizer nutrients applied were 120 kg N and 46 kg P ha^{-1} through urea and DAP. Entire DAP was applied as basal dose during last puddling operation and remaining N was applied through urea in two equal splits at 20 days interval from transplanting. Rice is normally broadcasted under rainfed farming but transplanting was preferred to shorten crop duration in the field by raising nursery 21 days before commencement of the monsoon. After cessation of rainfall in September, water requirement of rice was much higher than 120 mm irrigation applied in 6 m × 1 m nursery plots. Submergence by rainfall or runoff was allowed to a maximum of 0.1 m. Results of 2013 experiment showed superiority of puddling on crop performance and, therefore, only puddling experiment was repeated in 2014.

RESULTS AND DISCUSSION

Construction of Waterway and Pond

As per design, waterway was constructed to intercept and carry runoff from 3 m upstream of the cultivated land, running 50 m across the slope with minor bending to accommodate local land surface situations before turning downslope towards the pond (Figure 3). The waterway downslope followed one of the drains almost centrally dividing the lower catchment. Soil excavated from the first 50 m waterway running across the slope was placed only on the downstream side of the embankment (Figure 3). Embankment on the upstream was natural ground level to facilitate runoff interception. However, embankments were formed on both sides after turning downslope until next 120 m. Thereafter slope of the cultivated land on the right side was perpendicular to the waterway and, therefore, embankment was again made only on the left bank to facilitate direct runoff interception from the right side. The waterway depth was varying with slope. Entire runoff from the cultivated land joined the waterway in its lower 1/4 section. Provision for inflow of runoff from left side of the waterway was also made at appropriate locations.

The pond of designed size (60 m × 9 m × 1 m) was constructed adjacent to rice field on the upstream side. Pond embankment (dam) adjacent to rice field was of base width 4 m, top width 2.5 m and height 1.5 m from ground surface. The remaining 3 sides were undisturbed mango orchard land sloping towards the pond, which facilitated not only direct inflow of local runoff into the pond but also increased pond capacity with rising water level above the land surface. Soil excavated from the pond was used for construction of the dam after removing gravels and stones. Two orifices of diameter 0.125 m were installed at ends of the dam for safe release of excess runoff from pond into 1 m wide (base width) field channel constructed on the two sides of the rice field. The dam was repeatedly compacted by moving a 600 kg tractor driven roller. Thorough passage of tractor and human traffic was allowed over the dam for its safety and proper use of the farmland. Entire work was done manually by students (Figure 4). In case of emergency, provision for natural spillway was also made on eastern side of the embankment from where overflowing runoff could be diverted into another safety channel of 1.5 m base width made along irrigation channel in the rice field. Natural inflow of runoff from sides was also intercepted and stored in this safety channel before allowing into irrigation channels or rice plots.

Figure 3: Waterway to intercept runoff.

Figure 4: Students digging the pond.

Performance of Waterway and Pond

The waterway has been performing well for the last 3 years. Soil texture of the rice field was slowly transforming from loam to sandy loam and loamy sand in lower depths below 0.8 - 1 m. Pond was thus draining quickly after every rainfall. In 2013, pond received runoff to almost its full capacity twice but in 2014, it received only once due to lower storm rainfall. However, waterway, pond and rice field channels assured additional water availability to the crop and zero runoff from the watershed. The incoming colloidal sediments with runoff in the pond are expected to clog soil pores with time allowing water to stay for longer time. Part of the sediments received in the pond could also be used as manure in the rice field or elsewhere as necessary.

Development of Water Table and Soil Wetness

Seepage from Anseba river, runoff stored in the pond and irrigation channels and wetness in the rice plots raised groundwater table in the rice field from 3.25 m depth in June end to 1.4 m by 4th week of August, which receded down to 1.7 m by crop maturity (Figure 5). Rate of rise of water table was faster than receding. The water table rose by 1.85 m in about 48 days but receded only 0.3 m in next 75 days. The water table was oscillating within 1.5 ± 0.1 m for about 2 months during grand growth to reproductive stages of rice crop.

The shallow groundwater table greatly affected wetness in the root zone. Soil water content in the third week of September showed upward gradient (Figure 6). Whereas surface soil was dry forming cracks, soil wetness below 0.7 m depth was near field capacity and was increasing with depth (Figure 6). Roots of NERICA rice were observed down to 0.8 m [3] and, therefore, parts of the rice roots were receiving water by capillarity from the fluctuating water table and the crop was greener until October (Figure 7). The rising groundwater table due to combined effect of water level in the river and percolation from the pond and rice field may serve as a natural source of sub-irrigation to the rice crop. The crop was harvested in November first week.

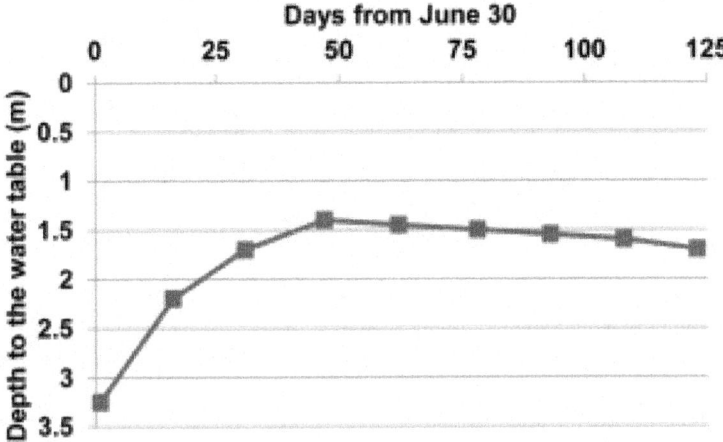

Figure 5: Two-year average water table fluctuation during the crop season.

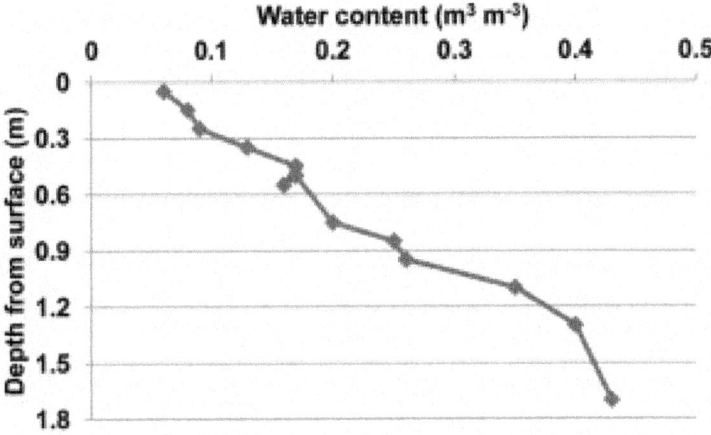

Figure 6: Soil moisture profile in the rice field.

Figure 7: Rice experimental crop on October 25, 2014.

Performance of Rainfed Rice

Rice yields in 2013 ranged from 782 kg·ha^{-1} under no puddling to 2307 kg·ha^{-1} under 4 puddlings (Table 1). Effect of 2 or 4 puddlings was statistically equal on crop performance. In 2014, yields increased from 850 kg·ha^{-1} under no puddling to 2438 kg·ha^{-1} under 2 puddlings. Yields in second year were also at par under 2 and 4 puddlings. Results thus show that 2 puddling operations by country plow would be sufficient for optimum rice yields in the loam soil of Hamelmalo under runoff farming. Similarly compaction, which had similar effect on percolation, improved rice yields from 690 kg·ha^{-1} under no compaction to 1980 kg·ha^{-1} under compaction by 4 passes of 600 kg roller. Yields were at par under 2 and 4 passes of roller. Results show that puddling was superior

Table 1: Effect of puddling and compaction on grain yield of NERICA 4 rice

Puddling by spade	Grain yield in 2013 (kg·ha^{-1})	Grain yield in 2014 (kg·ha^{-1})	Compaction by 600 kg roller	Grain yield in 2013 (kg·ha^{-1})
T$_0$ (No puddling)	782	850	T$_0$ (no compaction)	690
T$_1$ (1 puddling)	1181	1488	T$_1$ (1 pass roller)	1234
T$_2$ (2 puddling)	1811	2438	T$_2$ (2 pass roller)	1430
T$_3$ (3 puddling)	1921	2313	T$_3$ (3 pass roller)	1550
T$_4$ (4 puddling)	2307	2063	T$_4$ (4 pass roller)	1980
Mean	1600	1830	Mean	1369
LSD at 5%	552	572	LSD at 5%	605

to compaction in terms of crop performance under runoff farming conditions of Hamelmalo. These yields are as against highly poor sorghum stands due to recurrent floods and consequent wetness and deposition of eroded materials in such fields. Since successful rice crop has been harvested for the first time in Eritrea under runoff farming conditions more multilocational trails are necessary to optimize soil, water and nutrient requirements in addition to development of less than 4 month duration NERICA varieties suitable for different rainwater management conditions.

CONCLUSIONS

- Water harvested from >50% - 80% nonagricultural lands of all agricultural watersheds in Eritrea can significantly meet water requirements of rice.
- Earthen waterway and pond constructed to harvest runoff would be viable in arresting further land degradation.
- Rice yields exceeding 2000 kg·ha^{-1} could be harvested under runoff farming conditions.
- Soil puddling twice by local plow would be better option in loam soil but it should be verified for coarser texture soils.
- River or any natural drain on the downstream rice field and upstream pond could raise groundwater table significantly to serve as source of sub-irrigation directly into root zone.
- Efforts on rice cultivation could save hard currency to import rice for local consumption.

ACKNOWLEDGEMENTS

Authors are grateful to UNDP for providing funds from SGP/GEF grant for carrying out this research and to Mr. Tseneo Tsrusaki, Japanese International Cooperation Agency for supplying NERICA rice seed.

REFERENCES

1. MOA (Ministry of Agriculture) (2005) Area and Production by Zoba from 1992-2005. MOA, Asmara, Eritrea.

2. Tripathi, R.P. and Ogbazghi, W. (2010) Development and Management of a Hilly Watershed in Hamelmalo Agricultural College farm, as a Demonstration Site for Farmers and a Study Site for Students. Final Technical Report of the Project Financed by Eastern and Southern Africa Partnership Programme (ESAAP), Department of Land Resources and Environment, Hamelmalo Agricultural College, Keren, Eritrea, 59 p.

3. Tripathi, R.P., Ogbazghi, W., Amlesom, S. and Araia, W. (2014) Optimizing Tillage and Rain Water Conservation in the Soils of Hamelmalo Region of Eritrea for Arresting Soil Degradation and Achieving Sustainable High Crop Yields. Final Technical Report of the Project Financed by GEF/SGP, UNDP, Department of Land Resources and Environment, Hamelmalo Agricultural College, Keren, Eritrea, 112 p.

4. Temesgen, M., Rockstrom J., Savenije, H.H.C., Hoogmoed, W.B. and Alemu, D. (2008) Determination of Tillage Frequency among Smallholder Farmers in Two Semiarid Areas in Ethiopia. Physics and Chemistry of the Earth, 33, 183-191. http://dx.doi.org/10.1016/j.pce.2007.04.012

5. FAO (2005) Global Information and Early Warning System on Food and Agriculture World Food Programme. Special report FAO/WFP Crop and Food Supply Assessment Mission to Eritrea. FAO, Rome.

6. Elwell, H.A. and Stocking, M.A. (1988) Loss of Soil Nutrients by Sheet Erosion Is a Major Hidden Farming Cost. The Zimbabwe Science News, 22, 79-82.

7. Tesfai, M. (2001) Soil and Water Management in Spate Iirrigation Systems in Eritrea. Ph.D. Thesis, Department of Erosion and Soil and Water Conservation, Wageningen University and Research Center, The Netherlands, 211 p.

8. Ashouri, M. (2012) The Effect of Water Saving Irrigation and Nitrogen Fertilizer on Rice Production in Paddy Fields of Iran. International Journal of Bioscience, Biochemistry and Bioinformatics, 2, 56-59. http://dx.doi.org/10.7763/IJBBB.2012.V2.70

9. Goto, A., Nishimaki, R., Suzuki, S., Watanabe, F. and Takahashi, S. (2012) Terrace Development Applied as a Water Harvesting Technology for Stable NERICA Production in Uganda. Journal of Arid Land Studies, 22, 243-246.

10. Abu, S.T. and Malgwi, W.B. (2012) Effects of Irrigation Regime and Frequency on Soil Physical Quality, Water Use Efficiency, Water Productivity and Economic Returns of Paddy Rice. ARPN Journal of Agricultural and Biological Science, 7, 86-99.

11. Tripathi, R.P., Kushwaha, H.S. and Mishra, R.K. (1986) Irrigation Requirement of Rice under Shallow Water Table Conditions. Agricultural Water Management, 12, 127-136. http://dx.doi.org/10.1016/0378-3774(86)90011-9

12. Bajpai, R.K. and Tripathi, R.P. (2000) Evaluation of Non-Puddling under Shallow Water Tables and Alternative Tillage Methods on Soil and Crop Parameters in a Rice-Wheat System in Uttar Pradesh. Soil and Tillage

Research, 55, 99-106. http://dx.doi.org/10.1016/S0167-1987(00)00111-2

13. Tripathi, R.P., Kushwaha, H.S. and Agrawal, A. (1987) A Simple Non-Weighing LysimeterInstallation with Rain Shelter. Agricultural and Forest Meteorology, 41, 275-288.http://dx.doi.org/10.1016/0168-1923(87)90084-0

14. Tesfamichael, F. (2007) Integrated Effect of Irrigation and Nitrogen on Wheat Productivity under Shallow Water Table Condition in Eritrea. M.Sc. Thesis, Department of Land Resources and Environment, Hamelmalo Agricultural College, Eritrea, 91 p.

15. Tesfai, M. and Stroosnijder, L. (2000) The Eritrean Spate Irrigation System. Agricultural Water Management, 48, 512-560.

16. Amlesom, S. (2005) Review of Irrigation Development in Eritrea. In: Mehari, T. and Ghebru, B., Eds., Irrigation Development in Eritrea: Potentials and Constraints, Proceedings of the Workshop of the Association of Eritreans in Agricultural Sciences (AEAS) and the Sustainable Land Management Programme (SLM), Eritrea, 14-15 August 2003, 5-11.

17. Abraham, K., Dawit, F., Kibrom, B., Natsnet, O. and Selemawit, W. (2014) Effect of Puddling on Soil Properties and Yield of Rice. Senior Research Project, Department of Land Resources and Environment, Hamelmalo Agricultural College, Eritrea, 40 p.

18. Asmelash, M., Filmon, D., Amanuel, G., Selam, E. and Natsnet, K. (2014) Effect of Compaction on Soil Properties and Yield of Rice. Senior Reseach Project, Department of Land Resources and Environment, Hamelmalo Agricultural College, Eritrea, 50 p.

19. ASAE (American Society of Agricultural Engineers) (1989) Design, Layout, Construction and Maintenance of Terrace Systems. ASAE Standard S268.3, St. Joseph, MI.

Chapter 14

FACTOR-CLUSTER ANALYSIS AND EFFECT OF PARTICLE SIZE ON TOTAL RECOVERABLE METAL CONCENTRATION IN SEDIMENTS OF THE LOWER TENNESSEE RIVER BASIN

Paul S. Okweye[1], Karnita G. Garner[2], Anthony S. Overton[2], Elica M. Moss[2]

[1]College of Engineering, Technology & Physical Sciences, Department of Physics, Chemistry and Mathematics, Alabama A&M University, Normal, AL, USA

[2]College of Agricultural, Life and Natural Sciences, Department of Biology and Environmental Sciences, Alabama A&M University, Normal, AL, USA

ABSTRACT

Total recoverable concentration of five elements of concern: Aluminum, Iron, Manganese, Arsenic and Lead (Al, Fe, Mn, As, Pb) were measured by inductively coupled plasma atomic emission spectrometry, and mass spectrometry. The results show that sediment texture plays a controlling role in the concentrations and their spatial distribution. Principal Component Analysis and Cluster Analysis were used to analyze the grain sizes of the sediments. Result of texture analysis classified the samples into three main components in percentages: sand, silt, and clay. Significant differences among the element concentrations in the three groups were observed, and the concentrations of the elements in each group are reported in this study. Most of the elements have their highest concentrations in the fine-grained samples with clay playing an important role, in comparison with the sand component of the soil/sediment samples. There appears to be a strong correlation between samples with high silt, and clay content with the areas of elevated concentrations for Al, Fe, and Mn. There was a strong correlation between aluminum and lead with clay; lead with silt; and sand with manganese, aluminum, and lead. However, there was no strong relationship between the soil textures and iron or arsenic. All elements measured were statistically significant (at $P \leq 0.05$) by watershed. The upland areas, and depositional areas' spatial variation of element concentrations in the sediments were also observed, which was in line with the spatial distribution of the grain size and was thought to be related to the watersheds hydrological dynamics.

INTRODUCTION

Lower Tennessee River basin in Alabama includes the Flint Creek (FC) and Flint River (FR) watersheds, and the spatial distribution of the grain size of sediments at these watersheds is largely fine grained in the upper, middle to lower reaches of the rivers. To study the grain size effect on the total recoverable metal concentrations in sediments, samples within, and along the river, sites were collected in winter/spring of 2014. Sediments in riverbeds serve as depositories for most aquatic pollutants, including heavy metals. Hence sediments are considered to be an important indicator for environmental pollution [1] . Sediments act as sinks and sources of contaminants in aquatic systems because of their variable physical and chemical properties [2] . Heavy metals in sediments in the Flint Creek (FC) and the Flint River (FR) watersheds in the Tennessee River (TR) basin have received very little scientific attention. Okweye et al. [3] concluded that the surface water from the FC and the FR watersheds in the TR basin had been polluted with heavy metals from anthropogenic sources surrounding the watersheds. Furthermore, the concentrations of heavy metals in the soil/sediment of these watersheds exceeded the maximum contaminant level(MCL) allowed by USEPA in drinking water. This paper examines the concentrations of heavy metals (Al, Fe, Mn, As, and Pb) in relation to sediment grain size distribution on the watersheds. Particle size is important because the grain size of soil particles and their aggregate structures affect the ability of soil to transport and retain water and nutrients.

The purpose of this study was: 1) to determine the distribution of the particle size of the soil/sediment and heavy metals (Al, Fe, Mn, As, and Pb) at depositional and upland areas of each site; and 2) to identify the controlling factors by using Principal Component Analysis(PCA) and Cluster Analysis(CA) because multivariate analyses are useful for interpreting elements in spatial patterns, which might be related to similar input sources or transport pathways [4] .

The study tested the null hypothesis that soil/sediment particle size distribution and sampling location influenced the level of heavy metal concentration within the watersheds. The results from this study are expected to provide a framework for interpreting sediment toxicity and group elements with similar properties at these watersheds.

MATERIALS AND METHOD

Study Area and location of sampling Sites

Table 1: Showing the Flint Creek (FC) and Flint River (FR) Geographic Point Coordinates (GIS)

FR	Latitude	Longitude
WR-FR	N34°49'25.932"	W086°28'59.081"
BF-FR	N34°49'25.915"	W086°29'6.778"
HR-FR	N34°32'3.822"	W086°29'59.782"
FC	Latitude	Longitude
RB-FC	N34°30'30.264"	W086°57'30.264"
MB-FC	N34°27'49.038"	W086°58'41.844"
VB-FC	N34°29'50.894"	W087°01'3.679"

FR (codes: WR-FR = Winchester Road, BF-FR = Briar Fork, HR-FR = Hobbs Road); FC (Codes: RB-FC = Red Bank, MB-FC = Means Bridge, VB-FC = Vaughn Bridge).

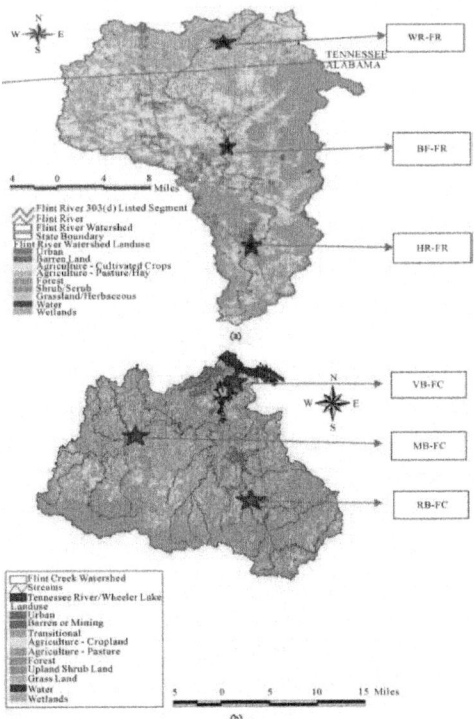

Figure 1: (a) Sampling sites: at the Flint River Watershed (Okweye, P., PhD dissertation, 2009); (b) Sampling sites: at the Flint Creek Watershed (Okweye, P., PhD dissertation, 2009).

Table 2. Soil Types for the FC and FR Watersheds (Courtesy: Joe Gardinski, NRCS)

FR	Code	Soil Types
	WR-FR	Ennis silt loam
	BF-FR	Bodine cherty silt loam
	HR-FR	Melvin silty clay loam
FC	**Code**	**Soil Types**
	RB-FC	Bruno loamy fine sand
	MB-FC	Lindside silty clay loam
	VB-FC	Muskingum stony fine sandy loam

Sample collection

Soil/sediment samples were collected in winter/spring 2014 from six sites (WR-FR, BF-FR, HR-FR, RB-FC, MB-FC, and VB-FC) as shown in Figure 1(a); Figure 1(b) and Table 1. Samples were transferred into sampling bags and placed in a cooler at 4°C, and then transported to Alabama A&M University laboratories for storage. About 1 kg of sample was obtained from each site and air dried before analysis. All the sampling locations were recorded with a GPS. Most of the samples were fine-grained clay and silt, and passed through a 1 mm metal sieve.

The soil/sediment sampling locations for each site included: 1) an in-stream/ Depositional area, 2) a Bank-side, and 3) an Upland in riparian zone. At each location, five samples were collected, composited, and well-mixed to obtain a representative sample (Figure 2). A stainless steel soil probe was used to collect soils from the banks and upland areas, and a 250-cm pole sediment sampler— Pakar [5] was used for collecting sediments from in- stream/depositional areas across the upper, middle, and lower sections of the streams, covering a distance of ~110 km. The sediment sampling was carried out in low flow conditions because trace metal concentrations tend to be highest during this period as metals accumulate in the sediments from water. Under high water discharge, erosion of riverbeds takes place. According to the UNEP/WHO [6], following peak discharge, the concentrations of metals in bed sediments increase as the water flow again decreases. A total of seventy-two (72) soil/ sediment samples were collected from the watersheds. In the laboratory, samples were stored in the freezer until they were processed and analyzed. In-situ measurements for physical and chemical characteristics of water and soil properties were conducted at the sites with a 6600 Extended Deployment System (EDS).

Analytical methods

USEPA Method 3050B [7] was used for the digestion of heavy metals in soil/sediment samples at the Environmental Testing and Consulting (ETC) laboratories in Memphis, Tennessee. This method is suitable for hot block digestion of soil, sediment, and waste samples and for analysis by inductively coupled plasma optical emission/atomic emission spectrometry (ICP-OES/AES). A 1.0 g subsample of each sample was digested in nitric acid and hydrogen peroxide. The digestate was then further refluxed with hydrochloric acid. High-tech instrument ICP-AES, with SW-846 USEPA method 6010B, was used for the elemental determination. The USEPA standard sediment samples were used to monitor the analyses. Al, Fe, Mn, and Pb was measured by inductively coupled plasma optical emission spectrometry (ICP-OES, Optima 2000DV; Perkin Elmer, Waltham, MA, USA; detection limit 0.001 - 0.030 mg/L) and inductively coupled plasma-mass spectrometry (ICP-MS, 7500a; Agilent Technologies, Santa Clara, CA, USA; detection limit 0.015 - 0.120 mg/L) was used for the analysis for Arsenic (As). Laboratory quality control consisted of analysis of sediment reference material (GBW 07302- 07312a; National Institute of Standard and Technology, MD, USA) and triplicate samples were used. The results were consistent with the reference values, and the differences were generally within 10% (most were within 5%).

The recoveries all fell within the range of 90% - 110%, and the relative standard deviation was less than 5%. All The reagents used for the analysis were AR grade and double distilled water was used for preparation of solutions. The analyzing laboratory - ETC asserted that the results from the average values of the concentrations of the elements detected by both ICP-OES/AES (Al, Fe, Mn, and Pb) and ICP-MS (As) were consistent. The quality control in this study was similar to that used in a previous study in the area [3] . ICP-MS was used for arsenic

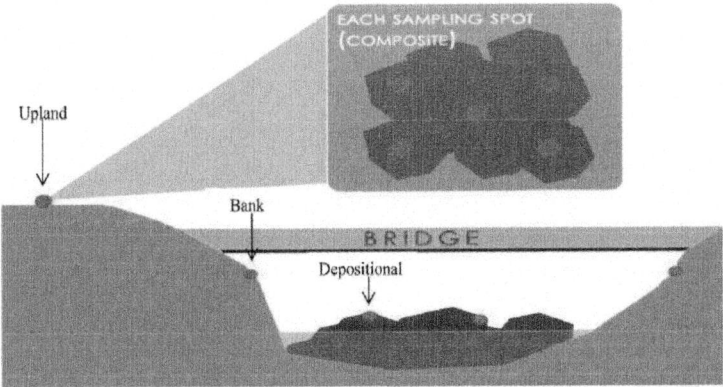

Figure 2: Diagram of a sampling site and spots sampled.

analysis in this research because it offers multi-element detection limits below parts per billion (ppb; 10^{-9}), sometimes down to parts per trillion (ppt; 10^{-12}), and can give a rapid throughput of samples [8] -[10] .

Soil/Sediment Texture Analysis

The relative proportions of sand (particle size between 0.06 and 2 mm), silt (0.002 and 0.06 mm), and clay (< 0.002 mm) in the soil/sediment samples were determined by means of the classical sieve-pipette technique. This method involves sieving of the sand fractions and pipette extraction of the clay fractions using settling tubes. The results may be due to the soil types of the various sites (Table 2)

DATA ANALYSIS

Results and Discussion

All the locations in both FC and FR watersheds had high percentages of the fine particle (clay) size and low percentages of the coarse (sand) particle sizes (Table 3). The samples were probably located at sites where there were low currents in the rivers during run-off periods and where only the fine sizes of the sediment were deposited and retained. Texture was not uniformly distributed along the sites; this may be due to soil types (Table 2); FC locations had lower percentages of silt particle sizes than FR locations (Table 3).

Table 3: Spatial distribution of particle sizes measured in depositional and upland areas of the FC and FR watersheds

Locations (Depositional)	Particle Sizes (%)			Locations (Upland)	Particle Sizes (%)		
	Sand	Silt	Clay		Sand	Silt	Clay
WR-FR-D	17.72	21.48	60.80	WR-FR-U	19.73	22.17	58.11
BF-FR-D	15.79	17.30	66.91	BF-FR-U	14.99	37.94	47.08
HR-FR-D	21.26	17.62	61.12	HR-FR-U	26.91	8.94	64.15
RB-FC-D	9.48	1.90	88.62	RB-FC-U	15.89	8.16	75.94
MB-FC-D	19.33	11.96	68.71	MB-FC-U	26.95	1.63	71.43
VB-FC-D	25.06	9.98	64.96	VB-FC-U	30.91	1.08	59.01

D = dispositional area, U = upland area.

Particle Size Effect

Table 4: Coefficient of determination (r²), SAS proc corr

Particle Sizes	Heavy Metals				
	Al	Fe	Mn	As	Pb
Sand	+ve ns	−ve *	−ve ns	−ve Ns	+ve ns
Silt	+ve ns	+ve **	+ve ns	+ve **	+ve ns
Clay	−ve ns	−ve Ns	+ve ns	−ve *	−ve ns

(a) +ve and −ve: positive and negative correlations: [*(orange), and **(red): significant and highly significant at 0.01 and 0.05 levels]; (b) ns: not significant.

Studies have shown that one of the most significant parameters influencing total recoverable metal levels in sediments is particle dimension [1] . Bio-available sediment-bound metals such as Al, Fe, Mn, As, and Pb, depend to a significant extent on the particle size fraction with which a metal is associated. This study showed that the highest concentrations of metals were associated with fine grained sediment particles. Table 4 shows that Fe was significant, but negative correlations with sand, and positive but highly significant correlations with silt. Arsenic (As) was highly significant. There were positive correlations with silt and negative but significant correlations with clay.

In the FC and the FR watersheds, the spatial distribution of the particle size of sediments was coarser from upland to depositional (Table 3). The results showed that texture played a controlling role on the concentrations and their spatial distribution. Principal component analysis and cluster analysis were carried out based on the particle sizes of the sediments, and the samples were classified into three fractions: sand, silt and clay. Significant differences among the element concentrations in the three groups were observed, and the concentrations of the elements in each group were reported in this study (Table 5). Most of the elements had their highest concentrations at the sites with high, fine particle (silt and clay) samples; in comparison with the sand samples, with clay minerals possibly playing an important role. The heavy metals being mainly present in the clay-silt particles with particle sizes less than 63 μm (<0.06 mm), may be due to the increase in specific surface properties of this fraction. An upland to depositional spatial variation of element concentrations in the sediments was observed. This is in line with the spatial distribution of the particle size and may be due to the water hydrological dynamics in the rivers.

Correlation Coefficients

The Factor Analysis for the data was developed in three stages:

Table 5: Average of heavy metals in combined soil/sediment samples

Heavy Metals in Combined Soil/Sediment (n = 72), µg/kg					
Sites	Al	Fe	Mn	As	Pb
		WR-FR			
Ave	10,342	29,625.0	1004.75	9.01	14.45
		BF-FR			
Ave	11,240	17,425.0	735.25	8.27	27.93
		HR-FR			
Ave	21,000	19,900.1	1463.75	6.67	20.75
		RB-FC			
Ave	11,452	12,442.5	862.9	3.93	13.29
		MB-FC			
Ave	10,148	10,725.0	9055.0	3.96	21.78
		VB-FC			
Ave	6622	7745.0	504.08	5.55	17.07

- A correlation matrix was generated for all the variables,
- Factors were extracted from the correlation matrix based on the correlation coefficients of the variables; and
- The factors were rotated in order to maximize the relationship between the variables and some of the factors.

For a better understanding of the relationships between the heavy metal log transformed concentrations and particle sizes, correlation coefficients were calculated and listed in Table 6(a) andTable 6(b).

The Pearson correlation coefficient was used to determine the linear association between two variables, using data that was transformed and normally distributed. The Fe and As elements inTable 7(a) had negative correlations with the sand (>0.06 mm). With the exception of Al, all the other elements had moderate to significantly positive correlations with silt (<0.06 mm) and clay fractions of <0.02 mm of the sediments. Mn is the only element with significantly positive correlations with the finest grained sizes of <0.02 - 0.05 mm, and sand fractions (>0.06 mm).

All the elements in Table 7(b) have significantly positive correlations with the silt (<0.06 mm) and sand fractions (>0.06 mm) of the sediments. Mn was the only element with significantly positive correlations with the finest particle sizes of <0.02 mm. However, Al, Fe, As, and Pb had moderately negative correlations with the coarse particle sizes of >0.06 mm. Therefore, all the elements in this study appear to be influenced by the particle size effect in some manner.

Sample Classification

Cluster Analyses was applied for sample classification based on the particle sizes and elements detected. In addition, PCA and correlations were applied to reduce the variables between the particle sizes and the elements; factor analysis uses the correlation matrix to determine which sets of variables cluster together. Before the multivariate analysis, tests were conducted to test the normality of the datasets. All of the elements and the particle sizes passed the normality tests at the significance level of $P \leq 0.05$. Therefore, all the raw data (real observed data points) were used without any transformation for the following multivariate analyses. The matrix of correlation coefficients and their significance levels are shown in Table 8(a) through Table 9(b).

According to Hatcher, L. [14] , confirmatory factor analysis (CFA) is a statistical approach used to examine the internal reliability of a measure; to investigate the theoretical constructs, or factors, that might be represented

Table 6: (a) Pearson correlation coefficients for depositional areas; (b) Pearson correlation coefficients (upland−reference area)

	Log-Sand	Log-Silt	Log-Clay	Log-Al	Log-Fe	Log-Mn	Log-As	Log-Pb
Log-Sand	1.000							
Log-Silt	−0.575**	1.000						
Log-Clay	0.517	−0.562**	1.000					
Log-Al	0.398	−0.452*	0.417	1.000				
Log-Fe	−0.462*	0.326	0.655	0.748	1.000			
Log-Mn	0.869	0.613	0.359	0.338	0.554	1.000		
Log-As	−0.417*	0.277	0.595	0.738	0.969**	0.510	1.000	
Log-Pb	0.707	0.517	0.528	0.748**	0.390	0.647**	0.397	1.000

(b)

	Log-Sand	Log-Silt	Log-Clay	Log-Al	Log-Fe	Log-Mn	Log-As	Log-Pb
Log-Sand	1.000							
Log-Silt	0.302	1.000						
Log-Clay	−0.615**	−0.764**	1.000					
Log-Al	0.560	0.660	−0.290	1.000				
Log-Fe	0.246	0.272	−0.452*	0.882**	1.000			
Log-Mn	0.251	0.893	0.828	0.851**	0.618**	1.000		
Log-As	0.357	0.354	−0.450*	0.259	0.554**	0.666	1.000	
Log-Pb	0.481	0.361	−0.256	0.537	0.355	0.966	0.651	1.000

*= Correlation is significant at the 0.05 level (2-tailed); **= Correlation is significant at the 0.01 level (2-tailed).

Table 7: (a) Correlation matrix for depositional areas; (b) Correlation matrix for upland areas

(a)

	Sand	Silt	Clay	AL	Fe	Mn	As	Pb
Sand	1.000							
Silt	−0.583	1.000						
Clay	−0.447	−0.466	1.000					
AL	0.496	−0.295	−0.216	1.000				
Fe	0.016	0.695	−0.782	0.340	1.000			
Mn	−0.110	−0.030	0.152	0.740	0.155	1.000		
As	0.074	0.638	−0.783	0.306	0.949	0.026	1.000	
Pb	0.251	0.091	−0.373	0.828	0.640	0.541	0.704	1.000

(b)

	Sand	Silt	Clay	AL	Fe	Mn	As	Pb
Sand	1.000							
Silt	−0.186	1.000						
Clay	−0.653	−0.623	1.000					
AL	−0.315	0.105	0.170	1.000				
Fe	0.087	0.039	−0.099	0.877	1.000			
Mn	−0.743	0.110	0.507	0.815	0.451	1.000		
As	0.549	0.088	−0.505	−0.357	−0.103	−0.461	1.000	
Pb	0.168	0.301	−0.366	−0.144	−0.267	0.059	0.487	1.000

Table 8: (a) Component score coefficient matrix for depositional area; (b) Component score coefficient matrix for upland area

(a)

Variables	Component		
	1	2	3
Sand	0.059	−0.091	0.572
Silt	0.213	−0.057	−0.362
Clay	−0.298	0.162	−0.224
AL	−0.024	.353	0.148
Fe	0.267	0.040	−0.068
Mn	−0.130	0.465	−0.222
As	0.279	0.003	−0.014
Pb	0.089	0.293	0.027

(b)

Variables	Component		
	1	2	3
Sand	−0.426	0.137	−0.165
Silt	0.096	0.035	0.500
Clay	0.265	−0.136	−0.254
AL	0.026	0.407	0.031
Fe	−0.192	0.504	−0.106
Mn	0.282	0.153	0.166
As	−0.223	−0.008	0.135
Pb	0.066	−0.109	0.436

Extraction Method: Principal Component Analysis; Rotation Method: Varimax with Kaiser Normalization.

Table 9: (a) Total variance explained (Depositional); (b) Total variance explained (Upland)

(a)

Component	Initial Eigenvalues			Extraction Sums of Squared Loadings			Rotation Sums of Squared Loadings		
	Total	Variance %	Cumulative %	Total	Variance %	Cumulative %	Total	Variance %	Cumulative %
1	3.794	47.419	47.419	3.794	47.419	47.419	3.421	42.763	42.763
2	2.324	29.045	76.464	2.324	29.045	76.464	2.394	29.927	72.690
3	1.493	18.661	95.126	1.493	18.651	95.126	1.795	22.435	95.126
4	0.337	4.212	99.338						
5	5.296E−02	0.662	100.000						
6	4.102E−17	5.128E−16	100.000						
7	−8.49E−17	−1.061E−15	100.000						
8	−9.47E−16	−1.184E−14	100.000						

(b)

Component	Initial Eigenvalues			Extraction Sums of Squared Loadings			Rotation Sums of Squared Loadings		
	Total	Variance %	Cumulative %	Total	Variance %	Cumulative %	Total	Variance %	Cumulative %
1	3.327	41.591	41.591	3.327	41.591	41.591	2.683	33.531	33.531
2	1.961	24.508	66.099	1.961	24.508	66.099	2.230	27.875	61.407
3	1.420	17.746	84.844	1.420	17.746	83.844	1.795	22.438	83.844
4	0.917	11.461	95.305						
5	0.376	4.695	100.000						
6	1.595E−17	1.994E−16	100.000						
7	−2.68E−17	−3.354E−15	100.000						
8	−4.94E−16	−6.171E−15	100.000						

Extraction Method: PCA by a set of variables, such as, given in this study (Al, Fe, Mn, As, Pb, sand, silt, and clay) and to assess the quality of individual variables. According to Cattell, R.B. [15], researchers typically use maximum likelihood in the CFA model to estimate factor loadings, and the most common approach to deciding the number of factors to generate a scree plot. The scree plot is a two dimensional graph with factors on the x-axis and eigenvalues on the y-axis. The scree plot was used to help determine the number of factors to keep in the analysis. Theoretically, an eight-item variable will have eight possible underlying factors, and each factor will have an eigenvalue that indicates the amount of variation in the items accounted for by each factor. It should be noted that this approach to selecting the number of factors involved a certain amount of subjective judgment. From the scree plot (Figure 3(a) and Figure 3(b)), there were three underlying factors for each plot (components 3, 4, 5 for depositional and components and 1, 2, and 3 for upland area). It is believed that the remaining of factors were due to unknown error variation.

The component plot in rotated space shows the first three principal components (PCs) for the depositional area (Figure 4(a)), and the upland area (Figure 4(b)). The first three PCs may reveal a clearer explanation of the particle-size and element factor controlling the distribution and accumulation of the metals and particle sizes. The significant relationships between the particle-size variables and the elemental components were presented by using FA as the indicators.

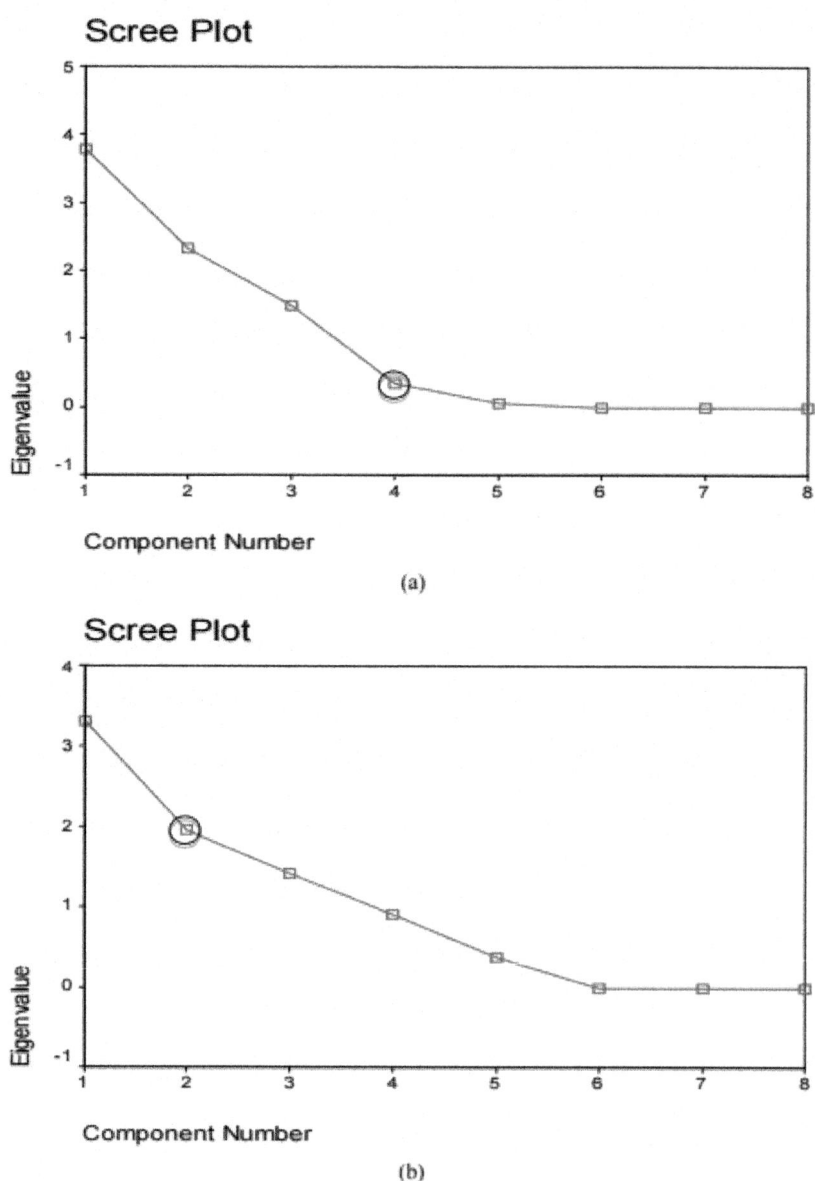

Figure 3: (a) Scree Plot for the Depositional Area; (b) Scree Plot for the Upland Area.

Component Plot in Rotated Space

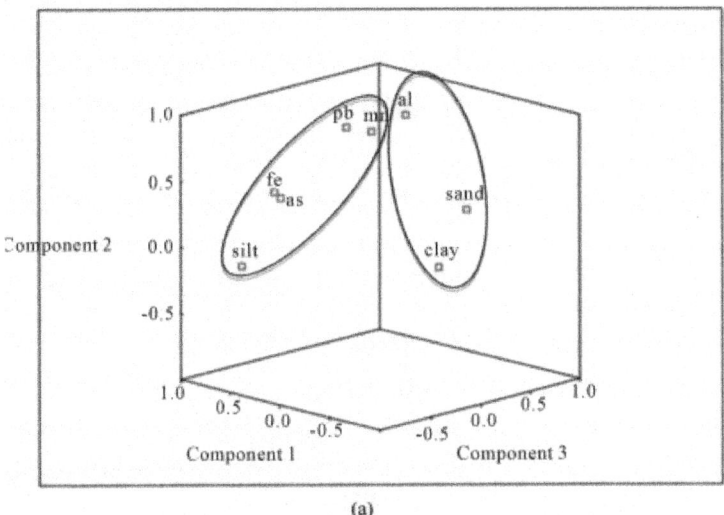

(a)

Component Plot in Rotated Space

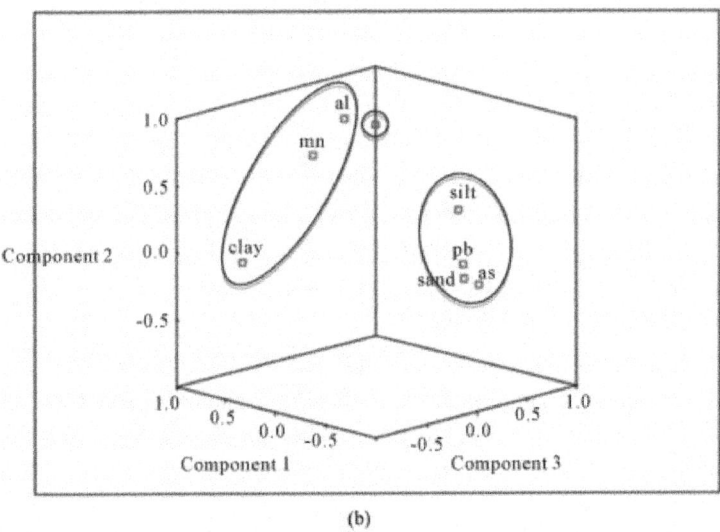

(b)

Figure 4: (a) FA in Rotated Depositional Space; (b) FA in Rotated Upland Space.

The first three PCs in the depositional area accounted for 95.12%, and the upland area accounted for 83.84% of the total data variance explained. They were selected for subsequent graphical displays and analysis. Figure 4(a) and Figure 4(b) show the PC loadings of the particle sizes and the elements on the first three principal components. In the depositional area the fine particle sizes of <0.05 and <0.02 mm had low negative loadings of PC2, and the coarse sizes of >0.06 mm had high positive loadings of PC3. In upland area the fine particle sizes of <0.05 mm had low negative loadings of PC2, and the coarse sizes of >0.06 mm had low negative loadings of PC3, but silt <0.02 mm had a high positive loadings in PC3. The summary scores on the two PCs are shown in Figure 6(a) and Figure 6(b). It can be seen that many samples were located on both sides of the positive and negative values of PC2, and some samples were located at the side of high values of PC3 and no values in PC1. It is expected that the multi-element concentrations among the three PC loading of samples should be different. Since the initial solution was obtained, the loadings are rotated. Rotation is a way of maximizing high loadings and minimizing low loadings so that the simplest possible structure is achieved.

For this study, oblique rotation was used because it derives factor loadings based on the assumption that the factors were correlated. This was the case for these measures (variables). The oblique rotation gave the correlation between the factors in addition to the loadings (See Table 10).

Cluster Analysis for Upland Area

Table 11, cluster membership displays the single solution cluster (between-groups linkage model option) to which each variable was assigned. Each stage (site) was assigned one predominant variable. The icicle plot, in Table 12, displays vertical information about how variables were combined into clusters at each iteration of the analysis. The agglomeration scheduled hierarchical CA identified relatively homogeneous groups of variables based on selected characteristics. It shows that Al was dominant in WR-FR; Fe in BF-FR; Mn in HR-FR; As in RB-FC; Pb in MB-FC, and all the metals were clustered in VB-FCs site sediments of upland areas.

The dendrograms plot below used the nearest neighbor model to assess the cohesiveness of the clusters formed. It shows that Mn, Pb, and As were clustered in all the particle sizes as shown, and the cluster of Al and Fe were observed mainly in the clay proportion of the sediments (see Figure 5(a) and Figure 5(b)).

Table 10: Summary of the Loadings from the oblique rotations for depositional and upland areas of the fc and FR watersheds

Variables from Depositional Area			Variables from Upland Area		
Comp-1	Comp-2	Comp-3	Comp-1	Comp-2	Comp-3
	Pb	Al		Al	Silt
	Mn	Sand		Mn	Pb
	As, Fe	Clay		Clay	Sand
	Silt				Fe, As

Table 11: Cluster membership

Case	3 Clusters
WU	1
BU	1
HU	1
RU	2
MU	2
VU	3

Table 12: Verticle icicle

Number of Clusters	Case										
	VU		MU		RU		HU		BU		WU
1	X	X	X	X	X	X	X	X	X	X	X
2	X	X	X	X	X		X	X	X	X	X
3	X		X	X	X		X	X	X	X	X
4	X		X	X	X		X		X	X	X
5	X		X		X		X		X	X	X

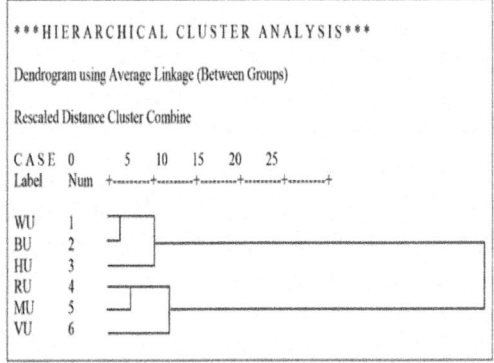

(a)

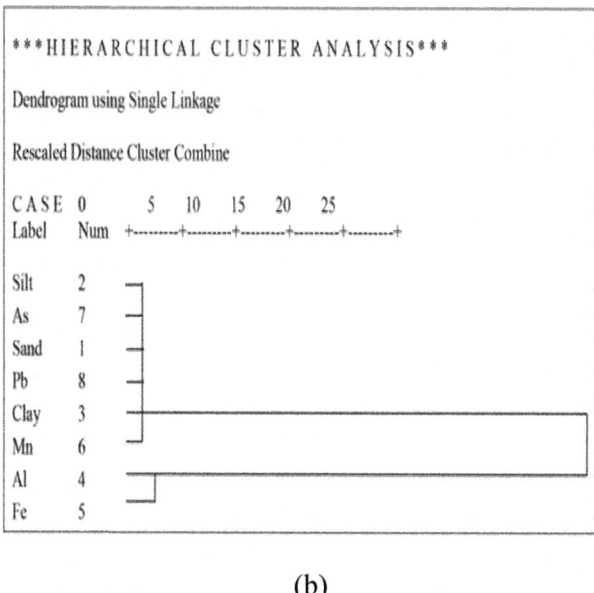

(b)

Figure 5: (a) Dendrogram using average linkage (between groups); (b) Den- drogram using single linkage.

Cluster Analysis for Depositional Area

Table 13(a) Cluster membership, shows the single solution cluster (between-groups linkage model option) also to which each variable is assigned. Here each depositional site on the watershed was assigned only one predominant variable. The icicle plot, in Table 13(b), displayed vertical information about how variables are combined into clusters at each iteration of the analysis. The agglomeration scheduled hierarchical cluster analysis identified relatively homogeneous groups of variables based on selected characteristics. It showed that MB-FC and VB-FC had similar clustering, BF-FR and RBFC had similar clustering, and that HR-FR and WF-FR had different clustering for all the metals in site sediments of depositional area.

The dendrograms plots below used the (nearest neighbor model) to assess the cohesiveness of the clusters formed. It shows that As and Pb were clustered in all the particle sizes as shown, and Mn was clustered in clay, but Al and Fe cluster were observed mainly in the clay proportion of the sediments (see Figure 6(a) and Figure 6(b)).

Concentrations

The arithmetic means of the particle size compositions and element concentrations in all the sediment samples consisting of sand, silt, and clay were calculated, and listed in the tables below. The depositional samples, as hypothesized, had high percentages of the grain sizes <0.02 mm. The sand samples, on the other hand, mainly consisted of the coarse sizes >0.1 and 0.05 - 0.1 mm. All of the particle-size groups were relatively evenly distributed in the silt-clay samples. Most of the elements under study were elevated in the clay samples, and depleted in the silt samples. Concentrations of the elements in the silt-clay samples were between the clay and silt samples.

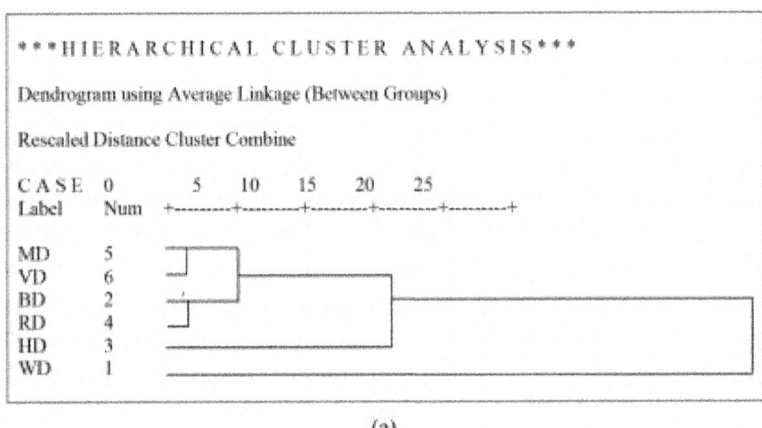

(a)

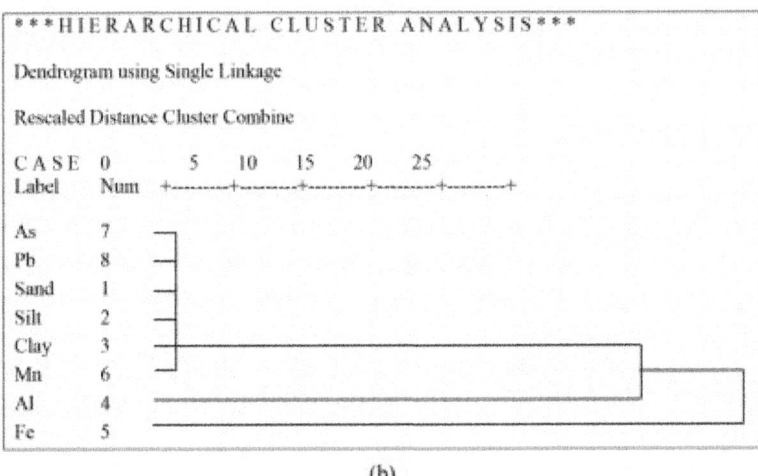

(b)

Figure 6: (a) Dendrogram using average linkage (Between groups); (b) Den- drogram using single linkage.

Table 13: (a) Cluster membership; (b) Verticle icicle

(a)

Case	3 Clusters
WD	1
BD	2
HD	3
RD	2
MD	2
VD	2

(b)

Number of Clusters	Case									
	HD		VD		MD		RD		BD	WD
1	X	X	X	X	X	X	X	X	X	X	X
2	X	X	X	X	X	X	X	X	X		X
3	X		X	X	X	X	X	X	X		X
4	X		X	X	X		X	X	X		X
5	X		X	X	X		X		X		X

Table 14(a) and Table 14(b) list the concentrations of the five elements Al, Fe, Mn, Pb, and As. Most of the element concentrations of Al, Fe, Mn, As and Pb were highest in the clay samples. However, Al in depositional and As in upland were elevated in the sand samples. The particle size effect on total metals has been widely studied. Clay minerals and other fine particle minerals play an important role in trapping and holding total metals. In this study, the same conclusion for the total metals was reached. The PC loadings of the particle sizes on the second and third PCs contain all of the Al, Fe, Mn, As and Pb. Further studies on the relationships between multi-element composition, and particle sizes are needed.

CONCLUSIONS

Sediments along rivers have different textures, as shown in the samples studied in these two watersheds, depending on whether the stream moves quickly or slowly [16] [17] . Fast-moving water leaves gravel, rocks, and sand. Without storm events, the FC and the FR have slow-moving water and tend to leave fine textured material (clay and silt) when sediments in the water settle out. The study evaluated spatial and compositional patterns in the FC and FR sediment contaminant data using the multivariate techniques—principal components analysis (PCA) and

Table 14: (a) Soil texture and heavy metals in depositional sediments; (b) Soil texture and heavy metals in upland soil samples. (SAS proc glm with Duncan's Multiple Range Test) (a, b, c, d = Average Means (Duncan Grouping) with the same letter are not significantly Different)

(a)

	WR-FR	BF-FR	HR-FR	RB-FC	MB-FC	VB-FC
	(%) Distribution of Particle sizes of Soil/Sediment and Heavy Metals					
Sand	9.79c	14.98c	31.91a	15.89b,c	26.95b,a	30.92a
Silt	34.28a	37.94a	3.95b	8.16b	1.63b	10.08b
Clay	55.93b,a	47.08b	64.15b,a	75.94a	71.43a	59.01b,a
	(µg/kg)					
Al	7465.00d	10683.00c	23800.00a	14478.00b	10665.00c	8366.00d,c
Fe	42433.00a	19713.00b	23250.00b	16331.00c,b	10939.00c,d	8701.00d
Mn	969.00b	675.50b	1640.00a	853.80b	995.50b	913.80b
As	11.67a	7.18b	7.62b	4.63c	3.93c	2.82c
Pb	13.17b	14.11b	20.33a	13.79b	12.09b	9.83b

(b)

	WR-FR	BF-FR	HR-FR	RB-FC	MB-FC	VB-FC
	(%) Distribution of Particle sizes of Soil/Sediment and Heavy Metals					
Sand	17.62c	15.79b,c	31.26a	9.48d	19.33b,c	25.07b,a
Silt	22.50a	17.29b,a	7.62b,c	1.90c	11.97b,a,c	9.98b,a,c
Clay	65.88c	69.23c,a	65.96c	88.62a	75.07b	71.17c,b
	(µg/kg)					
Al	13600.00b,a	13098.00b	17213.00a	8253.00c	9116.00c	4431.00d
Fe	17967.00a	17950.00a	18125.00a	7404.00b	9934.00b	6624.00b
Mn	946.70a	1118.50a	1074.30a	1029.00a	669.00b,a	74.5b
As	5.73b,a	9.59a	5.53b,a	2.94b	3.72b	9.03a
Pb	16.28b	42.93a	16.82b	13.15b	31.05b,a	26.29b,a

cluster analysis (CA). With the aid of PC and CA, the sediments from these watersheds were classified based on their texture into three fractions: sand, silt and clay. Significant differences were observed among the concentrations of the elements in the categories. Most of the elements detected were enriched in the fine-particle samples, where clay minerals were an important constituent [3] . On the other hand, elevated concentrations of Al and As were observed in the coarse (sand) samples. This may imply that the Al and As were present in the parent materials that weathered to sand. PCA indicated association among all heavy metals determined in the environmental matrices (soil/sediment) analyzed, revealing that their high concentrations were due to the discharge of liquid wastes to the FC and FR watersheds. Interestingly, some of the wastewater from the Decatur plant enters the river untreated [18] . According to Gray [19] , sewage treatment removes less than 100% of the metals from wastewater, so untreated wastewaters can be an important source of metals such as the ones analyzed in this study. Further, CA correlated all sampling sites impacted by contamination resulting from non-point sources, municipal wastes, sewages, and other sources. Metal concentrations from this study were higher than the USEPA's MCL or background concentration values for Al, Fe, Mn, As and Pb. This indicated that there was metal pollution at all the sampling sites. Significant relationships from the Pearson's correlation were also supported by the results of the statistical methods (PCA and CA) used.

In summary, contaminants displayed distinct groupings (elements with similar properties) and spatial patterns; and relationships between particle size factors and chemical contaminant concentrations were explained in part with the multivariate statistical approaches. However, this multivariate statistical approach was as effective as simple correlation for this study, where contaminant concentrations were relatively high and correlation patterns were strong. In addition, the study refuted the null hypotheses. There was enough evidence from the overall results of the study to support the theoretical notion that particle size distribution and sampling location influenced the level of heavy metal concentration within the watersheds. These results provide a framework for interpreting heavy metal distribution and sediment toxicity in biological communities at these watersheds.

ACKNOWLEDGEMENTS

We would like to acknowledge Alabama A&M University Interdisciplinary Center for Health Sciences and Health Disparities in the College of Engineering, Technology and Physical Sciences with funding provided through the Evans-Allen Grant, administered by the College of Agricultural, Life and Natural Sciences (CALNS).

REFERENCES

1. Greany, K.M. (2005) An Assessment of Heavy Metal Contamination in the Marine Sediments of Las Perlas Archipelago, Gulf of Panama. Master of Science Thesis, Heriot-Watt University, Edinburgh.

2. Pekey, H.P., Karakas, S., Ayberk, L., Tolun, L. and Bakoglu, M. (2006) Ecological Risk Assessment Using Trace Elements from Surface Sediments of Izmit Bay (Northeastern Marmara Sea) Turkey. Marine Pollution Bulletin, 48, 9-10, 946-953.

3. Okweye, P., Tsegaye, T.D. and Garner. K.F. (2007) Distribution of Heavy Metals in Surface Water of the Wheeler Lake Basin in Northern Alabama. Journal of Environmental Monitoring and Restoration, 33, 91-100.

4. Phillips, C.R. (2007) Bulletin of the Southern California Academy of Sciences, 106, 163-178, E Article 1.

5. Okweye, P. and Garner, K.F. (2009) Sediment Sampler (Patent Pending).

6.   United Nations Environment Program and the World Health Organization (1996) Sediment Measurements. Water Quality Monitoring—A Practical Guide to the Design and Implementation of Freshwater Quality Studies and Monitoring Programs Edited by Jamie Bartram and Richard Balance.

7.   U.S. Environmental Agency (USEPA) (1994) 822-R-94-001. Drinking Water Regulations and Health Advisories. EPA's Toxics Release Inventory, 1-3.

8.   Date, A.R. and Gray, A.L. (1989) Applications of Inductively Coupled Plasma Source Mass Spectrometry: Blackie, Glasgow.

9.   Platzner, I.T. (1997) Modern Isotope Ratio Mass Spectrometry. Chemical Analysis, 145, 187-189.

10.   Kennett, D.J., Neff, H., Glascock, M.D. and Mason, A.Z. (2001) A Geochemical Revolution: Inductively Coupled Plasma Mass Spectrometry. The Archaeological Record, 1, 22-26.

11.   Grimm, L.G. and Yarnold, P.R. (2000) Introduction to Multivariate Statistics. In: Grimm, L.G. and Yarnold, P.R., Eds., Reading and Understanding More Multivariate Statistics, American Psychological Association, Washington DC, 3-21.

12.   SCB (2007) Bulletin of the Southern California Academy of Sciences, 106, 163-178, E Article 1.

13.   Phillips, C.R. (2007) Bulletin of the Southern California Academy of Sciences, 106, 163-178, E Article 1.

14.   Hatcher, L. (1994) A Step-by-Step Approach to Using the SAS® System for Factor Analysis and Structural Equation Modeling. SAS Institute Inc, Cary.

15.   Cattell, R.B. (1966) The Scree Test for the Number of Factors. Multivariate Behavioral Research, 1, 245-276.http://dx.doi.org/10.1207/ s15327906mbr0102_10

16.   Alagarsamy, R. (2006) Distribution and Seasonal Variation of Trace Metals in Surface Sediments of the Mandovi Estuary West Coast of India. Estuarine, Coastal and Shelf Science, 67, 333-339. http://dx.doi. org/10.1016/j.ecss.2005.11.023

17.   Jonathan, M.P., Ram-Mohan, V. and Srinivasalu, S. (2004) Geochemical Variations of Major and Trace Elements in Recent Sediments, off the Gulf of Mannar, Southeast Coast of India. Environmental Geology, 45, 466-480.http://dx.doi.org/10.1007/s00254-003-0898-7

18.   (2008) ADEM Files Suit against Hanceville Water Board. Huntsville Times, 15 June 2008, p. A20.

19.   . Gray, D. (2004) Doing Research in the Real World. SAGE Publications, Various Paginations.

Chapter 15

ECOTOXICOGENOMIC APPROACHES FOR UNDERSTANDING MOLECULAR MECHANISMS OF ENVIRONMENTAL CHEMICAL TOXICITY USING AQUATIC INVERTEBRATE, DAPHNIA MODEL ORGANISM

Hyo Jeong Kim [1,2], Preeyaporn Koedrith [1,3],and Young Rok Seo [1,2]

[1]Institute of Environmental Medicine for Green Chemistry, Dongguk University Biomedi Campus 32, Dongguk-ro, Ilsandong-gu, Goyang-si, Gyeonggi-do 410-820, Korea

[2]Department of Life Science, Dongguk University Biomedi Campus 32, Dongguk-ro, Ilsandong-gu, Goyang-si, Gyeonggi-do 410-820, Korea

[3]Faculty of Environment and Resource Studies, Mahidol University, 999 Phuttamonthon 4 Rd., Phuttamonthon District, Nakhon Pathom 73170, Thailand

ABSTRACT

Due to the rapid advent in genomics technologies and attention to ecological risk assessment, the term "ecotoxicogenomics" has recently emerged to describe integration of omics studies (*i.e.*, transcriptomics, proteomics, metabolomics, and epigenomics) into ecotoxicological fields. Ecotoxicogenomics is defined as study of an entire set of genes or proteins expression in ecological organisms to provide insight on environmental toxicity, offering benefit in ecological risk assessment. Indeed, *Daphnia* is a model species to study aquatic environmental toxicity designated in the Organization for Economic Co-operation and Development's toxicity test guideline and to investigate expression patterns using ecotoxicology-oriented genomics tools. Our main purpose is to demonstrate the potential utility of gene expression profiling in ecotoxicology by identifying novel biomarkers and relevant modes of toxicity in *Daphnia magna*. These approaches enable us to address adverse phenotypic outcomes linked to particular gene function(s) and mechanistic understanding of aquatic ecotoxicology as well as exploration of useful biomarkers. Furthermore, key challenges that currently face aquatic ecotoxicology (e.g., predicting toxicant responses among a broad spectrum of phytogenetic groups, predicting impact of temporal exposure on toxicant responses) necessitate the parallel use of other model organisms, both aquatic and terrestrial. By investigating gene expression profiling in an environmentally important organism, this provides viable support for the utility of ecotoxicogenomics.

INTRODUCTION

Recently, the term "ecotoxicogenomics" has been introduced [1–5] to describe the toxicogenomic context into ecotoxicological field (Figure 1) [6]. The toxicogenomic approach with the use of Daphnia magna as an aquatic invertebrate model advances our knowledge and understanding of ecotoxicity because current mechanism of chemical toxicity in invertebrates as well as potential biomarkers still need to be determined.

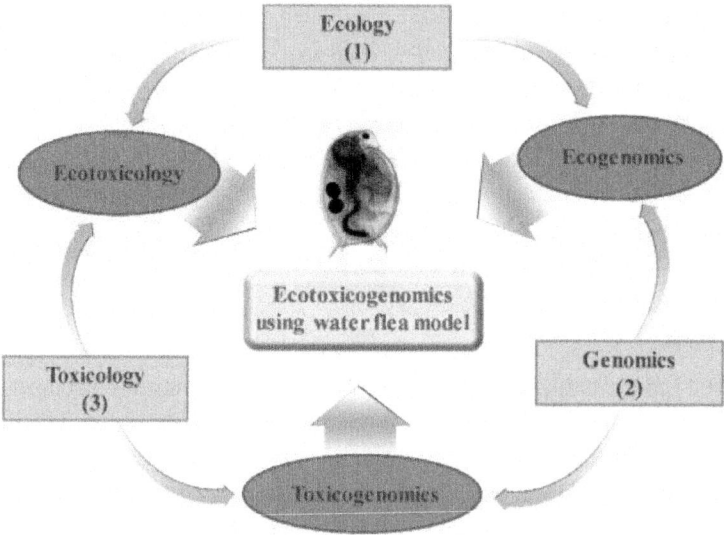

Figure 1: A scheme illustrating conceptual ecotoxicogenomics using Daphnia model. The thin arrows connect primary fields of study (in rectangles) to form interdisciplinary fields (in circles). The thick arrows indicate the tools or the knowledge that can be applied to integrated multidisplinary data sets, for instance: (1) ecological surveys; (2) genomic tools; (3) toxicity tests.

Basically, environmental pollutants may alter genomic expression profiles in an organism. In biomedicine, gene expression pattern is altered, directly or indirectly, owing to toxicant exposure in most cases [7]. Depending on the extent and period of the toxicant exposure, genomic alterations may be transient toxicological responses resulting in changes of "fitness" (survival and reproduction), or the "genotoxic disease syndrome" [8]. Previous studies have shown genotype-dependent effects in animals experienced to toxicants [9–11]. Daphnia is a keystone aquatic organism in the pelagic zone of most fresh water habitats (ranging from arctic and temperate lakes, lakes at high elevations, ephemeral ponds, to ponds in sand-dunes) and offers a key relationship between primary producer and higher trophic levels [12,13].

The establishment of ecology, phylogeny, toxicology, and physiology in Daphnia as well as its genome sequence accessibility (wfleabase.org) allow the development of genetic tools, such as genetic linkage map, [14], cDNA libraries and microarrays [6], and consequently investigate environmental impacts on gene functions in other model organisms that are difficult to be studied [15,16]. In particular, the availability of genetic linkage maps and the transferability of crossing panels across laboratories can facilitate the diagnosis of potential ecological and environmental traits via quantitative trait locus (QTL) analysis as well as the identification of heritable genotype-associated gene expression with the use of eQTL (expression QTL) approaches. Among freshwater organisms, daphnids have relatively high sensitivity to environmental contaminants [17].

Upon exposure to environmental stressors, daphnids exhibit significant reproductive decline [18], aberrant vertical mobility and behavioral pattern, and ultimately phenoplasticity [19–22]. These abiotic and biotic stressors in common include chemical substances, synthetic hormones, acidity, salinity, calcium levels, hypoxia, radiation, bacterial pathogens, predators and parasites [16]. Among Daphnia's closely related species, their multiple habitat transitions may be attributable to the extremely "eco-responsive genome" [16,23]. Among Daphnia closely related species complexes including Daphnia galeatamendotae, Daphnia longispina, and Daphnia pulex, their ecologically relevant traits are likely associated with the colonizing habitats under distinct environmental conditions [24–26]. The multiple lineages independently colonized and adapted to these freshwater habitats are distinguished in terms of extent of reproductive isolation and intraspecific genetic subdivision among populations [27,28]. In the field of molecular biology at post-genomic era, the availability of DNA sequence data combined with advent in genomic tools and technologies will promote the direct interrogation of gene expression at multiple levels in organisms experienced to various environmental stressors. Given that the genome sequencing tool has the potential to identify an increased number of ecologically relevant species in both vertebrates and invertebrates, this holds great promise to address the phenotypic and genotypic linkage based on fitness using a "bottom-up" approach from molecular to ecosystem level (Figure 2) [2].

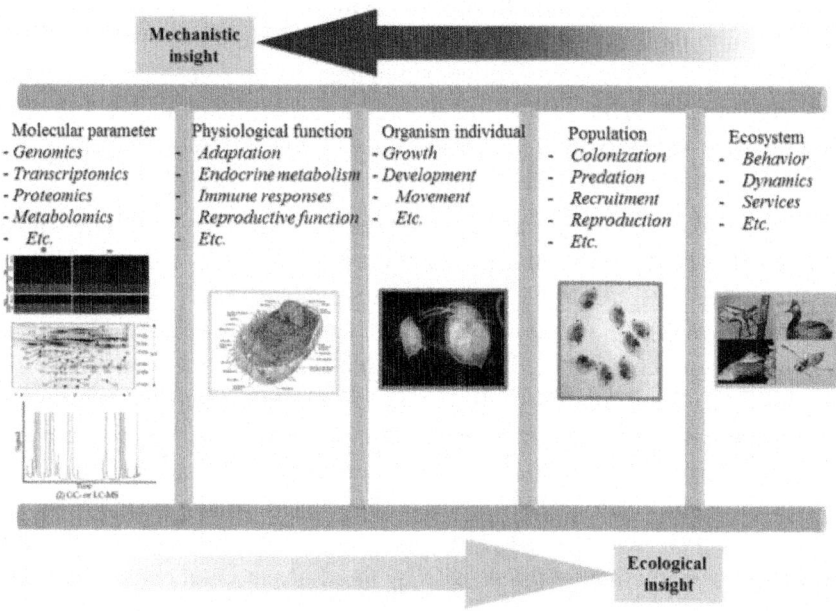

Figure 2: A scheme representing multilevel-framework for ecotoxicogenomic studies at multiple levels ranging from molecular, physiological, organismal, and population in ecosystem.

Genomic tools can facilitate ecology and evolutionary biology studies, allowing advance fundamental information and addressing the future issues related to chemical effects on environmental and human health. Purpose of this framework include: identification of ecological performance-regulated gene loci; functional analysis of ecological performance-related traits; evaluating individual, population, community, and ecosystem responses to the environment; examining the degree and significance of genetic variation among ecological performance-related traits [2]. Of ecological significance, omics technologies including microarrays enable rapid quantification of molecular variation among populations at multiple gene loci [2]. In the UK, the Natural Environment Research Council (NERC) Environmental Genomics Programme has been established in order to: examine the consequence of genetic variation on ecological performance; determine the degree of spatial and temporal variation at regulatory versus structural loci; investigate numbers of loci and their genomic distribution; and assess community structure and the influence of environmental alteration. At environmental contaminated sites, the omic tools are also useful for investigating important issues as follows: what are inducible genes and their functions; is there variation in gene expression in response to environmental change; is the variation adaptive; what are the consequences of the genetic variation-mediated molecular transformations at ecosystem-, community-, and population-level.

The omics-based technologies can facilitate comprehensive investigation to better understanding and improve the knowledge on how environmental stressors, such as heavy metals and persistent organic pollutants, cause toxicity in ecologically relevant organisms and adverse effects in ecosystem [29]. In this review, we mainly discuss three aspects: the commonly used omic technologies, including genomic (or mRNA-transcriptomic), proteomic, metabolomic analyses, and more recently emerged epigenetic technology; collective technologies (i.e., cDNA microarray, high-density oligonucleotide arrays, suppression subtractive hybridization PCR or high-throughput pyrosequencing, two-dimensional gel electrophoresis, fluorescence difference gel electrophoresis, ProteinChip, surface-enhanced laser desorption ionization (SELDI) mass spectrometry, and nuclear magnetic resonance) in conjunction with statistical testing or multivariate analysis; as well as development of bioinformatics tools and their application to aquatic ecotoxicology studies.

Practice of Ecotoxicogenomics in Genomic or Transcriptomic Responses

The DNA microarrays applied to ecotoxicology have ranged from nylon membranes that are custom spotted with a couple dozen selected cDNAs [30–32] to commercially available, high-density arrays consisting of thousands of oligonucleotides synthesized directly onto a solid support such as a glass slide [33–35]. To date, the availability of commercial, high-density oligonucleotide microarrays has been largely restricted to prominent model organisms for which considerable sequence information is available, such as zebrafish (Danio rerio), frog (Xenopus larvis), and fathead minnow (Pimphales promelas). Although the genomes of other species have not been completely sequenced, DNA microarrays have been developed and applied for studying a variety of additional model organisms for ecotoxicity testing. Coordinated efforts to share sequence information and other resources, such as the Daphnia Genomics Consortium and the Consortium for Genomic Research on All Salmon have greatly assisted the development of genomic tools for other model organisms. Up-to-date, high density customized microarrays of cDNA- or oligonucleotide-based chips containing thousands of targets have been dominantly developed for global analysis of transcripts associated with exposure to chemical or environmental stressors in ecotoxicology studies. The emergence of transcriptomics as a tool for ecosystem characterization is perhaps best shown by the fact that transcriptomics has been applied to a diverse group of species and stressors, representative of those traditionally examined in context of ecological risk assessment [29]. As a common test organism, the aquatic invertebrate, D. magna has also been used in a number of ecotoxicology-oriented transcriptomics studies [36–39]. To date, Daphnia and fish, particularly zebrafish and fathead minnow, appear to be the most common model organisms for ecotoxicogenomics studies employing DNA microarrays. As DNA sequencing technology becomes increasingly rapid and cost effective, the available sequence information for diverse taxa continues to grow, making it increasingly feasible to apply transcriptomic approaches to nearly any species of concern [40]. For example, the first version cDNA microarray of D. magna based on Suppression Subtractive Hybridisation PCR (SSH-PCR) 855 life stage-specific cDNAs has been successfully applied for elucidating mode of toxicity of propiconazole pesticide [41]. After 4 days of exposure to 1 μg/mL propiconazole, vitellogenin gene was repressed, suggesting that oocyte maturation was affected.

The vitellogenin mRNA might be considered as an early warning biomarker of chronic reproductive effects in aquatic invertebrates. Metal toxicity study was performed using a customized Daphnia magna cDNA microarray toward sublethal exposure to cadmium, copper, and zinc [37]. This finding revealed that distinct expression patterns toward each metal, and plausible exposure biomarkers, including putative metallothioneins and ferritin mRNA with a functional IRE were identified. Furthermore, using systems biology approach, transcriptomic patterns and phenotypic features were analyzed in D. magna model following exposure to ibuprofen, a nonsteroidal anti-inflammatory drug [36]. Result indicated that ibuprofen would affect reproductive capability at molecular-, organism-, and population-level in daphnids. Microarray data demonstrated ibuprofen-mediated early disturbance in crustacean eicosanoid metabolism via signal transduction, resulting in aberrant juvenile hormone metabolism and oogenesis. Recently, microarray analysis in combination with reproduction testing assay was conducted in D. magna system in order to determine bisphenol-A(BPA) toxicity since BPA has detrimental health impacts, particularly in reproduction, development, and organismal behavior [42]. Microarray results revealed significant change in expression levels of candidate genes homologous to animal nucleotide sequences, including cuticular protein, vitellogenin, protease, and ribosomal proteins. These genes reportedly discovered in other animal models might be recognized as novel biomarkers indicative of BPA exposure [42,43]. The application of genomic techniques into environmental toxicology holds great promise to identify exposure biomarkers and clarify the mode of toxicity of newly synthesized chemicals [44]. Using 15k oligonucleotide microarray, D. magna was also used as an aquatic model to investigate nanotoxicity, specific biomarkers, and effects of coating agents; for instance, citrate-coated and polyvinylpyrrolidone

(PVP)-coated silver nanoparticles (AgNPs) in comparison to bulk silver nitrate ($AgNO_3$) [44]. The microarray data showed distinct expression patterns toward AgNPs and $AgNO_3$, indicating distinct modes of toxicity. AgNPs affected to biological processes, especially protein metabolism and signal transduction whereas $AgNO_3$ suppressed sensory developmental processes. Only PVP-coated AgNPs could upregulate metal-and DNA repair-related genes. PVP-coated AgNPs-specific biomarkers, including metallothionein (MT) and DNA damage repair (REV_1) gene might be useful for the environmental detection. Regardless of the stressor(s) examined, the most prominent use of transcriptomics has improved the understanding of the mechanisms of action through which various stressors elicit or modulate adverse effects [29]. Mechanistic ecotoxicogenomic studies rely heavily on genome annotation to identify the specific transcripts modulated by exposure to a particular stressor and gain an understanding of the biological functions and pathways represented by the differentially expressed genes. Qualitative and quantitative analysis of "enriched" gene ontology (GO) terms tends to be one of the central analytical approaches used in mechanistic studies. Application of automated pathway analysis tools has been somewhat limited by the lack of standardized pathways for ecological model species (i.e., non-human, non-rodent). However, pathway-oriented analyses can still be conducted by hand [45,46] or by using bioinformatics approaches to identify human or rodent homologs and applying pathway tools developed for those species [40,47].

Ultimately most mechanistic transcriptomic studies have had to rely heavily on information available in the extant literature to make sense of the numerous, and often disparate, responses that are observed following exposure to a stressor. Transcriptomic studies have already produced a greater appreciation of the concept that exposure to stressors, even those traditionally thought to act through specific pathways, impacts on many fundamental cellular processes such as energy metabolism, protein metabolism, cell cycle, cytoskeletal organization, immune/inflammatory processes, and extracellular matrix development [34,48,49]. Microarray studies suggest that extensive crosstalk among pathways is perhaps the rule, more than the exception. Given the global nature of microarray analysis, one of the key challenges of mechanistic studies is differentiation of general stress responses from responses that are specific to a given mode or mechanism of action [50–52]. Furthermore, transcriptomic research appears to be enhancing appreciation for adaptive or compensatory responses to stressors as well [45,53,54]. The ability to differentiate adaptive responses from adverse ones is critical if transcriptomic responses are to be used as a basis for predictive ecological risk assessment. Toxicogenomics researchers have clearly recognized the need for phenotypic anchoring to link changes at the molecular level to outcomes at higher levels of organization. The majority of mechanistically oriented transcriptomic studies in ecotoxicology have included a variety of apical endpoints. For example, transcriptomic results for Daphnia magna were anchored to embryo abnormalities and carapax length in one study [55] and population growth rates in another [39]. An ecotoxicogenomic assessment of D. magna with the use of expressed sequence tags (ESTs) and the database has been conducted [56]. Based on this sequence information, an oligonucleotide-based DNA microarray has been developed in order to determine the acute toxicogenomic profiling of D. magna in response to various types of chemical stressors, including copper sulfate ($CuSO_4$), hydrogen peroxide (H_2O_2), pentachlorophenol (PCP), and β-naphthoflavone (βNF) as testing substances exerting distinct toxicities. The result showed that neonatal daphnids exposed to these compounds have distinct transcriptional changes toward each chemical. However, it is necessary to relate molecular mechanisms to toxicological outcomes when mechanistic ecotoxicogenomic studies in aspect of "proof-of-concept" are conducted. While ecotoxicogenomic studies have tended to focus on mechanisms, mechanistic research has not been conducted to the exclusion of fingerprinting approaches and biomarker discovery (Figure 3) [2]. Many in the field have used their transcriptomic data to both explore mechanisms and attempt to identify transcriptional fingerprints or biomarkers potentially indicative of specific types of exposure or effects. For example, Poynton et al. [37] identified distinct expression profiles in Daphnia magna for three different metals, found support for known mechanisms of metal toxicity, and postulated inhibition of chitinase activity by zinc as a novel mode of action [37]. Alternatively, techniques for high-throughput transcription analysis that are not relied on a priori information of DNA sequence database include suppression subtractive hybridization PCR and high-throughput pyrosequencing [57]. Nevertheless, the data interpretation somewhat depends on the knowledge of genome like that in microarray techniques. Several setbacks of the microarray techniques may thus be employed to the aforementioned techniques. This notwithstanding, researchers can predict that, except for genome model species, only a 5%–10% of the genome can be related to known genes and thus to defined functions. New functional data, novel developments in bioinformatics, and completion of sequencing projects will be needed to conciliate the spectrum of species and taxa required for an in depth analysis of environmental impacts, and the limitations of transcriptomic and functional analyses. Even in laboratory model organisms, much effort is still needed for gene functional analysis.

In the genetic organism models (e.g., Saccharomyces cerevisiae, Caenorhabditis elegans, Drosophila, and Mus musculus), gene functions can be analyzed on the phenotypic characterization of mutants, transgenic organisms using common techniques (i.e., knockdown or gene silencing, knockouts, and the reverse genetics method [58]. Some of these conventional genetics methods become available in D. magna [59,60], ecotoxicogenomics-based approaches would be complementary with the validation of results obtained by those techniques. Ecotoxicogenomics-oriented research in such small organisms like the water flea may help to overcome the difficulties in the higher levels of ecological organisms in terms of studying the whole organism.

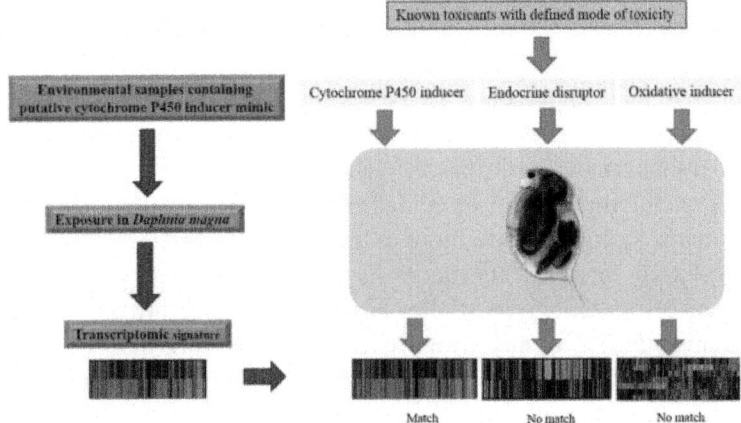

Figure 3: A scheme illustrating the utility of gene expression signatures to explore the possible mode of toxicity of an unknown environmental toxicant in an aquatic model organism (Daphnia spp.).

To date, proteomics has been the second most widely used of the omics approaches applied to ecosystem assessment [40]. Proteomic studies account for around 10% to 15% of published ecotoxicology-oriented "omics" research. Most of the proteomics studies have focused on fish and mussels. In contrast to transcriptomics-based ecotoxicology studies, which, to date, tend toward mechanistic investigation, proteomics-based ecotoxicology has skewed heavily toward identification of fingerprints, most often referred to as protein expression signatures (PES). There are two dominant practical approaches to PES generation. ProteinChip technology along with SELDI mass spectrometry, also known generally as retentate chromatography-mass spectrometry (RC-MS) [61], is the most common analytical technique used for PES identification. The second most prominent is two-dimensional gel electrophoresis and matrix-assisted laser deionization (MALDI)-based mass spectrometry has also been employed. Laboratory-based proof-of-principle type studies aimed at PES identification have shown some promise. Early on, this primarily consisted of work that employed two-dimensional gel electrophoresis (2-DE) and showed that exposure to different chemical treatments resulted in distinct spotting patterns on the gels [62–64]. While these studies provided basic support for the hypothesis that exposure to a date, proteomics has been the second most widely used of the omics approaches applied to ecosystem assessment [40]. Proteomic studies account for around 10% to 15% of published ecotoxicology-oriented "omics" research. Most of the proteomics studies have focused on fish and mussels. In contrast to transcriptomics-based ecotoxicology studies, which, to date, tend toward mechanistic investigation, proteomics-based ecotoxicology has skewed heavily toward identification of fingerprints, most often referred to as protein expression signatures (PES). There are two dominant practical approaches to PES generation. ProteinChip technology along with SELDI mass spectrometry, also known generally as retentate chromatography-mass spectrometry (RC-MS) [61], is the most common analytical technique used for PES identification. The second most prominent is two-dimensional gel electrophoresis and matrix-assisted laser deionization (MALDI)-based mass spectrometry has also been employed. Laboratory-based proof-of-principle type studies aimed at PES identification have shown some promise. Early on, this primarily consisted of work that employed two-dimensional gel electrophoresis (2-DE) and showed that exposure to different chemical treatments resulted in distinct spotting patterns on the gels [62–64]. While these studies provided basic support for the hypothesis that exposure to a biomarkers were successfully evaluated; meanwhile whole organism-based toxicity and biochemical testing were analyzed in D. magna exposed to citrate-coated silver nanoparticles (AgNPs) compared to bulk silver nitrate ($AgNO_3$) [68,69]. Levels of vitellogenins were increased upon exposure to both compounds, suggesting their functionally overlapped general stress response. Hemoglobin levels were increased toward AgNP exposure whereas 14-3-3 protein (a regulatory protein) carbonylation levels were decreased upon $AgNO_3$ exposure, indicating that both silver compounds has distinct impact on biological pathways possibly resulting in differential interactions with either natural or xenobiotic substances in the aquatic environment.

Practice of Ecotoxicogenomics in Metabolomic

Response Among the "omics" approaches, metabolomics has been the least widely applied to ecotoxicology [40]. Whereas transcriptomics and proteomics attempt to examine the entire complement of transcripts or proteins, respectively in a sample, metabolomics is concerned with the complement of small molecule metabolites found in biological samples [40]. Metabolomics studies account for approximately 5% of published ecotoxicogenomic studies. The majority of these have employed nuclear magnetic resonance (NMR) spectroscopy as the primary analytical technology. However, mass spectrometry (MS)-based environmental metabolomics studies are beginning to appear in the literature as well [70–72]. Similar to proteomics, the majority of ecometabolomics studies have focused on demonstrating that distinct metabolic profiles are observed for organisms exposed to different types of chemical stressors either in the laboratory or in the field [40]. Multivariate pattern-recognition analyses are typically used to assess complex differences between spectra. Thus, studies with a variety of species and stressors have demonstrated the ability to discriminate among groups using metabolite profiles, and to identify specific metabolites that may serve as biomarkers. However, to date, we are not aware of any concerted efforts to develop libraries of "metabolic fingerprints" for use in identifying or classifying various types of exposures or effects. When specific metabolites that were altered by a chemical exposure or differed among populations exposed from different (contaminated) environments can be identified, metabolomics studies also have the potential to yield mechanistic information [40].

Cautions have been raised that ad hoc attempts to infer mechanisms, based on metabolomic data in isolation, are not likely to be successful [73]. For example, an effective approach of high throughput, ultrahigh resolution mass spectrometry based metabolomics namely "direct infusion Fourier transform ion cyclotron resonance mass spectrometry (DI FT-ICR MS)" has been established as an exceptional tool in D. magna [74]. Copper was used as a testing chemical to validate this technique with an OECD 24 h acute toxicity testing in both univariate and multivariate models in order to screen and prioritize chemicals within tiered risk assessment. Later work with use of FT-ICR MS based metabolomic approach was successfully validated with D. magna toxicity testing to evaluate the acute metabolomic effects of chemicals and their mode of toxicity [75]. Test compounds with distinct toxicity modes including cadmium (oxidative stress inducer) [76], fenvalerate (sodium channel activator) [77], 2,4-dinitrophenol (DNP) [78], and propranolol (nonselective β-blocker) [79] were employed to evaluate whole-body metabolome relative to hemolymph metabolome with use of supervised multivariate modeling. The finding indicated that metabolomic patterns derived from whole-daphnids have discriminatory accountability to MOA of chemicals rather than hemolymphs as well as early metabolomic responses enable reflect discriminatory acute toxicities of chemicals. Furthermore, the integration of hemolymph metabolomics namely "FT ICR MS and NMR spectroscopy" and whole-daphnid transcriptomics namely "D. magna 44k oligonucleotide microarray" based method in conjugation with use of KEGG pathway database and gene ontology offers holistic insight on how cadmium at sublethal concentrations for 24 h interrupt nutrient uptake and metabolism.

This thus led to impaired energy production, resulting in chronic toxicity [80]. Recently, 1 H NMR-based metabolomics has been applied as viable platform for investigating metabolomic profiling and mode of toxicity of D. magna upon exposure to toxic metals including arsenic, copper, and lithium at sublethal concentrations for 48 h [81]. Metabolomic responses under all treatments were statistically compared using principal component analysis (PCA), and differentially expressed metabolites were quantitatively identified. Metabolomic data indicated that lithium exposure significantly exhibits an analogous mode of toxicity to copper as evident by disrupted energy reservoir and regulation, while arsenic exposure has a metabolic shift with non-significant changes.

Practice of Ecotoxicogenomics in Epigenetic Response Epigenetic effects can be defined as inheritable changes in phenotypes, by either mitotically or meiotically, without changes in DNA sequence [82]. Gene expression changes can be mediated via well-studied processes including DNA methylation, histone modifications, and RNA interference as well as less well-studied epigenetic processes (i.e., histone variation, nucleosome phasing, higher-order chromatin structure organization, and nuclear localization [83,84]. DNA methylation, which is occurred by either de novo or maintenance DNA methyltransferases, is related with transcriptional regulation, chromosome inactivation, and transposable element regulation [85]. Even though DNA methylation is present in various eukaryotic organisms, the degree of methylation and the chromatin structure organization are dependent on species and developmental stages [83]. DNA methylation interacts with other epigenetic processes including histone modifications at amino- or carboxyl-termini, thereby affecting chromosome coiling and accessibility to transcriptional machinery and ultimately resulting in gene expression changes [86,87]. Additionally, DNA methylation and histone modifications can interact with the RNA interference (RNAi) system that is involved in the generation of small noncoding RNA molecules (ncRNA) [88]. The nc RNAs, such as microRNA (miRNA) and short interfering RNA (siRNA) can form RNA-induced silencing complexes (RISC) that recruit DNA methyltransferases and histone modifying enzymes [89]. Epigenetic markers are regulated by environmental conditions (i.e., nutrients, chemical stressors, hypoxia, and developmental stages) [90]. For example, histone methylation status can be modified by hypoxia via interruption of Jumonji protein ($JMJD_2$) activity in aquatic system [91]. Indeed, the term "epigenetics" can be simply described to any environmentally modified process of DNA, such as DNA methylation or histone modification regardless to maternal inheritability. For example, previous studies revealed that transgenerational incidence is unlikely related to methylation status in either Daphnia [92] or Fundulus [93]. These epigenetic mechanisms in normal Daphnia development and their adaptations remain to be elucidated. Vandegehuchte et al. [94] have firstly discovered that D. magna is capable of methylating DNA as well as genes homologous to major vertebrate DNA methyltransferases (Dnmt1, Dnmt2, and Dnmt3A) with their activities confirmed [94,95]Using ultra-performance liquid chromatography (UPLC) and microarrays, DNA methylation and transcriptome profiling were respectively analyzed in D. magna exposed to several chemicals [96]. This finding showed that global or localized DNA methylation levels could be altered by 5-azacytidine, vinclozolin, genistein, and zinc but not by 5aza-2′-deoxycytidine, biochanin A, and cadmium [94,96]. The transgenerational impacts of methylation status were also determined [96]. With available accessibility of complete genome for D. pulex and D. magna, the advanced techniques, such as bisulfite sequencing, methylated DNA immunoprecipitation (meDip), or DNA methylation sensitive restriction enzyme digests preferably enable measurement of methylation status of particular genes, offering biologically meaningful information. Using these methods, D. magna's certain genes related to growth and reproduction were definitely determined since body length, brood size and sex determination as well as helmet and neck-teeth are affected upon exposure to toxicant [92,94,96–100]. Regarding to Daphnia epigenetics in DNA methylation process, CpG methylation is occurred at relatively low level but it is sensitive to developmental stage, as evident by the 2-fold increase in percentage of CpG dinucleotides in adults at 32-day age compared to that at 7-day age [101]. In epigenetic aspects in other core processes (i.e., histone modification or noncoding RNA), or the impact of these epigenetic mechanisms on either normal development or the well-known predator-induced epigenetic polyphenisms, this information is still lacking. For example, previous investigation reported that both histone H_3 and H4 modifications were occurred in embryonic cells. Interestingly, histone H_3 dimethylated at lysine 4 ($H_3K_4me_2$) was non-uniformly present in a cell-cycle-specific manner in D. magna gastrula cells but was absent in oocytes.

SUGGESTION OF PROMISING BIOMARKERS OF ENVIRONMENTAL TOXICITY OR EXPOSURE

The toxicogenomic studies using Daphnia have potential advantages not only to identifying interlinked crosstalks and biological processes in response to environmental toxicants, but also to suggest promising biomarkers that are indicative of certain types of environmental stressor's effect or exposure. Indeed, this recognizes the need for phenotypic anchoring to link changes at the molecular level to outcomes at higher levels of organization. The majority of mechanistically oriented transcriptomic studies in ecotoxicology have included a variety of apical endpoints. In various studies about toxicity using

Daphnia as an aquatic test model to metals, endocrine disruptors, drugs and so on, data enables us to understanding the symptoms induced by those toxicants. Although the detailed pathways and mechanisms of the toxic effects have not been clearly elucidated, Daphnia-customized microarray data, in particular, reveals a number of biomolecules including genes involved in ion transport and chelating (ferritin, putative metallothionein), metamorphosis (vitellogenin and chitinase), invertebrate immune system (eicosanoid), glycolytic and proteolytic process (amylase, cellulose, esterase, and serine protease), cellular anti-oxidative defense (glutathione-S-transferase, catalase, and peroxiredoxin), and stress response (heat shock proteins) that have been predominantly recognized as suggestive biomarkers in response to environmental stressor's exposure or effect (Figure 4) [37–39,102,103].

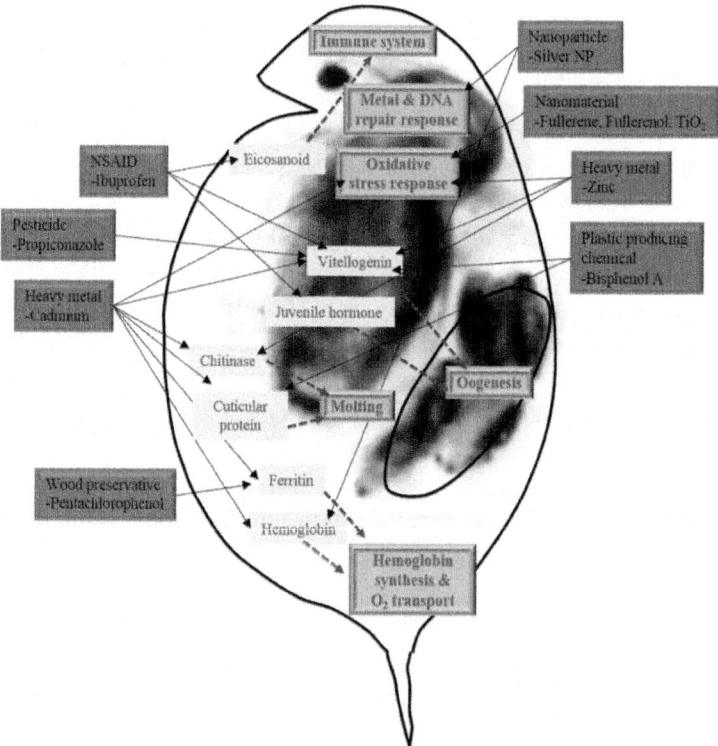

Figure 4: A scheme showing overview of biomarkers and biological interactions in Daphnia in response to stressors (as indicated 3-D red square). These candidate genes might be considered as potential biomarkers (as indicated in pink square), and their products are involved in important biological processes (as indicated in 3-D orange square) via interference with certain biomolecules (as indicated in 2-D pink square)

Heckmann et al. [104] revealed the profiles of critical genes in daphnia caused by ibuprofen exposure using microarray and temporal real-time quantitative PCR. In acute ibuprofen exposure, several genes, such as Lip (triacylglycerollipase), Ltb_4dh (leukotriene B_4 12-hydroxydehydrogenase), $FABP_3$ (fatty acid binding protein 3), DmagVTG1 (vitellogenin 1), dmHb (hemoglobin), JHE (juvenile hormone esterase), VMO_1 (vitelline outer layer membrane protein 1), and Cht (chitinase) are up- and down-regulated, inducing or suppressing the metabolisms and cellular processes. Lip and Ltb4dh are linked to eicosanoid metabolism and play a vital role in the invertebrate immune system, and $FABP_3$ induce Peroxisome Proliferator Activated Receptor (PPAR) signaling pathway in the endocrine system. A temporal expressed gene, Cht is a key enzyme to molting fluid enzyme secreted during apolysis [105].

JHE, $DmagVTG_1$ and VMO_1 is related to oogenesis of daphnia, and separately expressed at different times in the developing oocyte [104]. Cadmium exposure microarray data are also analogous to ibuprofen exposure data, inducing glycolytic, proteolytic, homeostatistic, heat shock protein genes, and oxidative stress responses. Up-regulated Cht gene encoding chitinase and CYP450 gene encoding Cytochrome P450 monooxygenase, down-regulated LDLa encoding low-density lipoprotein receptor domain class A, and DD_5 gene encoding a cuticular protein are involved in endocrine system of molting. In addition, up-regulated ferritin 1-like protein A and hemoglobin 2 (BJ9311841) and down-regulated hemoglobin1, 2 (BJ932805), 3 are linked to iron storage and iron absorption. Up-regulated ferritin genes in D. magna have previously been associated with cadmium exposure [37,102]. The general stress response gene HSP70 is also up-regulated [39]. The concentrations and time periods have an effect on different gene expression profiles. Soetaert et al. [102] stated that alteration of gene expression resulted from high concentration of cadmium. According to performed microarray data, 266 genes changed their expression after cadmium exposure.

The genes are related to several cell processes including digestion, oxygen transport, vitellogenin, cuticular metabolism, immune response, acid-base balance, visual-sensory preption and signal transduction. The up-regulated genes, cellulase (DW724578), alpha-amylase (DW985556) and alpha-esterase (DW724473), induce digestive processes of carbohydrates and lipids. Hemoglobin (dhb1) (DW724693) and di-domain hemoglobin precursor were down-regulated, resulting in transport of less O2. Furthermore four genes encoding vitellogenin, two of which are well documented in Daphnia magna [vitellogenin1 (DY037262) and vitellogenin2 (DY037256)], were also down-regulated. Seven genes related to cuticula metabolism showed altered expression [102]. In the case of cadmium and copper exposure, D. magna ferritin gene (AJ292556) was up-regulated. Putative metallothionein1 (MT1) cDNAs, DV437826 and DV437799, were up-regulated by cadmium exposure and copper and cadmium co-exposure [37]. MT proteins are involved in detoxification of heavy metals and have been used as biomarkers of metal exposure for many years [106].

For $CuSO_4$ or H_2O_2 exposure, the genes had similar expression pattern. Both chemicals are known to induce oxidative stress [107–110]. $CuSO_4$ and H_2O_2 induced gene expression of glutathione-S-transferase (GST), serine proteinase inhibitor (Serpins), and alcohol dehydrogenase (ADH). $CuSO_4$ also up-regulate lysosomal thiol reductase and expressed sequence tag, the flamingo homolog was up-regulated by H_2O_2 exposure. Pentachlorophenol (PCP) exposure induced ferritin-encoding gene, and beta-naphthoflavone induced gene encoding cathepsin L-like protease [38]. GST is also the biomarker of exposures to nanoparticle, fullerene. Catalase (CAT) is also the biomarker of nanoparticles.

These two proteins involved in the detoxification processes during oxidative stress, are commonly used biomarkers in toxicity tests and indicate the early alteration of response to pollutants. The GST and CAT levels were induced in fullerene nC60, fullerenol $C_{60}(OH)_{24}$, and titanium dioxide (TiO_2) exposure by the daphnids. Fullerene related to functionalization of nanoparticles and TiO_2 is the core structure of the nanoparticles. Both functionalization and core structure affect GSH and CAT expression [103,111,112]. In a study focusing on Zn exposure, large numbers of genes were up- and down-regulated. Transcription- and translation-related genes and all vitellogenin-related genes were down-regulated while oxidative stress responsive genes (such as glutathione-S-transferase and peroxiredoxin) and other different types of metabolism-related genes were up-regulated, in comparison to the control [92]. Ribosomal proteins were both up- and down-regulated [92]. In a study of BPA toxicity using microarray techniques and Blast2GO, Jeong et al. [42] elucidated the molecular mechanisms by altered genomic responses, the molecular functions, and biological processes. They selected several candidate biomarkers which were related to BPA exposure. The reproductive activity of D. magna has been related to the function of molting and chitinase activity [105]. Genes related to chitin activity, chitin deacetylase 2 isoform A (NP_001096047), chitin binding domain-containing protein (ABU80624), chitin binding protein 4 (BAI44118), and encoding cuticle protein (cuticle protein 1b) were up-regulated. Serine protease encoding gene, serine protease 13 (ABZ04021), serine protease H82 (EFA11957), and serine proteinase stubbles (EFN83787, EGI65883), were also up-regulated. Various proteolytic enzymes, such as serine protease can be affected molting cycle of Daphnia causing degradation of exoskeleton [102,113]. When vitellogenin is fused with superoxide dismutase, it is down-regulated and influences embryo development and reproduction of daphnia by providing nourishment for growing embryos [114].

ADVERSE OUTCOMES FROM ENVIRONMENTAL STRESSOR EXPOSURE

Various toxicants such as endocrine disrupting chemicals, heavy metals, pharmaceutical drugs, and organic compounds that are widely distributed in environment can perturb the ecologically relevant organisms. These toxicants disturb ecosystems by influencing organisms at the bottom of the ecological pyramid. When the organisms are affected, the entire ecological pyramid can collapse. Therefore the base of ecosystem is crucial for the entire web of living things, including humans. Studies on the toxic effect using the daphnia systems have been carried out recently. Daphnia is a model species to study aquatic environmental toxicity designated in OECD toxicity test guideline [115,116].

Endocrine disrupting chemicals (EDCs) interact with various receptors, such as the estrogen receptor (ER), androgen receptor and aryl-hydrocarbon receptor [117,118]. EDCs have an effect on reproduction by reducing the proportion of juveniles and by producing the male daphnia [119–121]. Heavy metal contaminants usually have negative effects on the development of population growth rate, longevity, and reproduction [6]. Organic compounds and drugs also have detrimental effects on reproduction and development impacting fecundity, abnormality of embryogenesis, and molting [104,122]. Table 1 summarizes studies of the effects of toxicants (i.e., EDCs, heavy metals, pharmaceutical drugs, organic compounds, and nanoparticles) on daphnia system.

Table 1: Toxicological studies using daphnia system in response to potential environmental stressors including endocrine disrupting chemicals (EDCs), heavy metals, organic compounds, pharmaceutical drugs, and nanoparticles

Chemicals	Species	Chemical Exposure	Effect	Reference
EDCs	*Daphnia magna*	Nonylphenol at 0.024 mg/L for 48 h	Abnormal proportion of juveniles	[119]
	Daphnia magna	Juvenile hormone at 330 ng/L for 21 days	Production of male offspring	[121]
	Ceriodaphnia dubia	Styrene at 0.04–1.7 μg/L for 7 days	Mortality and reduced fertility	[120]
	Daphnia magna	Bisphenol A at 6.67 and 10 mg/L for 21 days	Reduced offspring production	[42]
	Daphnia magna	Cadmium at 6, 20, and 37 μg/L for 24 h	Reduced survival and somatic growth	[39]
Heavy metals	*Daphnia magna*	Cadmium chloride at 71 μg/L for 24 h	Increased mortality, reduced survival, depleted glutathione level, and induced oxidative stress	[75]
	Ceriodaphnia dubia and *Daphnia carinata*	Copper, lead, zinc (at 1.3, 1.1, 13.0 mg/L., respectively) for 48 h or 7 days	Increased mortality and reduced reproduction ability	[123]
	Daphnia pulex	Organic selenium at 0.025 mg/L for 48 h	Death, immobility, and molting delay	[122]
Organic compounds	*Daphnia magna*	Propiconazole (pesticide) at 1 μg/L for 4 and 8 days	Impaired adult growth, decreased offspring development, impaired oocyte maturation, and interrupted resulting reproduction	[41]
	Daphnia magna	Fenvalerate (insecticide) at 0.6 μg/L for 24 h	Increased mortality, reduced survival, increased arginine phosphate level, and disrupted amino sugar metabolism	[75]
	Daphnia magna	Uncoupler of oxidative phosphorylation 2,4-dinitrophenol at 1.5 mg/L for 24 h	Increased mortality, reduced survival, and increased arginine phosphate level	[75]
	Daphnia magna	Alkylpolyglucosides (GCP 650, GCP 600, GCP 215) at IC_{50} (29, 14, 111 mg/L, respectively) for 24 h	Increased immobility	[124]

Chemicals	Species	Chemical Exposure	Effect	Reference
Drugs	*Daphnia magna*	Ibuprofen (non-steroidal anti-inflammatory drug or NSAID) at 20, 40, 80 mg/L for 8 days	Reduced fecundity and arrested early embryogenesis	[104]
	Daphnia magna	Beta-blocker anti-hypertension drug propanolol at 1.4 mg/L for 24 h	Increased mortality, reduced survival, and disrupted fatty acid metabolism and eicosanoid biosynthesis	[75]
	Daphnia magna	Mefenamicacid at EC_{50} (17.16 mg/L) for 48 h and 1 mg/L for 21 days	Increased immobility and reduced offspring production	[125]
Nanoparticles	*Daphnia magna*	Silver nanoparticles with surface coating at LC_{50} (0.88 μg/L) for 48 h	Increased mortality and reduced survival	[126]
	Daphnia magna	Polyvinylpyrrolidone-coated silver nanoparticles at LC_{50} (0.18 mg/L) for 24 h	Increased mortality, reduced survival, disrupted proteolysis and cell cycle	[44]
	Daphnia magna	Coated silver nanoparticles (Ag-GAs, Ag-PEGs, and Ag-PVPs) at LC_{50} (3.41, 3.16, 14.81 μg/L, respectively) for 48 h	Increased mortality	[127]
	Daphnia magna	Collargol (protein-coated nano Ag) and $AgNO_3$ nanoparticles at EC_{50} (20–27 ppb) for 48 h	Increased immobility	[128]
	Daphnia magna	Ag and CuO nanoparticles at EC_{50} (3.8 and 2.6 mg/L, respectively) for 24 h	Increased immobility	[129]
	Daphnia magna	Titanium dioxide nanoparticles at 0.01–10 mg/L for 48 h	Reduced survival offspring production, and digestion ability	[130]
	Daphnia magna	Nano and bulk titanium dioxide at 20 g/L for 48 h	Undetectable toxicity	[131]
	Daphnia magna	Bulk CuO, nano CuO and $CuSO_4$ at $L(E)C_{50}$ (165, 3.2, 0.17 mg/L, respectively) for 48 h	Increased immobility	[131]
	Daphnia magna	Bulk ZnO, nano ZnO and $ZnSO_4 \cdot 7H_2O$ at $L(E)C_{50}$ (1.8, 1.9, 1.1 mg/L, respectively) for 48 h	Increased immobility	[131]

PERSPECTIVES

Daphnia offers advantages for toxicological investigation of multiple stressors owing to high growth rate, high sensitivity to environmental changes, wide spatial distribution, parthenogenetic life cycle, and availability of omics-based tools [122]. In particular, its unique parthenogenesis facilitates the study of epigenetic effects without profound genetic differences [132,133]. Epigenetically regulated-sex determination and sexual reproduction are undergone toward harsh environmental conditions. Mapping the phenotype to the genotype in Daphnia system is considerable challenge. Genomic responses to genetic and environmental stressors as well as environmental impact on the phenotype have been pursued [134–136]. Integration of genome sequencing with the functional genomic tools provides a better insight on phenotypic evolution, mechanistic elucidation of the evolutionary novel traits, evolutionary adaptation, and regulatory pathways underlying to adaptive evolution. In the context of ecological risk assessment regarding complex mixtures, genomic tools and fingerprinting approaches offer exceptional platform [40]. Several transcriptomic research projects that use either simple mixtures of particular chemicals or complex mixtures like entire effluents have emerged to evaluate these capabilities, suggesting that the microarray approach was informative [137,138]. A mixture consisting of four chemicals was used to examine the hypothesis that salient transcriptomic responses of individual chemicals could be retained in mixtures, including compounds with dissimilar modes of toxicity [139,140]. This is one of the concepts to applying transcriptome signatures/fingerprints in environmental exposure diagnosis. Metabolomic analyses might complement with phenotypic data or transcriptomic data, while systemic ontologies and pathway analysis software for metabolomics has been developed even less in comparison to transcriptomics and proteomics. However, metabolomics likely has as much potential as transcriptomics or proteomics for utility to ecological risk assessment. Understanding of genetic and epigenetic mechanisms underlying phenotypic responses to environment (i.e., sex determination, sexual reproduction, helmets, and neckteeth) offers great benefit. In addition to DNA methylation, extensive researches on histone modification, RNAi, and changes in DNA methylation toward environmental stressors throughout developmental stages are necessary. Using Daphnia model, the epigenetic discriminatory between sexual and asexual as well as stressor-exposed and non-exposed daphnid individuals will be warranted with potential applications in the area of evolutionary and developmental biology

ACKNOWLEDGMENTS

This research was supported by a grant from "The Ecoinnovation Project" (412-112-011), by the Korea Ministry of Environment. This research project was also supported by a grant from Mahidol University, Thailand, and by a grant from the Faculty of Environment and Resource Studies, Mahidol University, Thailand.

REFERENCES

1.   Bartosiewicz, M.; Penn, S.; Buckpitt, A. Applications of gene arrays in environmental toxicology: Fingerprints of gene regulation associated with cadmium chloride, benzo(a)pyrene, and trichloroethylene. Environ.

Health Perspect. 2001, 109, 71–74

2. Snape, J.R.; Maund, S.J.; Pickford, D.B.; Hutchinson, T.H. Ecotoxicogenomics: The challenge of integrating genomics into aquatic and terrestrial ecotoxicology. Aquat. Toxicol. 2004, 67, 143–154

3. Miracle, A.L.; Ankley, G.T. Ecotoxicogenomics: Linkages between exposure and effects in assessing risks of aquatic contaminants to fish. Reprod. Toxicol. 2005, 19, 321–326

4. Iguchi, T.; Watanabe, H.; Katsu, Y. Application of ecotoxicogenomics for studying endocrine disruption in vertebrates and invertebrates. Environ. Health. Perspect. 2006, 114, 101–105

5. Watanabe, H.; Iguchi, T. Using ecotoxicogenomics to evaluate the impact of chemicals on aquatic organisms. Mar. Biol. 2006, 149, 107–115.

6. Altshuler, I.; Demiri, B.; Xu, S.; Constantin, A.; Yan, N.D.; Cristescu, M.E. An Integrated multi-disciplinary approach for studying multiple stressors in freshwater ecosystems: Daphnia as a model organism. Integr. Comp. Biol. 2011, 51, 623–633

7. Nuwaysir, E.F.; Bittner, M.; Trent, J.; Barrett, J.C.; Afshari, C.A. Microarrays and toxicology: The advent of toxicogenomics. Mol. Carcinog. 1999, 24, 153–159.

8. Kurelec, B. The genotoxic disease syndrome. Mar. Environ. Res. 1993, 35, 341–348.

9. Oakshott, J.G. Selection at the Adh locus in Drosophila melanogaster imposed by environmental ethanol. Genet. Res. 1976, 26, 265–274.

10. Hawkins, A.J.S.; Rusia, J.; Bayne, B.L.; Day, A.J. The metabolic/physiological basis of genotype-dependent mortality during copper exposure in Mytilus edulis. Mar. Environ. Res. 1989, 28, 139–144.

11. Schat, H.; Ten Bookum, W.M. Genetic control of copper tolerance in Silene vulgaris. Heredity 1992, 68, 219–229.

12. Hebert, P.D.N. The population biology of Daphnia (Crustacea, Daphnidae). Biol. Rev. 1978, 53, 387–426.

13. Lampert, W. Daphnia: Model herbivore, predator and prey. Pol. J. Ecol. 2006, 54, 607–620.

14. Cristescu, M.E.; Colbourne, J.K.; Radivojac, J.; Lynch, M. A microsatellite-based linkage map of the water flea Daphnia pulex: On the prospect of crustacean genomics. Genomics 2006, 88, 415–430

15. Eads, B.D.; Andrews, J.; Colbourne, J.K. Ecological genomics in Daphnia: Stress responses and environmental sex determination. Heredity 2008, 100, 184–190

16. Colbourne, J.K.; Hebert, P.D.N.; Taylor, D.J. Evolutionary origins of phenotypic diversity in daphnia. In Molecular Evolution and Adaptive Radiation; Givnish, T.J., Sytsma, K.J., Eds.; Cambridge University Press: Cambridge, UK, 1997; pp. 163–188.

17. Schindler, D.W. Detecting ecosystem responses to anthropogenic stress. Can. J. Fish. Aquat. Sci. 1987, 44, 6–25.

18. Hebert, P.D.N.; Crease, T. Clonal diversity in populations of Daphnia pulex reproducing by obligate parthenogenesis. Heredity 1983, 51, 353–369.

19. Stich, H.B.; Lampert, W. Growth and reproduction of migrating and non-migrating Daphnia species under simulated food and temperature conditions of diurnal vertical migration. Oecologia 1984, 61, 192–196.

20. Dawidowicz, P.; Loose, C.J. Metabolic costs during predator-induced diel vertical migration of Daphnia. Limnol. Oceanogr. 1992, 37, 1589–1595.

21. Gerhardt, A.; de Bisthoven, B.L.; Soares, A.M.V. Evidence for the stepwise stress model: Gambusia holbrooki and Daphnia magna under acid mine drainage and acidified reference water stress. Environ. Sci. Technol. 2005, 39, 4150–4158

22. Tollrian, R. Neckteeth formation in Daphnia pulex as an example of continuous phenotypic plasticity: Morphological effects of Chaoborus kairomone concentration and their quantification. J. Plankton Res. 1993, 15, 1309–1318.

23. Tautz, D. Not just another genome. BMC. Biol. 2011, 9, 8

24. Fryer, G. Functional morphology and the adaptive radiation of the Daphnidae (Branchiopoda: Anomopoda). Philos. Trans. R. Soc. B. 1991, 331, 1–99.

25. Hebert, P.D.N. The Daphnia of North America: An Illustrated Fauna; University of Guelph: Guelph, ON, Canada, 1995.

26. Wellborn, G.A.; Skelly, D.K.; Werner, E.E. Mechanisms creating community structure across a freshwater habitat gradient. Annu. Rev. Ecol. Syst. 1996, 27, 337–363.

27. Crease, T.J.; Lee, S.K.; Yu, S.L.; Spitze, K.; Lehman, N.; Lynch, M. Allozyme and mtDNA variation in populations of the Daphnia pulex complex from both sides of the Rocky Mountains. Heredity 1997, 79, 242–251.

28. Pfrender, M.E.; Spitze, K.; Lehman, N. Multi-locus genetic evidence for rapid ecologically based speciation in Daphnia. Mol. Ecol. 2000, 9, 1717–1735

29. Boverhof, D.R.; Gollapudi, B.B. Applications of Toxicogenomics. In Safety Evaluation and Risk Assessment, 1st ed.; Wiley: Singapore, 2011.

30. Larkin, P.; Villeneuve, D.L.; Knoebl, I.; Miracle, A.L.; Carter, B.J.; Liu, L.; Denslow, N.D.; Ankley, G.T. Development and validation of a 2000-gene microarray for the fathead minnow (Pimephales promelas). Environ. Toxicol. Chem. 2007, 26, 1497–1506

31. Morgan, M.B.; Edge, S.E.; Snell, T.W. Profiling differential gene expression of corals along a transect of waters adjacent to the Bermuda municipal dump. Mar. Pollut. Bull. 2005, 51, 524–533

32. Park, H.R.; Yang, H.; Kim, G.D.; Son, G.W.; Park, Y.S. Microarray analysis of gene expression in 3-methylcholanthrene-treated human endothelial cells. Mol. Cell. Toxicol. 2014, 10, 19–27.

33. Carney, S.A.; Chen, J.; Burns, C.G.; Xiong, K.M.; Peterson, R.E.; Heideman, W. Aryl hydrocarbon receptor activation produces heart-specific transcriptional and toxic responses in developing zebrafish. Mol. Pharmacol. 2006, 70, 549–561

34. Santos, E.M.; Paull, G.C.; van Look, K.J.; Workman, V.L.; Holt, W.V.; van Aerle, R.; Kille, P.; Tyler, C.R. Gonadal transcriptome responses and physiological consequences of exposure to estrogen in breeding zebrafish (Danio rerio). Aquat. Toxicol. 2007, 83, 134–142

35. Wang, R.L.; Bencic, D.; Biales, A.; Lattier, D.; Kostich, M.; Villeneuve, D.; Ankley, G.T.; Lazorchak, J.; Toth, G. DNA microarray-based ecotoxicological biomarker discovery in a small fish model species. Environ. Toxicol. Chem. 2008, 27, 664–675

36. Heckmann, L.H.; Connon, R.; Hutchinson, T.H.; Maund, S.J.; Sibly, R.M.; Callaghan, A. Expression of target and reference genes in Daphnia magna exposed to ibuprofen. BMC Genomics 2006, 7, 175

37. Poynton, H.C.; Varshavsky, J.R.; Chang, B.; Cavigiolio, G.; Chan, S.; Holman, P.S.; Loguinov, A.V.; Bauer, D.J.; Komachi, K.; Theil, E.C.; et al. Daphnia magna ecotoxicogenomics provides mechanistic insights into metal toxicity. Environ. Sci. Technol. 2007, 41, 1044–1050

38. Watanabe, H.; Takahashi, E.; Nakamura, Y.; Oda, S.; Tatarazako, N.; Iguchi, T. Development of a Daphnia magna DNA microarray for evaluating the toxicity of environmental chemicals. Environ. Toxicol. Chem. 2007, 26, 669–676

39. Connon, R.; Hooper, H.L.; Sibly, R.M.; Lim, F.L.; Heckmann, L.H.; Moore, D.J.; Watanabe, H.; Soetaert, A.; Cook, K.; Maund, S.J.; et al. Linking molecular and population stress responses in Daphnia magna exposed to cadmium. Environ. Sci. Technol. 2008, 42, 2181–2188

40. Garcia-Reyero, N.; Griffitt, R.J.; Liu, L. Construction of a robust microarray from a non-model species (largemouth bass) using pyrosequencing technology. J. Fish. Biol. 2008, 72, 2354–2376

41. Soetaert, A.; Moens, L.N.; van der Ven, K.; van Leemput, K.; Naudts, B.; Blust, R.; de Coen, W.M. Molecular impact of propiconazole on Daphnia magna using a reproduction-related cDNA array. Comp. Biochem. Physiol. C Toxicol. Pharmacol. 2006, 142, 66–76

42. Jeong, S.W.; Lee, S.M.; Yum, S.S.; Iguchi, T.; Seo, Y.R. Genomic expression responses toward bisphenol-A toxicity in Daphnia magna in terms of reproductive activity. Mol. Cell. Toxicol. 2013, 9, 149–158.

43. Dong, Y.; Zhai, L.; Zhang, L.; Jia, L.; Wang, X. Bisphenol A impairs mitochondrial function in spleens of mice via oxidative stress. Mol. Cell. Toxicol. 2013, 9, 401–406.

44. Poynton, H.C.; Lazorchak, J.M.; Impellitteri, C.A.; Blalock, B.J.; Rogers, K.; Allen, H.J.; Loguinov, A.; Heckman, J.L.; Govindasmawy, S. Toxicogenomic responses of nanotoxicity in Daphnia magna exposed to silver nitrate and coated silver nanoparticles. Environ. Sci. Technol. 2012, 46, 6288–6296

45. Villeneuve, D.L.; Larkin, V.P.; Knoebl, I.; Miracle, A.L.; Kahl, M.D.; Jensen, K.M.; Makynen, E.A.; Durhan, E.J.; Carter, B.J.; Denslow, N.D.; et al. A graphical systems model to facilitate hypothesis-driven ecotoxicogenomics research on the teleost brain-pituitary-gonadal axis. Environ. Sci. Technol. 2007, 41, 321–330

46. Villeneuve, D.L.; Knoebl, I.; Larkin, P.; Miracle, A.L.; Carter, B.J.; Denslow, N.D.; Ankley, G.T. Altered gene expression in the brain and liver of female fathead minnows Pimephales promelas Rafinesque exposed to fadrozole. J. Fish. Biol. 2008, 72, 2281–2340.

47. Denslow, N.D.; Kocerha, J.; Sepúlveda, M.S.; Gross, T.; Holm, S.E. Gene expression fingerprints of largemouth bass (Micropterus salmoides) exposed to pulp and paper mill effluents. Mutat. Res. 2004, 552, 19–34

48. Williams, T.D.; Diab, A.M.; George, S.G.; Godfrey, R.E.; Sabine, V.; Conesa, A.; Minchin, S.D.; Watts, P.C.; Chipman, J.K. Development of the GENIPOL European flounder (Platichthys flesus) microarray and determination of temporal transcriptional responses to cadmium at low dose. Environ. Sci. Technol. 2006, 40, 6479–6488

49. Yeung, L.W.Y.; Guruge, K.S.; Yamanaka, N.; Miyazaki, S.; Lam, P.K.S. Differential expression of chicken hepatic genes responsive to PFOA and PFOS. Toxicology 2007, 237, 111–125

50. Krasnov, A.; Koskinen, H.; Pehkonen, P.; Rexroad, C.E.; Afanasyev, S.;

Mölsä, H. Gene expression in the brain and kidney of rainbow trout in response to handling stress. BMC Genomics 2005, 6, 3

51. Krasnov, A.; Koskinen, H.; Pehkonen, P.; Rexroad, C.; Afanasyev, S.; Mölsä, H.; Oikari, A. Transcriptome responses to carbon tetrachloride and pyrene in the kidney and liver of juvenile rainbow trout (Oncorhynchus mykiss). Aquat. Toxicol. 2005, 74, 70–81

52. Kassahn, K.S.; Crozier, R.H.; Ward, A.C.; Stone, G.; Caley, M.J. From transcriptome to biological function: Environmental stress in an ectothermic vertebrate, the coral reef fish Pomacentrus moluccensis. BMC Genomics 2007, 8, 358

53. Lam, S.H.; Winata, C.L.; Tong, Y.; Korzh, S.; Lim, W.S.; Jan Spitsbergen, V.K.; Mathavan, S.; Miller, L.D.; Liu, E.T. Transcriptome kinetics of arsenic-induced adaptive response in zebrafish liver. Physiol. Genomics 2006, 27, 351–361

54. Volz, D.C.; Hinton, D.E.; Law, J.M.; Kullman, S.W. Dynamic gene expression changes precede dioxin-induced liver pathogenesis in medaka fish. Toxicol. Sci. 2006, 89, 524–534

55. Soetaert, A.; van der Ven, K.; Moens, L.N.; Vandenbrouck, T.; van Remortel, P.; de Coen, W.M. Daphnia magna and ecotoxicogenomics: Gene expression profiles of the anti-ecdysteroidal fungicide fenarimol using energy-, molting- and life stage-related cDNA libraries. Chemosphere 2007, 67, 60–71

56. Watanabe, H.; Tatarazako, N.; Oda, S.; Nishide, H.; Uchiyama, I.; Morita, M.; Iguchi, T. Analysis of expressed sequence tags of the water flea Daphnia magna. Genome 2005, 48, 606–609

57. Margulies, M.; Egholm, M.; Altman, W.E.; Attiya, S.; Bader, J.S.; Bemben, L.A.; Berka, J.; Braverman, M.S.; Chen, Y.J.; Chen, Z.T.; et al. Genome sequencing in microfabricated high-density picolitre reactors. Nature 2005, 437, 376–380

58. Hardy, S.; Legagneux, V.; Audic, Y.; Paillard, L. Reverse genetics in eukaryotes. Biol. Cell 2010, 102, 561–580

59. Martyniuk, C.J.; Trudeau, V.L. Fish endocrinology meets functional genomics: What exactly is the message? Gen. Comp. Endocrinol. 2009, 164, 132–134

60. Kato, Y.; Kobayashi, K.; Watanabe, H.; Iguchi, T. Environmental sex determination in the branchipod crustacean Daphnia magna: Deep conservation of a double-sex gene in the sex-determining pathway. PLoS Genet. 2011, 7, e1001345

61. Wetmore, B.A.; Merrick, B.A. Toxicoproteomics: Proteomics applied to toxicology and pathology. Toxicol. Pathol. 2004, 32, 619–642

62. Shepard, J.L.; Olsson, B.; Tedengren, M.; Bradley, B.P. Protein expression signatures identified in Mytilus edulis exposed to PCBs, copper and salinity stress. Mar. Environ. Res. 2000, 50, 337–340.

63. Rodríguez-Ortega, M.J.; Grøsvik, B.E.; Rodríguez-Ariza, A.; Goksøyr, A.; López-Barea, J. Changes in protein expression profiles in bivalve molluscs (Chamaelea gallina) exposed to four model environmental pollutants. Proteomics 2003, 3, 1535–1543

64. Shrader, E.A.; Henry, T.R.; Greeley, M.S., Jr.; Bradley, B.P. Proteomics in zebrafish exposed to endocrine disrupting chemicals. Ecotoxicology 2003, 12, 485–458

65. Hogstrand, C.; Balesaria, S.; Glover, C.N. Application of genomics and proteomics for study of the integrated response to zinc exposure in a non-model fish species, the rainbow trout. Comp. Biochem. Physiol. B Biochem. Mol. Biol. 2002, 133, 523–535.

66. Bjørnstad, A.; Larsen, B.K.; Skadsheim, A.; Jones, M.B.; Andersen, O.K. The potential of ecotoxicoproteomics in environmental monitoring: Biomarker profiling in mussel plasma using Protein Chip array technology. J. Toxicol. Environ. Health A 2006, 69, 77–96

67. Le, T.H.; Lim, E.S.; Hong, N.H.; Lee, S.K.; Shim, Y.S.; Hwang, J.R.; Kim, Y.H.; Min, J. Proteomic analysis in Daphnia magna exposed to As(III), As(V) and Cd heavy metals and their binary mixtures for screening potential biomarkers. Chemosphere 2013, 93, 2341–2348

68. Rainville, L.C.; Carolan, D.; Varela, A.C.; Doyle, H.; Sheehan, D. Proteomic evaluation of citrate-coated silver nanoparticles toxicity in Daphnia magna. Analyst 2014, 7, 1678–1686

69. Choi, Y.S.; Lee, M.Y.; David, A.E.; Park, Y.S. Nanoparticles for gene delivery: Therapeutic and toxic effects. Mol. Cell. Toxicol. 2014, 10, 1–8.

70. Ralston-Hooper, K.; Hopf, A.; Oh, C.; Zhang, X.; Adamec, J.; Sepúlveda, M.S. Development of GCxGC/TOF-MS metabolomics for use in ecotoxicological studies with invertebrates. Aquat. Toxicol. 2008, 2, 48–52

71. Yoo, G.; Jeong, S.H.; Ryu, W.I.; Lee, H.; Kim, J.H.; Bae, H.C.; Son, S.W. Gene expression analysis reveals a functional role for the Ag-NPs-induced Egr-1 transcriptional factor in human keratinocytes. Mol. Cell. Toxicol. 2014, 10, 149–156.

72. Park, Y.H.; Bae, H.C.; Jang, Y.; Jeong, S.H.; Lee, H.N.; Ryu, W.I.; Yoo,

M.G.; Kim, Y.R.; Kim, M.K.; Lee, J.K.; et al. Effect of the size and surface charge of silica nanoparticles on cutaneous toxicity. Mol. Cell. Toxicol. 2013, 9, 67–74.

73. Bundy, J.G.; Keun, H.C.; Sidhu, J.K.; Spurgeon, D.J.; Svendsen, C.; Kille, P.; Morgan, A.J. Metabolic profile biomarkers of metal contamination in a sentinel terrestrial species are applicable across multiple sites. Environ. Sci. Technol. 2007, 41, 4458–4464

74. Taylor, N.S.; Weber, R.J.M.; Southam, A.D.; Payne, T.G.; Hrydziuszko, O.; Arvanitis, T.N.; Viant, M.R. A new approach to toxicity testing in Daphnia magna: Application of high throughput FT-ICR mass spectrometry metabolomics. Metabolomics 2009, 5, 44–58.

75. Taylor, N.S.; Weber, R.J.; White, T.A.; Viant, M.R. Discriminating between different acute chemical toxicities via changes in the daphnid metabolome. Toxicol. Sci. 2010, 118, 307–317

76. Stohs, S.J.; Bagchi, D. Oxidative mechanisms in the toxicity of metal ions. Free Radic. Biol. Med. 1995, 18, 321–336.

77. Ray, D.E.; Fry, J.R. A reassessment of the neurotoxicity of pyrethroid insecticides. Pharmacol. Ther. 2006, 111, 174–193

78. Drysdale, G.R.; Cohn, M. Mode of action of 2,4-dinitrophenol in uncoupling oxidative phosphorylation. J. Biol. Chem. 1958, 233, 1574–1577. [PubMed]

79. Huggett, D.B.; Brooks, B.W.; Peterson, B.; Foran, C.M.; Schlenk, D. Toxicity of select beta adrenergic receptor-blocking pharmaceuticals (b-blockers) on aquatic organisms. Arch. Environ. Contam. Toxicol. 2002, 43, 229–235

80. Poynton, H.C.; Taylor, N.S.; Hicks, J.; Colson, K.; Chan, S.; Clark, C.; Scanlan, L.; Loguinov, A.V.; Vulpe, C.; Viant, M.R. Metabolomics of microliter hemolymph samples enables an improved understanding of the combined metabolic and transcriptional responses of Daphnia magna to cadmium. Environ. Sci. Technol. 2011, 45, 3710–3717

81. Nagato, E.G.; D'eon, J.C.; Lankadurai, B.P.; Poirier, D.G.; Reiner, E.J.; Simpson, A.J.; Simpson, M.J. 1H NMR-based metabolomics investigation of Daphnia magna responses to sub-lethal exposure to arsenic, copper and lithium. Chemosphere 2013, 93, 331–337

82. Ho, D.H.; Burggren, W.W. Epigenetics and transgenerational transfer: A physiological perspective. J. Exp. Biol. 2010, 213, 3–16

83. Vandegehuchte, M.B.; Janssen, C.R. Epigenetics and its implications for ecotoxicology. Ecotoxicology 2011, 20, 607–624

84. Jaenisch, R.; Bird, A. Epigenetic regulation of gene expression: How the genome integrates intrinsic and environmental signals. Nat. Genet. 2003, 33, 245–254

85. Santos, K.F.; Mazzola, T.N.; Carvalho, H.F. The prima donna of epigenetics: The regulation of gene expression by DNA methylation. Braz. J. Med. Biol. Res. 2005, 38, 1531–1541

86. Fuks, F. DNA methylation and histone modifications: Teaming up to silence genes. Curr. Opin. Genet. Dev. 2005, 15, 490–495

87. Lennartsson, A.; Ekwall, K. Histone modification patterns and epigenetic codes. Biochim. Biophys. Acta 2009, 1790, 863–868

88. Lippman, Z.; Martienssen, R. The role of RNA interference in heterochromatic silencing. Nature 2004, 431, 364–370

89. Carthew, R.W.; Sontheimer, E.J. Origins and Mechanisms of miRNAs and siRNAs. Cell 2009, 136, 642–655

90. Reamon-Buettner, S.M.; Mutschler, V.; Borlak, J. The next innovation cycle in toxicogenomics: Environmental epigenetics. Mutat. Res. 2008, 659, 158–165

91. Krieg, A.J.; Rankin, E.B.; Chan, D.; Razorenova, O.; Fernandez, S.; Giaccia, A.J. Regulation of the histone demethylase JMJD1A by hypoxia-inducible factor 1 alpha enhances hypoxic gene expression and tumor growth. Mol. Cell. Biol. 2010, 30, 344–353

92. Vandegehuchte, M.B.; Vandenbrouck, T.; Coninck, D.D.; de Coen, W.M.; Janssen, C.R. Can metal stress induce transferable changes in gene transcription in Daphnia magna? Aquat. Toxicol. 2010, 97, 188–195

93. Aluru, N.; Karchner, S.I.; Hahn, M.E. Role of DNA methylation of AHR1 and AHR2 promoters in differential sensitivity to PCBs in Atlantic Killifish, Fundulus heteroclitus. Aquat. Toxicol. 2011, 101, 288–294

94. Vandegehuchte, M.B.; Kyndt, T.; Vanholme, B.; Haegeman, A.; Gheysen, G.; Janssen, C.R. Occurrence of DNA methylation in Daphnia magna and influence of multigeneration Cd exposure. Environ. Int. 2009, 35, 700–706

95. Youngson, N.A.; Whitelaw, E. Transgenerational epigenetic effects. Annu. Rev. Genomics Hum. Genet. 2008, 9, 233–257

96. Vandegehuchte, M.B.; Lemi'ere, F.; Vanhaecke, L.; Vanden Berghe, W.; Janssen, C.R. Direct and transgenerational impact on Daphnia magna of chemicals with a known effect on DNA methylation. Comp. Biochem. Physiol. C Toxicol. Pharmacol. 2010, 151, 278–285

97. Vandegehuchte, M.B.; Vandenbrouck, T.; de Coninck, D.; de Coen,

W.M.; Janssen, C.R. Gene transcription and higher-level effects of multigenerational Zn exposure in Daphnia magna. Chemosphere 2010, 80, 1014–1020

98. Eads, B.D.; Colbourne, J.K.; Bohuski, E.; Andrews, J. Profiling sex-biased gene expression during parthenogenetic reproduction in Daphnia pulex. BMC Genomics 2007, 8, 464

99. Agrawal1, A.A.; Laforsch, C.; Tollrian, R. Transgenerational induction of defences in animals and plants. Nature 1999, 401, 60–63.

100. Miyakawa, H; Imai, M.; Sugimoto, N.; Ishikawa, Y.; Ishikawa, A.; Ishigaki, H.; Okada, Y.; Miyazaki, S.; Koshikawa, S.; Cornette, R.; et al. Gene up-regulation in response to predator kairomones in the water flea, Daphnia pulex. BMC Dev. Biol. 2010, 10, 45.

101. Vandegehuchte, M.B.; Lemière, F.; Janssen, C.R. Quantitative DNA-methylation in Daphnia magna and effects of multigeneration Zn exposure. Comp. Biochem. Physiol. C Toxicol. Pharmacol. 2009, 150, 343–348

102. Soetaert, A.; Vandenbrouck, T.; van der Ven, K.; Maras, M.; van Remortel, P.; Blust, R.; de Coen, W.M. Molecular responses during cadmium-induced stress in Daphnia magna: Integration of differential gene expression with higher-level effects. Aquat. Toxicol. 2007, 83, 212–222

103. Klaper, R.; Crago, J.; Barr, J.; Arndt, D.; Setyowati, K.; Chen, J. Toxicity biomarker expression in daphnids exposed to manufactured nanoparticles: Changes in toxicity with functionalization. Environ. Pollut. 2009, 157, 1152–1156

104. Heckmann, L.H.; Sibly, R.M.; Connon, R.; Hooper, H.L.; Hutchinson, T.H.; Maund, S.J.; Hill, C.J.; Bouetard, A.; Callaghan, A. Systems biology meets stress ecology: Linking molecular and organismal stress responses in Daphnia magna. Genome. Biol. 2008, 9, R40

105. Merzendorfer, H.; Zimoch, L. Chitin metabolism in insects: Structure, function and regulation of chitin synthases and chitinases. J. Exp. Biol. 2003, 206, 4393–4412

106. Hennig, H.F. Metal-binding proteins as metal pollution indicators. Environ. Health Perspect. 1986, 65, 175–187. [PubMed]

107. Gaetke, L.M.; Chow, C.K. Copper toxicity, oxidative stress, and antioxidant nutrients. Toxicology 2003, 189, 147–163.

108. Wang, Y.J.; Lee, C.C.; Chang, W.C.; Liou, H.B.; Ho, Y.S. Oxidative stress and liver toxicity in rats and human hepatoma cell line induced

by pentachlorophenol and its major metabolite tetrachlorohydroquinone. Toxicol. Lett. 2001, 122, 157–169.

109. Xie, F.; Koziar, S.A.; Lampi, M.A.; Dixon, D.G.; Norwood, W.P.; Borgmann, U.; Huang, X.-D.; Greenberg, B.M. Assessment of the toxicity of mixtures of copper, 9,10-phenanthrenequinone, and phenanthrene to Daphnia magna: Evidence for a reactive oxygen mechanism. Environ. Toxicol. Chem. 2006, 25, 613–622

110. Kim, D.K.; Song, J.W.; Park, J.D.; Choi, B.S. Copper induces the accumulation of amyloid-beta in the brain. Mol. Cell. Toxicol. 2013, 9, 57–66.

111. Park, H.G.; Yeo, M.K. Effects of TiO2 nanoparticles and nanotubes on zebrafish caudal fin regeneration. Mol. Cell. Toxicol. 2013, 9, 375–383.

112. Park, H.G.; Kim, J.I.; Kang, M.; Yeo, M.K. The effect of metal-doped TiO2 nanoparticles on zebrafish embryogenesis. Mol. Cell. Toxicol. 2014, 10, 293–301.

113. Ote, M.; Mita, K.; Kawasaki, H.; Daimon, T.; Kobayashi, M.; Shimada, T. Identification of molting fluid carboxypeptidase A (MF-CPA) in Bombyx mori. Comp. Biochem. Physiol. B Biochem. Mol. Biol. 2005, 141, 314–322

114. Kato, Y.; Tokishita, S.; Ohta, T.; Yamagata, H. A vitellogenin chain containing a superoxide dismutase-like domain is the major component of yolk proteins in cladoceran crustacean Daphnia magna. Gene 2004, 334, 157–165

115. OECD/OCDE 202, OECD Guideline for Testing of Chemicals; Daphnia sp. Acute Immobilisation Test. Available online: http://www.oecd-ilibrary.org/environment/test-no-202-daphnia-sp-acute-immobilisation-test_9789264069947-en (accessed on 28 May 2015).

116. OECD/OCDE 211, OECD Guideline for Testing of Chemicals; Daphnia magna Reproduction Test. Available online: http://www.oecd-ilibrary.org/environment/test-no-211-daphnia-magna-reproduction-test_9789264185203-en (accessed on 28 May 2015).

117. McLachlan, J.A. Environmental signaling: What embryos and evolution teach us about endocrine disrupting chemicals. Endocr. Rev. 2001, 22, 319–341

118. Iguchi, T.; Sumi, M.; Tanabe, S. Endocrine disruptor issues in Japan. Congen. Anorm. 2002, 42, 106–119.

119. Comber, M.H.I.; Williams, T.D.; Stewart, K.M. The effects of nonylphenol on Daphnia magna. Water Res. 1993, 27, 273–276.

120. Tatarazako, N.; Takao, Y.; Kishi, K.; Onikura, N.; Arizono, K.; Iguchi, T. Styrene dimers and trimers affect reproduction of daphnid (Ceriodaphnia dubia). Chemosphere 2002, 48, 597–601.

121. Tatarazako, N.; Oda, S.; Watanabe, H.; Morita, M.; Iguchi, T. Juvenile hormone agonists affect the occurrence of male Daphnia. Chemosphere 2003, 53, 827–833.

122. Schultz, T.W.; Freeman, S.R.; Dumont, J.N. Uptake, depuration, and distribution of selenium in Daphnia and its effects on survival and ultrastructure. Arch. Environ. Contam. Toxicol. 1980, 9, 23–40

123. Cooper, N.L.; Bidwell, J.R.; Kumar, A. Toxicity of copper, lead, and zinc mixtures to Ceriodaphniadubia and Daphnia carinata. Ecotoxicol. Environ. Saf. 2009, 72, 1523–1528

124. Jurado, E.; Fernández-Serrano, M.; Núñez Olea, J.; Lechuga, M.; Jiménez, J.L.; Ríos, F. Acute toxicity of alkylpolyglucosides to Vibrio fischeri, Daphnia magna and microalgae: A comparative study. Bull. Environ. Contam. Toxicol. 2012, 88, 290–295

125. Collard, H.J.; Ji, K.; Lee, S.; Liu, X.; Kang, S.; Kho, Y.; Ahn, B.; Ryu, J.; Lee, J.; Choi, K. Toxicity and endocrine disruption in zebrafish (Danio rerio) and two freshwater invertebrates (Daphnia magna and Moina macrocopa) after chronic exposure to mefenamic acid. Ecotoxicol. Environ. Saf. 2013, 94, 80–86

126. Zhao, C.M.; Wang, W.X. Importance of surface coatings and soluble silver in silver nanoparticles toxicity to Daphnia magna. Nanotoxicology 2012, 6, 361–370

127. Newton, K.M.; Puppala, H.L.; Kitchens, C.L.; Colvin, V.L.; Klaine, S.J. Silver nanoparticle toxicity to Daphnia magna is a function of dissolved silver concentration. Environ. Toxicol. Chem. 2013, 32, 2356–2364

128. Blinova, I.; Niskanen, J.; Kajankari, P.; Kanarbik, L.; Käkinen, A.; Tenhu, H.; Penttinen, O.P.; Kahru, A. Toxicity of two types of silver nanoparticles to aquatic crustaceans Daphnia magna and Thamnocephalus platyurus. Environ. Sci. Pollut. Res. Int. 2013, 20, 3456–3463

129. Jo, H.J.; Choi, J.W.; Lee, S.H.; Hong, S.W. Acute toxicity of Ag and CuO nanoparticle suspensions against Daphnia magna: The importance of their dissolved fraction varying with preparation methods. J. Hazard. Mater 2012, 227–228, 301–308

130. Fouqueray, M.; Dufils, B.; Vollat, B.; Chaurand, P.; Botta, C.; Abacci, K.; Labille, J.; Rose, J.; Garric, J. Effects of aged TiO2 nanomaterial from sunscreen on Daphnia magna exposed by dietary route. Environ. Pollut. 2012, 163, 55–61

131. Heinlaan, M.; Ivask, A.; Blinova, I.; Dubourguier, H.C.; Kahru, A. Toxicity of nanosized and bulk ZnO, CuO and TiO2 to bacteria Vibrio fischeri and crustaceans Daphnia magna and Thamnocephalus platyurus. Chemosphere 2008, 71, 1308–1316

132. Lubbock, J. An account of the two methods of reproduction in Daphnia, and of the structure of the ephippium. Philos. Trans. R. Soc. Lond. 1857, 147, 79–100.

133. Zaffagnini, F.; Peters, R.H.; de Bernardi, R. Reproduction in Daphnia. Mem. Ist. Ital. Idrobiol. 1987, 45, 245–284.

134. Kleiven, O.T.; Larsson, P.; Oikos, A.H. Sexual reproduction in Daphnia magna requires three stimuli. OIKOS 1992, 65, 197–206.

135. Thompson, J.D. Phenotypic plasticity as a component of evolutionary change. Trends Evol. Ecol. 1991, 6, 246–249.

136. Depledge, M.H. Genotypic toxicity: Implications for individuals and populations. Environ. Health Perspect. 1994, 102, 101–104

137. Shugart, L.R.; Theodorakis, C. Genetic ecotoxicology: The genotypic diversity approach. Comp. Biochem. Physiol. 1996, 113, 273–276.

138. Moens, L.N.; van der Ven, K.; van Remortel, P.; del-Favero, J.; de Coen, W.M. Expression profiling of endocrine-disrupting compounds using a customized Cyprinus carpio cDNA microarray. Toxicol. Sci. 2006, 93, 298–310

139. Moens, L.N.; Smolders, R.; van der Ven, K.; van Remortel, P.; del-Favero, J.; de Coen, W.M. Effluent impact assessment using microarray-based analysis in common carp: A systems toxicology approach. Chemosphere 2007, 67, 2293–2304

140. Koedrith, P.; Boonprasert, R.; Kwon, J.Y.; Kim, I.S.; Seo, Y.R. Recent toxicological investigations of metal or metal oxide nanoparticles in mammalian models in vitro and in vivo: DNA damaging potential, and relevant physicochemical characteristics. Mol. Cell. Toxicol. 2014, 10, 107–126.

CITATION

CHAPTER 1

Rebelo, M. , Santos, G. and Silva, R. (2014) A Methodology to Develop the Integration of the Environmental Management System with Other Standardized Management Systems. Computational Water, Energy, and Environmental Engineering, 3, 170-181. doi: 10.4236/cweee.2014.34018.

CHAPTER 2

Hardisty, P.E.; Sivapalan, M.; Brooks, P. The Environmental and Economic Sustainability of Carbon Capture and Storage. doi:10.3390/ijerph8051460

CHAPTER 3

Assessment and Management Perspectives, Emerging Pollutants in the Environment - Current and Further Implications, Prof. Marcelo Larramendy (Ed.), ISBN: 978-953-51-2160-2, InTech, DOI: 10.5772/60205

CHAPTER 4

Masachika Suzuki (2013). What Are the Roles of National and International Institutions to Overcome Barriers in Diffusing Clean Energy Technologies in Asia?: Matching Barriers in Technology Diffusion with the Roles of Institutions, Environmental Change and Sustainability, Dr. Steven Silvern (Ed.), ISBN: 978-953-51-1094-1, DOI: 10.5772/54124.

CHAPTER 5

Volker Schneider and Jana K. Ollmann (2013). Punctuations and Displacements in Policy Discourse: The Climate Change Issue in Germany 2007-2010, Environmental Change and Sustainability, Dr. Steven Silvern (Ed.), ISBN: 978-953-51-1094-1, InTech, DOI: 10.5772/54302.

CHAPTER 6

Merayyan, S. and Mrayyan, S. (2014) Jordan's Water Resources: Increased Demand with Unreliable Supply.Computational Water, Energy, and Environmental Engineering, 3, 48-56. doi: 10.4236/cweee.2014.32007.

CHAPTER 7

Uddin, M. , Alam, J. , Khan, Z. , Jahid Hasan, G. and Rahman, T. (2014) Two Dimensional Hydrodynamic Modelling of Northern Bay of Bengal Coastal Waters. Computational Water, Energy, and Environmental Engineering, 3, 140-151. doi: 10.4236/cweee.2014.34015.

CHAPTER 8

Mohd Saudi, A. , Juahir, H. , Azid, A. , Amri Kamarudin, M. , Toriman, M. and Abdul Aziz, N. (2014) Flood Risk Pattern Recognition Using Chemometric Technique: A Case Study in Muda River Basin. Computational Water, Energy, and Environmental Engineering, 3, 102-110. doi: 10.4236/cweee.2014.33011

CHAPTER 9

Saada, N. (2015) Simulation of Long Term Characteristics of Annual Rainfall in Selected Areas in Saudi Arabia.Computational Water, Energy, and Environmental Engineering, 4, 18-24. doi: 10.4236/cweee.2015.42003.

CHAPTER 10

Makaka, G. (2014) Influence of Fly Ash on Brick Properties and the Impact of Fly Ash Brick Walls on the Indoor Thermal Comfort for Passive Solar Energy Efficient House. Computational Water, Energy, and Environmental Engineering, 3, 152-161. doi: 10.4236/cweee.2014.34016.

CHAPTER 11

Hadgu, L. , Nyadawa, M. , Mwangi, J. , Kibetu, P. and Mehari, B. (2014) Application of Water Quality Model QUAL2K to Model the Dispersion of

Pollutants in River Ndarugu, Kenya. Computational Water, Energy, and Environmental Engineering, 3, 162-169. doi: 10.4236/cweee.2014.34017.

CHAPTER 12

Chauhan, M. , Dikshit, P. and Dwivedi, S. (2015) Modeling of Discharge Distribution in Bend of Ganga River at Varanasi. Computational Water, Energy, and Environmental Engineering, 4, 25-37. doi:10.4236/cweee.2015.43004.

CHAPTER 13

Ramesh Prasad Tripathi, Woldeselassie Ogbazghi, SemereAmlsom,Simon Measho, (2016) Runoff Harvesting and Storage for Rice Crop at Hamelmalo, Semiarid Region of Eritrea. Computational Water, Energy, and Environmental Engineering,05,1-9. doi: 10.4236/cweee.2016.51001

CHAPTER 14

Paul S.Okweye,Karnita G.Garner,Anthony S.Overton,Elica M.Moss, (2016) Factor-Cluster Analysis and Effect of Particle Size on Total Recoverable Metal Concentration in Sediments of the Lower Tennessee River Basin. Computational Water, Energy, and Environmental Engineering,05,10-26. doi: 10.4236/cweee.2016.51002

CHAPTER 15

Hyo Jeong Kim Preeyaporn Koedrith and Young Rok Seo ; Ecotoxicogenomic Approaches for Understanding Molecular Mechanisms of Environmental Chemical Toxicity Using Aquatic Invertebrate, Daphnia Model Organism; doi:10.3390/ijms160612261

INDEX